Springer Theses

Recognizing Outstanding Ph.D. Research

For further volumes:
http://www.springer.com/series/8790

Aims and Scope

The series "Springer Theses" brings together a selection of the very best Ph.D. theses from around the world and across the physical sciences. Nominated and endorsed by two recognized specialists, each published volume has been selected for its scientific excellence and the high impact of its contents for the pertinent field of research. For greater accessibility to non-specialists, the published versions include an extended introduction, as well as a foreword by the student's supervisor explaining the special relevance of the work for the field. As a whole, the series will provide a valuable resource both for newcomers to the research fields described, and for other scientists seeking detailed background information on special questions. Finally, it provides an accredited documentation of the valuable contributions made by today's younger generation of scientists.

Theses are accepted into the series by invited nomination only and must fulfill all of the following criteria

- They must be written in good English.
- The topic should fall within the confines of Chemistry, Physics, Earth Sciences and related interdisciplinary fields such as Materials, Nanoscience, Chemical Engineering, Complex Systems and Biophysics.
- The work reported in the thesis must represent a significant scientific advance.
- If the thesis includes previously published material, permission to reproduce this must be gained from the respective copyright holder.
- They must have been examined and passed during the 12 months prior to nomination.
- Each thesis should include a foreword by the supervisor outlining the significance of its content.
- The theses should have a clearly defined structure including an introduction accessible to scientists not expert in that particular field.

Corinna Reisinger

Epoxidations and Hydroperoxidations of α,β-Unsaturated Ketones

An Approach Through Asymmetric Organocatalysis

Doctoral Thesis accepted by the
University of Cologne, Germany

Research was carried out at the Max-Planck-Institut für Kohlenforschung, Mülheim an der Ruhr, Germany

Springer

Author
Dr. Corinna Reisinger
Max-Planck-Institut für Kohlenforschung
Mülheim an der Ruhr
Germany

Supervisor
Prof. Dr. Benjamin List
Max-Planck-Institut für Kohlenforschung
Mülheim an der Ruhr
Germany

ISSN 2190-5053 ISSN 2190-5061 (electronic)
ISBN 978-3-642-28117-4 ISBN 978-3-642-28118-1 (eBook)
DOI 10.1007/978-3-642-28118-1
Springer Heidelberg New York Dordrecht London

Library of Congress Control Number: 2012932956

Printed on acid-free paper

Springer is part of Springer Science+Business Media (www.springer.com)

Parts of this thesis have been published in the following journal articles:

"Catalytic Asymmetric Hydroperoxidation of α,β-Unsaturated Ketones: An Approach to Enantiopure Peroxyhemiketals, Epoxides, and Aldols": C. M. Reisinger, X. Wang, B. List, *Angew. Chem., Int. Ed.* **2008**, *47*, 8112–8115; *Angew. Chem.*, **2008**, *120*, 8232–8235.

"Catalytic Asymmetric Epoxidation of Cyclic Enones": X. Wang, C. M. Reisinger, B. List, *J. Am. Chem. Soc.* **2008**, *130*, 6070–6071.

"Preparation of chiral α,β-epoxy ketones": B. List, C. M. Reisinger, X. Wang, PCT Int. Appl. WO 2009112014 **2009**; Ger. Offen. DE 102008013962 **2009**.

Supervisor's Foreword

The enantioselective catalysis of olefin epoxidations has traditionally defined the state of the art in asymmetric synthesis. Remarkably though, despite the wonderful advancements in this area, which led to many name reactions such as those by Sharpless, Jacobsen-Katsuki, Julia-Colonna, and Shi, among others, simple enones have remained elusive substrates for highly enantioselective and general epoxidation catalysis. With this thesis, Corinna Reisinger has now delivered a close to perfect solution to this problem.

Corinna Reisinger has joined my research group at the Max-Planck-Institut für Kohlenforschung as a Ph.D. candidate after having completed her studies in chemistry at the University of Ulm with first-class honors. Her academic achievements had already been rewarded with an undergraduate fellowship from the Fonds der Chemischen Industrie and a diploma award of the Dr. Barbara Mez-Starck Foundation in 2006. During her stay in my group, she was supported by a highly prestigious Kekulé fellowship for graduate research from the Fonds der Chemischen Industrie.

My group focuses on the development of new organocatalytic methods for asymmetric synthesis. Among these, methods for generating highly enantioenriched epoxides are of paramount importance, due to the versatility of epoxides as building blocks in academic and industrial organic synthesis. Corinna Reisinger's work has focused on the development of a general and efficient catalytic asymmetric epoxidation of α,β-unsaturated ketones and aldehydes via iminium ion catalysis with primary amine salts. She has developed the first highly enantioselective epoxidation of cyclic enones in collaboration with another group member (*J. Am. Chem. Soc.* **2008**, 6070–6071). To the best of my knowledge, this work constitutes a breakthrough in the field of asymmetric epoxidation chemistry, standing to this date as the single most efficient method in the literature to generate the corresponding cyclic epoxyketones with exquisitely high enantioselectivities. It is a testament to her success that her seminal paper has collected over 70 citations only during the first 2 years after publication. Subsequently, she went on and implemented a highly enantioselective, primary amine salt-catalyzed hydroperoxidation of linear aliphatic enones to provide enantiomerically pure 3-hydroxy-1,2-dioxolanes. Corinna first identified these peroxy hemiketals as by-products of the epoxidation process and immediately recognized their synthetic potential. Indeed, she demonstrated that the products could be converted to the corresponding epoxides and to β-hydroxy ketones in a one-pot sequence without loss of enantioselectivity (*Angew. Chem. Int. Ed.* **2008**, 8112–8115).

Very recently, solely based on Corinna's encouragement, another graduate student of mine together with Corinna extended her epoxidation method to α-branched enals. This work provides the solution to yet another entirely unsolved problem in asymmetric epoxidation catalysis and was just published (*J. Am. Chem. Soc.* **2010**, 10227–10229).

Corinna's work could have far-reaching applications in asymmetric catalysis and a patent covering her oxidation method is currently pending. Indeed, the use of hydrogen peroxide as a safe inorganic oxidant that generates only water as a by-product, combined with the operational simplicity of this reaction (inexpensive cinchona alkaloids as catalysts, no requirement for inert atmosphere or scrupulously dried solvents) could well lead to commercial applications.

Based on her wonderful achievements, she has already won several awards, including being selected among the prestigious DSM Science and Technology awardees of 2010, and the Otto-Hahn-Medal of the Max-Planck-Society. Reading this thesis has been delightful and reminded me of the great privilege it has been to work with her.

Mülheim an der Ruhr Prof. Dr. Benjamin List

Contents

Abbreviations

Ac	Acetyl
ACDC	Asymmetric counteranion-directed catalysis
AcO	Acetate
Al, al	Alkyl, aliphatic
app.	Apparent
Ar, ar	Aryl, aromatic
aq.	Aqueous
calcd	Calculated
BINAM	2,2′-diamino-1,1′-binaphthalene
BINOL	1,1′-bi-2-naphthol
Bn	Benzyl
Boc	*tert*-butyloxycarbonyl
Bu	Butyl
C	Cinchonine
calcd	Calculated
CAN	Ceric ammonium nitrate
cat.	Catalyst/catalytic
CD	Cinchonidine
CHP	Cumyl hydroperoxide
conv.	Conversion
Cy	Cyclohexyl
cycl.	Cyclic
DACH	1,2-diaminocyclohexane
DET	Diethyl tartrate
DFT	Density functional theory
DHP	Dihydropyran
DHQ	Dihydroquinine
DIAD	Diisopropyl azadicarboxylate
DKR	Dynamic kinetic resolution
DMF	Dimethylformamide
DMSO	Dimethylsulfoxide

DNBA	2,4-dinitro benzoic acid
DPEN	1,2-diphenylethylenediamine
DPPA	Diphenylphosphoryl azide
DPPOH	Diphenyl phosphoric acid
dr	Diastereomeric ratio
E	Electrophile
EI	Electron impact
ee	Enantiomeric excess
EM	Exact mass
ent	Enantiomer(ic)
epo	Epoxide
equiv	Equivalent(s)
er	Enantiomeric ratio
Et	Ethyl
ESI	Electrospray ionization
FDP	Fructose-1,6-diphosphate
GC (GC-MS)	Gas chromatography (gas chromatography coupled with mass detection)
HBTU	*O*-(benzotriazol-1-yl)-tetramethyluronium hexafluorophosphate
HOMO	Highest occupied molecular orbital
HPLC	High performance liquid chromatography
HRMS	High resolution mass spectrometry
i	Iso
KR	Kinetic resolution
LA	Lewis acid
LDA	Lithium diisopropylamide
Lit.	Literature
LUMO	Lowest unoccupied molecular orbital
m	*Meta*
M	(alkali) Metal
M	Molar (concentration)
MCPBA	*Meta*-chloroperbenzoic acid
Me	Methyl
Mes	Mesityl (2,4,6-trimethylphenyl)
MS	Mass spectrometry, molecular sieves
Ms	Methylsulfonyl
MTBE	Methyl *tert*-butyl ether
MW	Molecular weight
m/z	Atomic mass units per charge
n	Normal
N	Normal (concentration)
n.a.	Not available
n.d.	Not determined
9-NH_2-Q	9-amino(9-deoxy)quinine
9-NH_2-*epi*Q	9-amino(9-deoxy)*epi*quinine

9-NH_2-*epi*QD	9-amino(9-deoxy)*epi*quinidine
NMP	*N*-methyl pyrrolidine
NMR	Nuclear magnetic resonance spectroscopy
NOE	Nuclear Overhauser effect
Nu-H/Nu	Nucleophile
Ns	Nosyl
o	*Ortho*
ol	Olefinic
P	Product
p	*Para*
PCC	Pyridinium chloro chromate
Ph	Phenyl
PHK	Peroxyhemiketal
Pr	Propyl
Py	Pyridine
Q	Quinine
QD	Quinidine
quin.	Quinuclidine
quint	Quintet
rac.	Racemic
r.t.	Room temperature
S	Substrate
sept	Septet
SOMO	Singly occupied molecular orbital
t	*Tert*
TEMPO	2,2,6,6-tetramethylpiperidine N-oxyl
TEP	Triethylphosphite
TBHP	*Tert*-butyl hydroperoxide
TBME	*Tert*-butylmethylether
TBS	Tetrabutyldimethylsilyl
TCA	Trichloroacetic acid
TFA	Trifluoroacetic acid
THF	Tetrahydrofuran
Thp	Tetrahydropyran-2-yl
TLC	Thin layer chromatography
TMS	Trimethylsilyl
TRIP	3,3′-bis(2,4,6-triisopropylphenyl)-1,1′-binaphthyl-2,2′-diyl hydrogenphosphate
Ts	*Para*-toluenesulfonyl
TsOH	*Para*-toluene sulfonic acid
UHP	Urea hydrogen peroxide
wt	Weight

Chapter 1
Introduction

Although the origin of chirality is still obscure [1], it is the source of diverse phenomena at the macro- and microscopic level, governing our environment and the existence of living organisms. The principles of molecular chirality were established over a century ago by van't Hoff and LeBel, but awareness of how they affect the biological activity of molecules is much more recent [2, 3].

The biological machinery made up from the basic building blocks of life—chiral amino acids, sugars, and lipids—is susceptible to enantioselective interactions. Biological systems are thus commonly capable of differentiating between enantiomeric forms of chiral molecules, including odorants, pheromones, agrochemicals, environmental pollutants and, most importantly, drug compounds. The two different enantiomers of a compound may have distinctly different effects on a given biological system. For instance, naturally occurring α-ionone of violets exhibits (*R*)-configuration and, differs significantly from the (*S*)-enantiomer in its olfactory properties (Fig. 1.1) [4].

Due to increased interest in the consequences of chirality on physical, biological and pharmacological properties of molecules [5, 6], the preparation of pure stereoisomers has become a topic of great importance, and methods of supplying optically pure materials are being intensively pursued. Nature has provided a variety of enantiomerically pure compounds referred to as the chiral pool, which can be subjected to further transformations. Besides this there are two general methods for obtaining enantiopure compounds: resolution of racemic mixtures and asymmetric synthesis. Among the stereoselective synthetic methodologies, catalytic processes hold a prominent place. In 1996, Nicolaou and Sorensen [7] stated that "in a catalytic asymmetric reaction, a small amount of an enantiomerically pure catalyst, either an enzyme or a synthetic, soluble transition metal complex, is used to produce large quantities of an optically active compound from a precursor that may be chiral or achiral". However, during the last decade, organocatalysis,

C. Reisinger, *Epoxidations and Hydroperoxidations of α,β-Unsaturated Ketones*,
Springer Theses, DOI: 10.1007/978-3-642-28118-1_1,

Fig. 1.1 Enantiomers of α-ionone

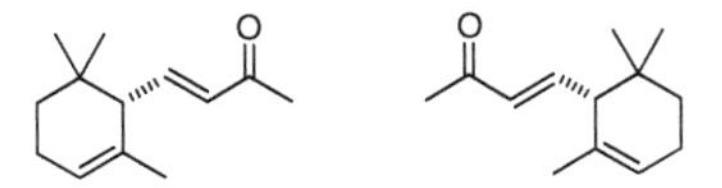

catalysis by low-molecular weight organic molecules,[1] has garnered increased importance within the realm of asymmetric catalysis by allowing a broad and diverse array of equally efficient and selective transformations, thereby complementing metal and biocatalysis [8–12].

Organocatalysis can even offer certain advantages over metal-catalyzed and biocatalytic methods: in particular, its operational simplicity and practicability represent major benefits. Most organocatalysts are readily available from inexpensive starting materials and are typically bench-stable and robust (inert towards moisture and oxygen). Thus, organocatalytic reactions generally do not require inert reaction conditions. They can regularly be conducted in wet solvents under an aerobic atmosphere. Indeed, the presence of water is often beneficial to the reaction rate and selectivity. In addition, organocatalytic—and thus commonly metal-free—methods are especially attractive in food and drug related contexts for the preparation of pharmaceutical products and agrochemicals that do not tolerate residual heavy metal impurities, not even in trace amounts.

This thesis deals with the utilization of organocatalysis for the synthesis of chiral, enantiopure epoxides. The importance of epoxides in organic synthesis arises partly from the occurrence of the strained three-membered ring unit in a number of interesting natural products, but more so because the stereospecific ring opening of epoxides allows straightforward elaboration to useful new functionality [13]. Nature's chiral pool has not proven to be a useful direct source of optically active epoxides for synthetic applications. Instead, enantioenriched epoxides have been accessed in indirect fashion via multistep procedures. These, however, tend to be inherently inefficient, and the range of epoxides available by this approach is also quite limited. As a consequence, the preparation of enantioenriched epoxides has long stood as a particularly significant endeavor in asymmetric synthesis. In particular, the identification of catalytic asymmetric olefin epoxidation methods has been an area of active research for several decades [14].

In the following chapters, both the history of organocatalysis and catalytic asymmetric epoxidation will be briefly outlined. This serves as a prelude for our own journey through the development of new organocatalytic asymmetric epoxidations and hydroperoxidations of α,β-unsaturated ketones—with its fair share of reaction design, rational analysis, and serendipitous discovery.

[1] Organocatalysts may contain metals if these play exclusively a structure-defining role and are not involved in the working principle.

References

1. Klussmann M, Blackmond DG (2009) Chemical evolution II: from the origins of life to modern society. American Chemical Society, Washington, p 133
2. van't Hoff JH (1874) Arch Neerl Sci Exactes Nat 9:445
3. Le Bel JA (1874) Bull Soc Chim France 22:331
4. Brenna E, Fuganti C, Serra S, Kraft P (2002) Eur J Org Chem 2002:967
5. Mellin GW, Katzenstein M (1962) New Engl J Med 267:1184
6. Blaschke G, Kraft HP, Fickentscher K, Köhler F (1979) Arzneimittel-Forsch 29:1640
7. Nicolaou KC, Sorensen EJ (1996) Classics in total synthesis: targets, strategies, methods. Wiley-VCH, Weinheim
8. Dalko PI (ed) (2007) Enantioselective organocatalysis, reactions and experimental procedures. Wiley-VCH, Weinheim
9. Berkessel A, Gröger H (2005) Asymmetric organocatalysis, from biomimetic concepts to applications in asymmetric synthesis. Wiley-VCH, Weinheim
10. List B (ed) (2009) Asymmetric organocatalysis. Top Curr Chem 291:77–144
11. List B (ed) (2007) Chem Rev 107(Special Issue: Organocatalysis): 5413
12. Dondoni A, Massi A (2008) Angew Chem Int Ed 47:4638
13. Gorzynski Smith J (1984) Synthesis 8:629
14. Xia QH, Ge HQ, Ye CP, Liu ZM, Su KX (2005) Chem Rev 105:1603

Chapter 2
Background

2.1 Asymmetric Organocatalysis

For a long time, the realm of asymmetric catalysis was dominated by metal and biocatalysis. Yet, at the beginning of this century, List's discovery of the (*S*)-proline-catalyzed direct asymmetric intermolecular aldol reaction [1] together with the development of an asymmetric Diels–Alder reaction catalyzed by a chiral imidazolidinone salt by MacMillan et al. [2] have raised awareness of the potential of purely organic molecules as efficient catalysts for a variety of asymmetric transformations and brought to life the term "organocatalysis" to address this research field (Scheme 2.1).

2.1.1 Historical Development

Organocatalysis has a rich background as it is suggested that extraterrestrial, enantiomerically enriched amino acids such as (*S*)-alanine and (*S*)-isovaline played a decisive role in the prebiotic formation of key building blocks such as sugars by promoting the self-aldol reaction of glycolaldehydes in water [3]. In this way, the reactions might have led to the introduction and widespread of homochirality in the living world. The historic roots of organocatalysis date back to the mid of the nineteenth century with von Liebig's accidentally discovery of what is today considered the first organocatalytic reaction, the transformation of dicyan into oxamide in the presence of an aqueous solution of acetaldehyde [4]. In the early 1900s, Bredig's pioneering studies on the use of natural Cinchona alkaloids as enantioselective catalysts were motivated by searching the chemical origin of enzyme activity observed in living organisms [5]. In this context, he developed the first asymmetric organocatalytic reaction by adding hydrogen cyanide to benzaldehyde in the presence of catalytic amounts of either quinine (**3**) or quinidine (**4**) (Scheme 2.2). These studies were conceptually groundbreaking, although the

C. Reisinger, *Epoxidations and Hydroperoxidations of α,β-Unsaturated Ketones*,
Springer Theses, DOI: 10.1007/978-3-642-28118-1_2,

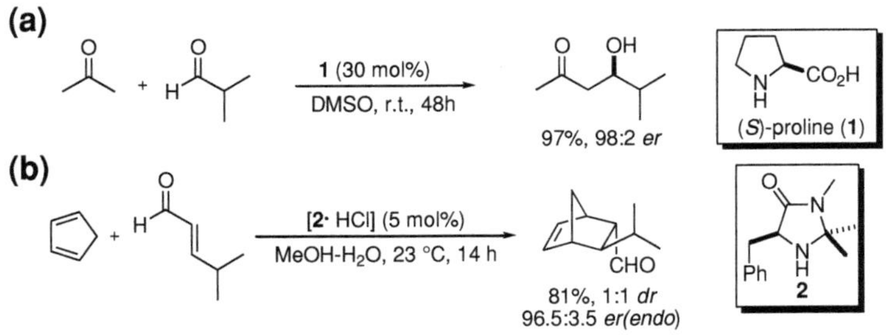

Scheme 2.1 **a** (*S*)-proline (**1**)-catalyzed direct asymmetric intermolecular aldol reaction. **b** Imidazolidinone (**2**) salt-catalyzed enantioselective Diels–Alder reaction

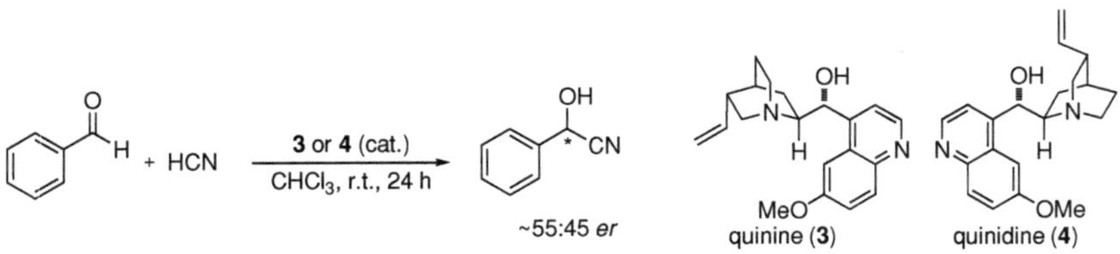

Scheme 2.2 First enantioselective organocatalytic reaction: asymmetric hydrocyanation of benzaldehyde by Bredig

enantiomeric ratio remained rather low (~55:45 *er*), and initiated a line of research which had been continued by Pracejus [6, 7] (~60:40 *er*) and others.

Another key event in the history of organocatalysis was the use of (*S*)-proline as an aldolization catalyst by Hajos and Parrish at Hoffmann-La Roche [8, 9] and Eder et al. [10, 11] at Schering in the early 1970s (Scheme 2.3). Their parallel studies were decisively inspired by the seminal work of Knoevenagel in the late nineteenth century [12].

Nonetheless, the real potential of this chemistry was only revealed three decades later by the groups of List and MacMillan who have demonstrated that aminocatalysis, the activation of carbonyl groups via enamine and iminium ion intermediates, is indeed a **general** concept for (asymmetric) catalysis (cf. Scheme 2.1). Evolving from a small collection of chemically unique and unusual, mechanistically poorly understood reactions, organocatalysis has advanced at a truly breathtaking pace since its birth in 2000 and has during the last ten years grown into a thriving area of research, which today represents the third pillar of asymmetric catalysis besides metal and biocatalysis [13–17].

2.1.2 Classification

Organocatalysis can be divided into four areas: Lewis base, Lewis acid, Brønsted base and Brønsted acid catalysis [18]. The corresponding (simplified) catalytic cycles are depicted in Scheme 2.4. Accordingly, Lewis base catalysts (B:) initiate

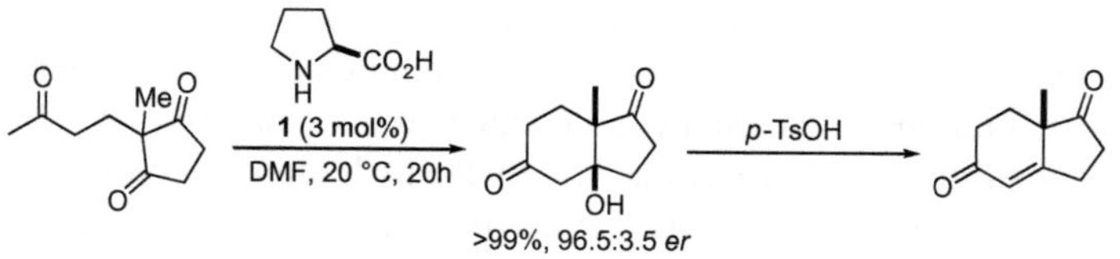

Scheme 2.3 Hajos–Parrish–Eder–Sauer–Wiechert reaction catalyzed by (*S*)-proline (**1**)

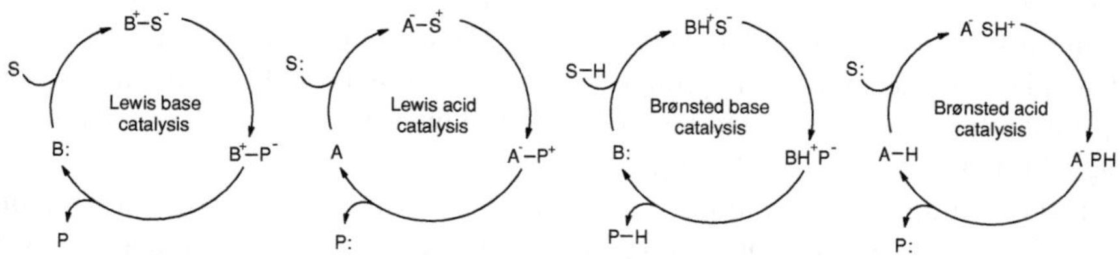

Scheme 2.4 (Simplified) catalytic cycles of Lewis base, Lewis acid, Brønsted base, and Brønsted acid catalysis

the catalytic cycle via nucleophilic addition to the substrate (S). The resulting complex undergoes a reaction and then releases the product (P) and the catalyst for further turnover. Lewis acid catalysts (A) activate nucleophilic substrates (S:) in a similar manner. Brønsted base and acid catalytic cycles are initiated via a (partial) deprotonation or protonation, respectively.

The majority of organocatalysts are N-, C-, O-, P-, and S-based Lewis bases such as amines, carbenes, formamides, phosphanes, phosphoramides, and sulfides. Of those, nitrogen-based systems account for the largest share, partly due to their abundance in the chiral pool. Lewis base organocatalysts operate through diverse mechanisms and convert the substrates into either activated nucleophiles or electrophiles. Typical reactive intermediates are iminium ions, enamines, acyl ammonium ions [19], 1-, 2-, or 3-ammonium enolates [20], among others. The underlying principle of the work discussed within this thesis is the activation of carbonyl compounds as enamine or iminium ion, which represents the basis of what is often called aminocatalysis.

2.1.3 Aminocatalysis

The roots of modern aminocatalysis trace back to the pioneering work of Knoevenagel who, at the turn of the nineteenth century, found that primary and secondary amines, as well as their salts, catalyze the aldol condensation of β-ketoesters or malonates with aldehydes or ketones [21–24]. Remarkably, Knoevenagel himself suggested a possible role of imine formation with the amine catalyst in the course of the reaction [22]. In the first half of the nineteenth century, Kuhn and Fischer and

Marschall discovered that amines and amine salts also catalyze aldol addition and condensation reactions [25–28]. The first iminium-catalyzed conjugate addition reaction to α,β-unsaturated carbonyl compounds was reported in 1937 by Langenbeck using piperidinium acetate as catalyst for the hydration of crotonaldehyde [29]. In the history of iminium catalysis, further important landmarks are: (a) the discovery of iminium-catalyzed transimination by Cordes and Jencks [30], (b) the reports of Baum and Viehe, and more recently Jung et al. [31, 32], on the acceleration of Diels–Alder reactions provided by α,β-unsaturated iminium ions, (c) the proline-catalyzed deracemization of a thianone intermediate in the synthesis of erythromycin by Woodward and co-workers [33], and most importantly (d) the use of alkali metal and ammonium proline salts in the conjugate addition of malonates to α,β-unsaturated aldehydes and ketones in the pioneering work of Yamaguchi and Taguchi between 1991 and 1996 [34–37].

Today's aminocatalytic transformations of carbonyl compounds via iminium ion and enamine intermediates using chiral primary and secondary amines (as well as their salts) as organocatalysts, hinges upon four distinct activation modes and is categorized accordingly in: (a) enamine, (b) iminium ion, (c) dienamine, and (d) SOMO catalysis, which involves the formation of enamine radical cations [38–41].

All distinct activation modes have in common that the initial step constitutes the reversible formation of an iminium ion by condensation of the amine (salt) catalyst (**A**) with the carbonyl compound (Scheme 2.5). The formation of iminium ion **B** effectively lowers the LUMO energy of the system (a)—as does the coordination of a Lewis acid (b). As a result, both nucleophilic additions and α-deprotonation become more facile. α-Deprotonation leads to the generation of enamine **C**, which is more nucleophilic than the enol form of the parent carbonyl compound. The formation of the enamine corresponds to a raise in the HOMO energy of the system and allows for subsequent reaction with a range of electrophiles to obtain α-functionalized carbonyl compounds. For conjugated π-systems, the electronic redistribution induced by the formation of iminium ion **D** facilitates nucleophilic additions (c), including conjugate additions as well as pericyclic reactions. γ-Deprotonation, finally, generates dienamine intermediates (*vide infra*).

In all aspects, the principle of aminocatalytic activation emulates the mechanism of the activation of carbonyl compounds by Lewis acids. In reactions of α,β-unsaturated ketones the steric and electronic similarity of the two carbonyl substituents does not generally permit high levels of discrimination between the free electron lone pairs in the metal-association step, which is essential for attaining high stereocontrol in the following transformation. In aminocatalysis by contrast, iminium ion formation overcomes the necessity of discriminating between the free electron lone pairs.

An overview of the divergent reaction pathways and functionalizations amenable to (a) aldehydes and ketones or (b) enals and enones, respectively, by means of aminocatalysis is provided in Scheme 2.6.

More recently, the formation of dienamines **F** from α,β-unsaturated iminium ions by γ-deprotonation has been synthetically exploited to give rise to γ-functionalized enals upon reaction with electrophiles (Scheme 2.6) [42, 43].

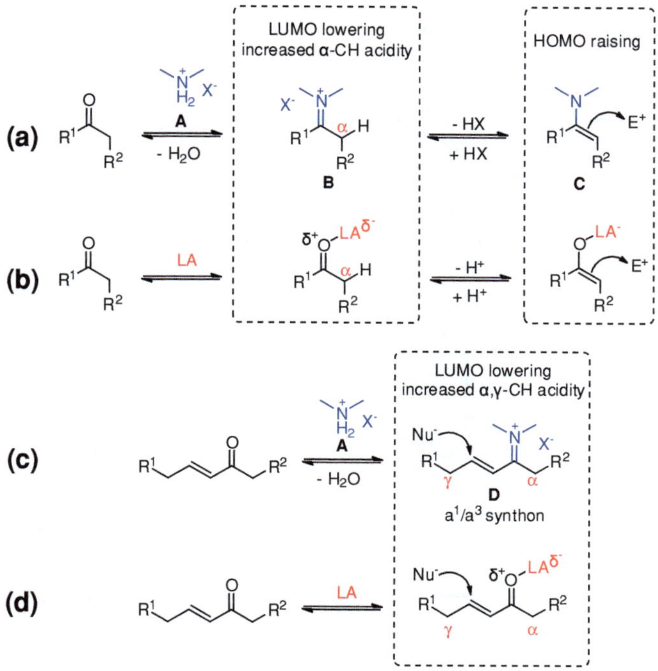

Scheme 2.5 Activation of carbonyl compounds by aminocatalysis (**a**, **c**) versus Lewis acid catalysis (**b**, **d**). (*LA* Lewis acid, *E* electrophile, *Nu* nucleophile)

Their use as electron rich dienes in Diels–Alder reaction has also been reported [44]. A fourth synthetic tool constitutes the SOMO activation of carbonyl compounds. Oxidation of the initially formed enamine **C** affords a radical cation **E** which allows the use of a group of reagents that has previously not been applicable to aminocatalysis such radicals (e.g. TEMPO), π-nucleophiles (e.g. allylsilanes, styrenes), or halides (e.g. LiCl) [45–48].

Enamine and iminium ion catalysis are two divergent reaction modes in organocatalysis, though sharing a common origin. Enamine catalysis proceeds via iminium ion formation and almost always results in iminium ion formation (cf. Scheme 2.6). In a complementary fashion, conjugate addition of a nucleophile to an iminium ion generates an enamine intermediate **G** which can in turn react with another electrophile [49]. The interdependency of those two catalytic intermediates (cf. Scheme 2.7a) combined with the ability of most amine catalysts to promote several types of transformations based on different activation modes side by side makes aminocatalysis the perfect platform for domino and tandem or cascade reactions and a powerful tool in the construction of complex molecular skeletons in a highly stereocontrolled manner as illustrated with numerous elegant examples in the recent literature [50, 51].

One example which presumably operates through the same mechanistic rationale, is the amine-catalyzed asymmetric epoxidation of α,β-unsaturated

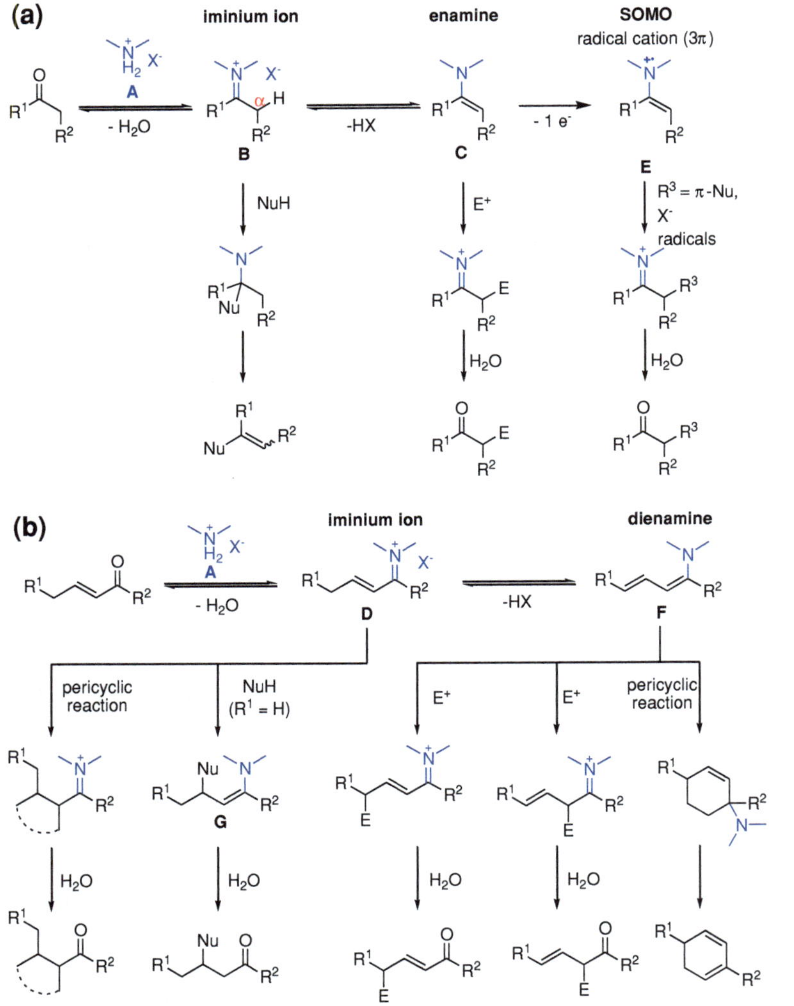

Scheme 2.6 Aminocatalytic activation modes and divergent reaction pathways with **a** aldehydes and ketones, **b** enals and enones

carbonyl compounds with alkyl hydroperoxides or hydrogen peroxide as studied within this thesis. In the course of this reaction, the hydroperoxide serves as both the nucleophile and the electrophile (Scheme 2.7b).

2.1.3.1 Asymmetric Counteranion-Directed Catalysis (ACDC)

Most chemical reactions proceed via charged intermediates or transition states. Such "polar reactions" can be influenced by the counterion, especially if

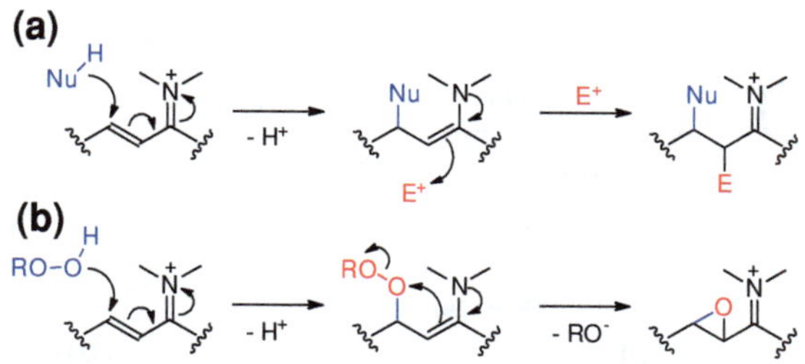

Scheme 2.7 **a** General picture of the iminium ion-enamine interdependency. **b** Iminium ion-enamine interplay in the nucleophilic epoxidation reaction. (*Nu-H* nucleophile, *E* electrophile; R = H, Alk)

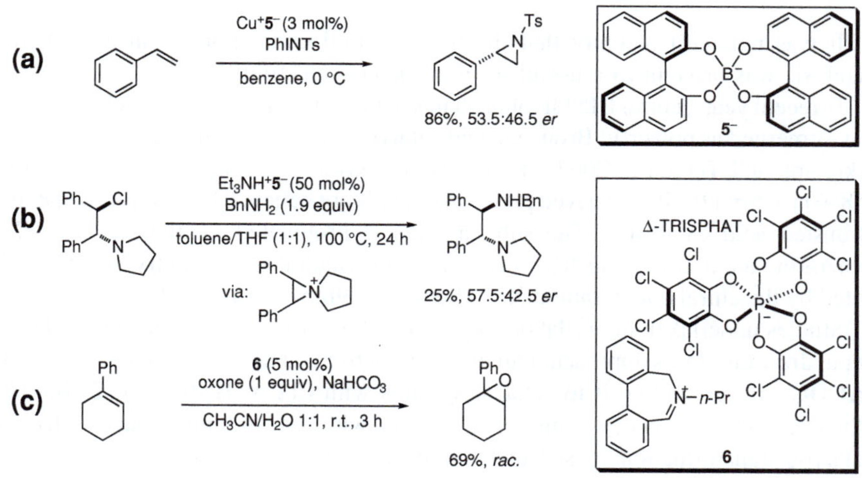

Scheme 2.8 Preliminary attempts toward the use of chiral counteranions in asymmetric catalysis

conducted in organic solvents, where ion pairs are ineffectively separated by the solvent. Although efficient asymmetric catalytic transformations involving anionic intermediates with chiral, cationic catalysts have been realized in the context of phase-transfer catalysis [52], analogous versions of inverted polarity attaining reasonable enantioselectivity have been elusive until recently. Attempts toward this end have been undertaken by Arndtsen and Nelson and co-workers (Scheme 2.8, eq. a and b) [53–55]. Both groups applied chiral borate anions such as BINOL-derived borate **5** to several catalytic transformations including copper-catalyzed aziridination and cyclopropanation of olefins and ring opening reactions of *meso*-aziridinium ions, albeit with only moderate success. The highest enantioselectivity observed in all those reactions was 67:33 *er* along with a yield of 3% of the desired product by using a chiral tartrate-derived borate anion [53]. Lacour et al. [56, 57] studied the influence of chiral TRISPHAT anions on enantioselective olefin epoxidation reactions catalyzed by iminium ion **6**. However, only racemic product was obtained (eq. c).

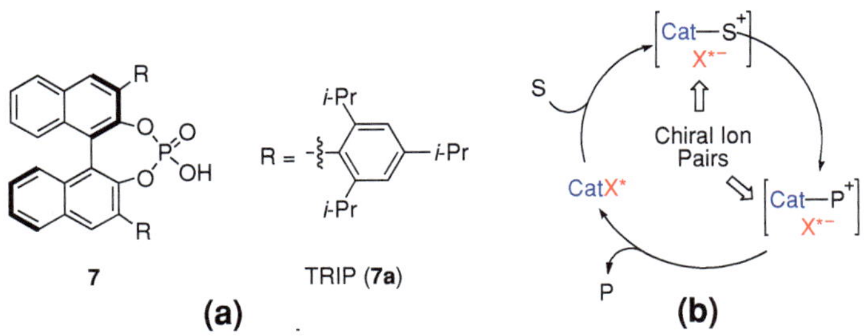

Scheme 2.9 **a** Chiral BINOL-derived phosphoric acids [e.g. TRIP (**7a**)]. **b** General concept of asymmetric counteranion-directed catalysis (ACDC) (with Cat = H: special case of Brønsted acid catalysis). (*S* substrate, *P* product, *CatX* catalyst, X^{*-}: chiral anion)

It was not until recently that the use of chiral counteranions in asymmetric catalysis was brought to a useful level of enantioselectivity.

In recent years, chiral BINOL phosphates of the general structure **7** (Scheme 2.9a) have emerged as powerful Brønsted acid catalysts triggered by the seminal works of Akiyama and Terada in 2004 on their use in asymmetric Mannich-type reactions [58–61]. Chiral BINOL-derived phosphoric acids are believed to function as specific Brønsted acid catalysts.[1] The substrate (S) is protonated by the catalyst HX* generating a chiral ion pair $[(S-H)^+ X^{*-}]$, in which the asymmetry is communicated by the chiral counteranion X^{*-} (Scheme 2.9b; with Cat = H).

Studies undertaken in our laboratory pushed forward the generalization and conceptualization of this approach, namely asymmetric counteranion-directed catalysis (**ACDC**), by expanding it to catalyst systems with Cat $\neq$ H (Scheme 2.9b). The efficiency of this concept was initially illustrated with various highly enantioselective catalytic transformations based on iminium ion catalysis (Scheme 2.10) [62–64]. Remarkably, the highest enantioselectivities were consistently observed when using the chiral BINOL-based phosphoric acid **TRIP** (**7a**), bearing sterically demanding 2,4,6-triisopropylphenyl substituents at the 3,3′-positions of the binaphthyl scaffold, as chirality inducing counteranion (Scheme 2.9a) [65]. Later, the ACDC concept was further expanded to Lewis acid and transition-metal catalysis, by our [66, 67] and other research groups [68–70]. The asymmetry transfer from the chiral counteranion X^{*-} to the activated substrate $(Cat-S)^+$ within a tight ion pair $[(Cat-S)^+ X^{*-}]$ (Scheme 2.9b; Cat $\neq$ H) is most likely the result of a cooperative effect of electrostatic and coordinative interactions conceivably assisted by hydrogen bonding.

The ACDC approach to asymmetric catalysis offers a manifest advantage compared with traditional strategies for catalyst design and optimization. Combinatorial libraries of ACDC catalysts are readily accessible by interchanging the two catalyst components separately. In addition, when using two **chiral** components matched combinations commonly allows further improvement of the asymmetric induction as has been demonstrated in the asymmetric transfer

[1] Specific Brønsted acid catalysis relies on the use of strong Brønsted acids as catalysts.

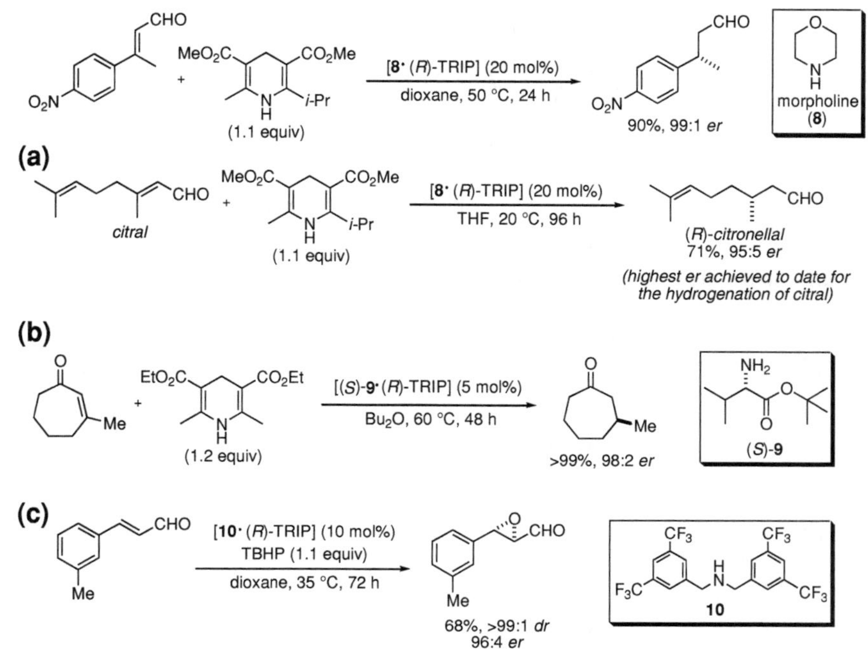

Scheme 2.10 The **ACDC** concept in iminium ion catalysis: application to the transfer hydrogenation of **a** enals, and **b** enones as well as, **c** the epoxidation of enals

hydrogenation of ketones catalyzed by the (*S*)-valin-*tert*-butyl ester (**9**) (*R*)-TRIP salt [(*S*)-**9** · (*R*)-TRIP] (Scheme 2.10b). This strategy will also find application in the work described within this thesis.

2.1.3.2 Primary Amine Salt Catalysts

At the outset, the focus of modern aminocatalysis was on the use of chiral secondary amine catalysts, especially substituted pyrrolidines, proline, or imidazolidinone derivatives, for the activation and functionalization of carbonyl compounds. In contrast, chiral primary amine catalysts have been largely neglected, possibly due to unfavorable imine-enamine equilibria [71]. This is surprising in view of the fact that early studies on aminocatalysis have already included the use of primary amino acids, amines and amine salts as catalysts [26, 72–75], and even more so since primary amine catalysis is effectively exploited by natural enzymes such as type I aldolases and decarboxylases, both containing catalytically active lysine residues (Fig. 2.1) [76].

Only recently has primary amine catalysis emerged as a powerful tool for the iminium ion activation of challenging classes of unsaturated carbonyl compounds such as sterically demanding α,β-unsaturated ketones and α-branched enals which are difficult to activate with conventional secondary amine catalysts [77–79]. Owing to reduced steric requirements, the use of primary amines overcomes the inherent difficulties of

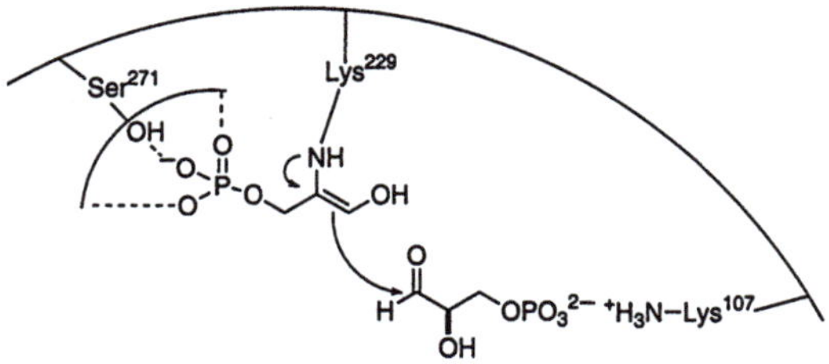

Fig. 2.1 Active site of type I rabbit muscle FDP (fructose-1,6-diphosphate) aldolase [76]

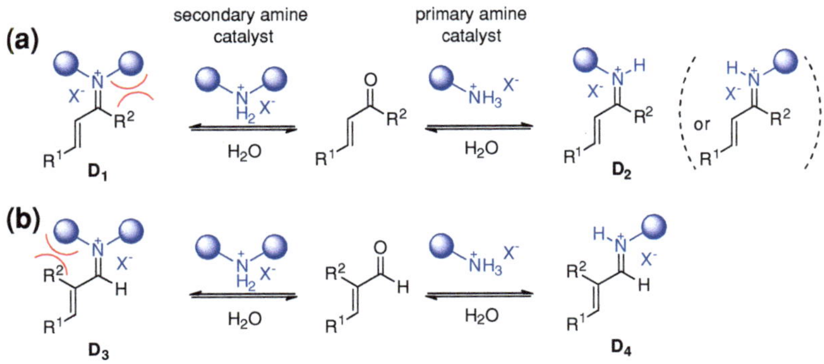

Scheme 2.11 Iminium ion activation of **a** α,β-unsaturated ketones, and **b** α-branched α,β-enals: secondary versus primary amine catalysts

secondary amine catalysts in generating congested covalent iminium ion intermediates, and thus enables transformations of sterically demanding carbonyl compounds which have previously not been realized through secondary amine catalysis (**D$_1$** and **D$_3$**, Scheme 2.11). Primary amine catalysts may provide higher equilibrium concentrations of the requisite iminium ion (**D$_2$** and **D$_4$**), and thus account for increased reaction rates.

The past years since 2005 have witnessed the rapid development of catalytic asymmetric transformations employing chiral primary amine catalysts [45]. Selected examples are depicted in Scheme 2.12.

Ishihara et al. [80–82] succeeded in activating α-substituted acroleins, in particular α-(acyloxy)acroleins, for Diels–Alder and [2 + 2] cycloaddition reactions by using novel primary amine **11** together with an acid co-catalyst. First evidence of the potential of primary amines as iminium ion activators of α,β-unsaturated ketones was provided by Chin and co-workers in 2006. Their synthesis of warfarin, a widely prescribed anticoagulant, via conjugate addition of 4-hydroxycoumarin to benzylideneacetone was catalyzed by (*S*,*S*)-DPEN (1,2-diphenylethylenediamine; **12**) in the presence of acetic acid (eq. b) [53–55]. In the same year, List and Martin disclosed the use of (*S*)-valin-*tert*-butyl ester (**9**) (*R*)-TRIP salt [(*S*)-**9** · (*R*)-TRIP] as highly efficient catalyst in the asymmetric transferhydrogenation of cyclic and acyclic α,β-enones (cf. Scheme 2.10b) [64]. Moreover, primary amines derived from naturally abundant Cinchona alkaloids were shown to provide an effective and general catalyst platform for

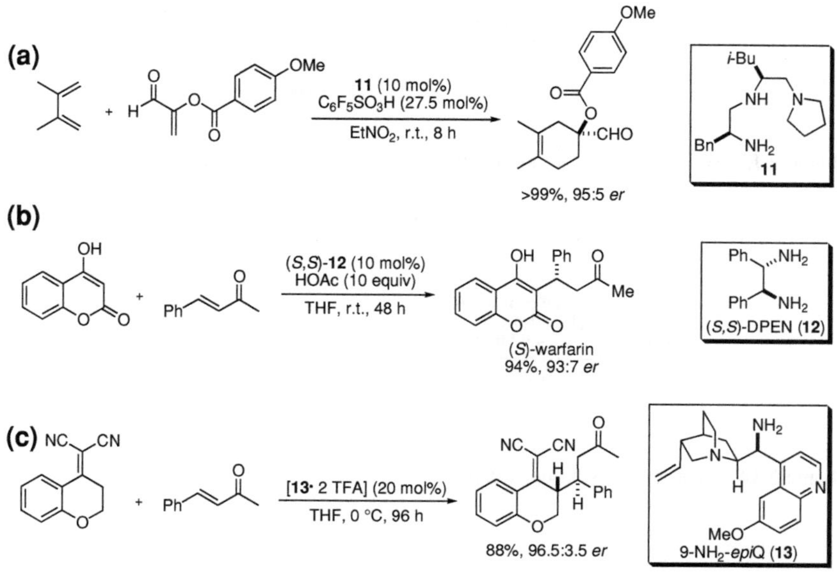

Scheme 2.12 Iminium ion catalysis with primary amine salt catalysts: seminal reports

asymmetric transformations of α,β-unsaturated ketones, including conjugate additions and pericyclic reactions [83]. The seminal report on their use as iminium ion activator of α,β-unsaturated ketones was disclosed by Chen and Deng et al. [84, 85] in 2007. Amino(9-deoxy)*epi*quinine (**13**) as its TFA salt efficiently mediated the enantioselective vinylogous conjugate addition of α,α-dicyanoalkenes to benzylideneacetone and derivatives (Scheme 2.12c). 9-Amino(9-deoxy)*epi*quinine (**13**) and analogues, whose chemistry has been pioneered by Brunner and co-workers [86, 87], have rapidly gained immense popularity and have found widespread application in primary amine catalysis. This is testified by not less than 50 publications since 2007 on their use as chiral organocatalysts from various research groups all around the world including the groups of Melchiorre, Connon, Deng and our group among others [77, 78, 87–91].

2.2 Catalytic Asymmetric Epoxidation of Electron-Deficient Olefins

2.2.1 Milestones in Catalytic Asymmetric Olefin Epoxidations

Olefin epoxidation holds a venerable place in the history of catalytic asymmetric synthesis. The pioneering work by Sharpless in the early 1980s on the titanium-tartrate-catalyzed asymmetric epoxidation of allylic alcohols (Scheme 2.13), paved the way for much of today's catalytic asymmetric synthesis [92].

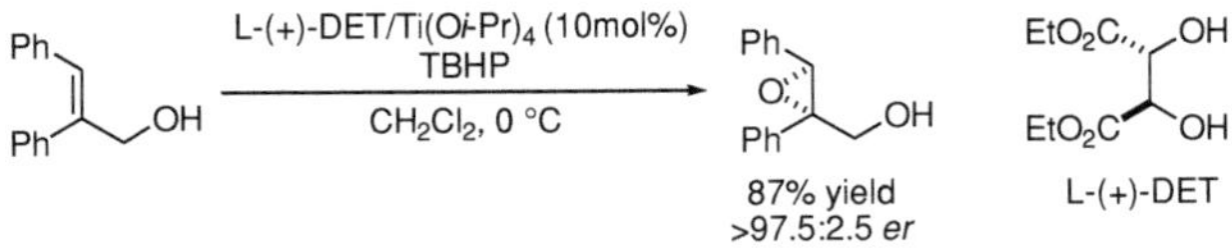

Scheme 2.13 Sharpless' catalytic asymmetric epoxidation of allylic alcohols

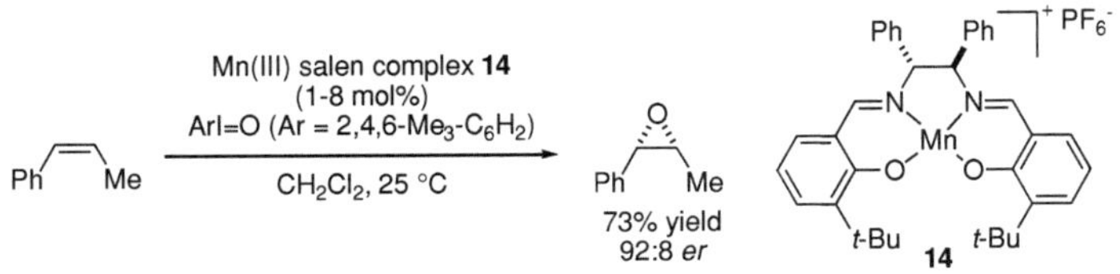

Scheme 2.14 Jacobsen–Katsuki epoxidation

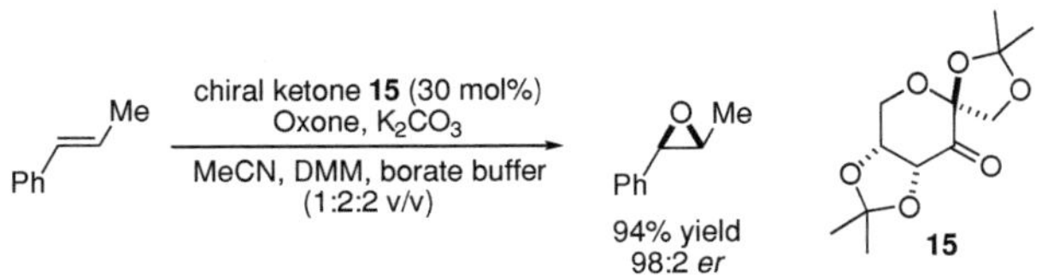

Scheme 2.15 Asymmetric epoxidation catalyzed by chiral ketones

Following this discovery, much progress has been made towards the asymmetric epoxidation of other classes of olefins. Jacobsen and Katsuki introduced manganese-salen complexes as valuable catalysts for the catalytic asymmetric epoxidation of unfunctionalized, and particularly (*Z*)-disubstituted olefins (Scheme 2.14) [93, 94].

More recently, the work of several groups, in particular Shi and co-workers, established dioxiranes, generated *in situ* from chiral ketones and Oxone, as asymmetric epoxidation reagents for a range of alkenes, especially (*E*)-disubstituted olefins (Scheme 2.15) [95].

The advances made in this field have increased greatly the number of enantiomerically enriched epoxides available for use in organic synthesis. Although a few different methods exist for the asymmetric epoxidation of electron-deficient olefins (*vide infra*), no system has gained widespread popularity amongst synthetic organic chemists [96–98]. However, due to the value of the corresponding optically active epoxides as versatile synthetic intermediates, much effort is devoted to the development of a general, highly enantioselective method applicable to a wide range of electron-deficient olefins.

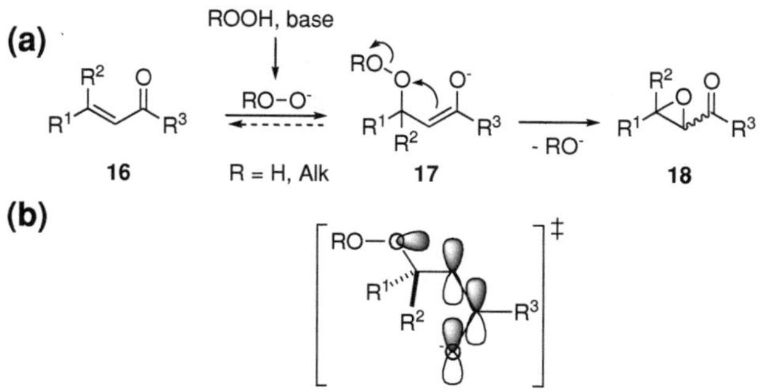

Scheme 2.16 **a** Two-step mechanism of the Weitz–Scheffer epoxidation: conjugate addition of a hydro- or alkyl peroxide followed by epoxide ring closure, **b** putative transition state for epoxide ring closure

2.2.2 Weitz–Scheffer Epoxidation

The general approach for the epoxidation of electron-deficient olefins is their treatment with alkaline hydroperoxides, well-known as the Weitz–Scheffer epoxidation [99, 100]. It is well-established that epoxide formation occurs via a two-step mechanism (Scheme 2.16), which was first proposed by Bunton and Minkoff [101]. Conjugate addition of a peroxyanion generated from hydrogen peroxide or alkyl hydroperoxides under basic conditions at the β-position of an α,β-unsaturated ketone **16** affords β-peroxyenolate **17**. Subsequent intramolecular nucleophilic displacement at the proximal oxygen atom breaks the weak O–O single bond with concomitant ejection of a leaving group—hydroxide or alkoxide—to furnish epoxide **18**.

The non-stereospecificity of these reactions is a strong indication for the formation of an intermediate **17** in the course of the reaction. Unlike for epoxidations with peracids, the alkene geometry is not necessarily retained in the epoxide. For example, the epoxidation of both *trans*- and *cis*-configured α,β-unsaturated ketones with alkaline hydrogen peroxide in methanol furnishes predominantly the corresponding *trans*-epoxide. The non-stereospecificity further implies that hydroperoxide addition is fast, reversible, and causes isomerisation by rotation about the vinylic bond in the peroxyenolate **17** [102].

Electron-deficient olefins exhibit low reactivity toward common electrophilic oxidants such as *meta*-chloroperbenzoic acid (MCPBA). However, under Weitz–Scheffer conditions they are selectively epoxidized in the presence of electron-rich olefins.

Extensive research has been devoted to the development of asymmetric versions of the Weitz–Scheffer epoxidation owing to the synthetic importance of enantiomerically enriched α,β-epoxyketones as chiral building blocks and products of pharmaceutical and biological interest (cf. Fig. 2.2, *vide infra*).

In the mid-1970s, Wynberg et al. [103–105] disclosed the first asymmetric Weitz–Scheffer epoxidation of *trans*-chalcone and derivatives attaining enantioselectivities of up to 77:23 *er* (Scheme 2.17). Cinchona alkaloid-derived quaternary ammonium

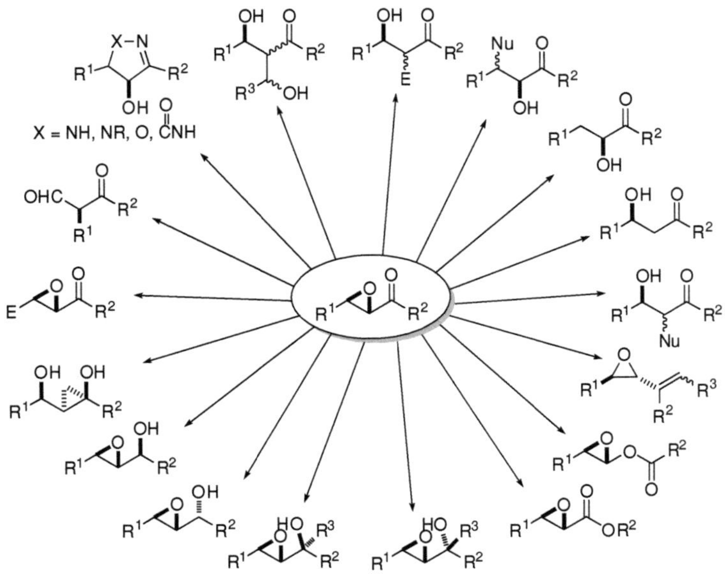

Fig. 2.2 Selected one-step transformations of α,*β*-epoxy ketones. (*Nu* nucleophile, *E* electrophile)

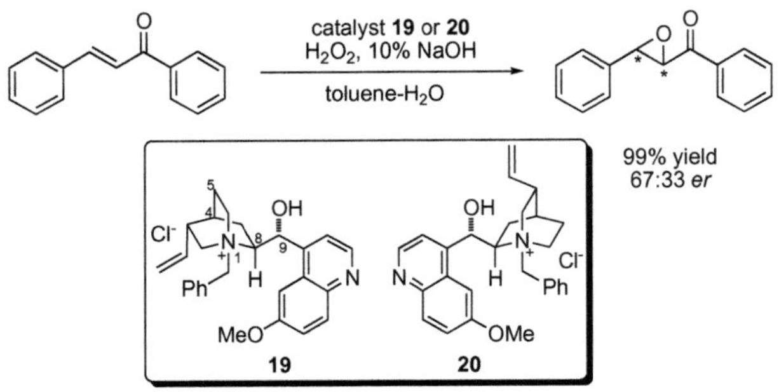

Scheme 2.17 First asymmetric Weitz–Scheffer epoxidation by Wynberg using *N*-benzylquininium or –quinidinium chloride **19** or **20** as chiral phase-transfer catalyst

salts were applied as chiral phase-transfer catalysts with either hydrogen peroxide, *tert*-butyl hydroperoxide, or sodium hypochlorite as the oxidant under biphasic reaction conditions. Pseudoenantiomeric[2] [106] catalysts **19** and **20** derived from quinine and quinidine, respectively, produced opposite epoxide enantiomers.

---

[2] Pseudoenantiomeric catalysts are diastereomers that afford antipodal products in catalytic asymmetric transformations. The configuration of catalysts **19** and **20** is inverted at C-8 and C-9 whereas it is identical at *N*-1, C-4, and C-5. Likewise, the starting materials, the naturally abundant cinchona alkaloids quinine and quinidine, are pseudoenantiomers themselves.

The phase-transfer catalyzed nucleophilic asymmetric epoxidation for the first time provided access to enantiomerically enriched chalcone- and naphthoquinone-derived epoxides, and moreover, this approach represents one of the first methods in general for the synthesis of optically active epoxides.

2.2.3 *Synthetic Versatility of α,β-Epoxy Ketones*

Enantiomerically enriched α,β-epoxy ketones are of particular value due to their dense functionalization [107]. Such optically active epoxides may be transformed into numerous products (i.a. pharmaceuticals, agricultural chemicals, and fragrances), where it is distinctly advantageous to use chiral starting materials of high optical purity. Both the epoxide and the adjacent ketone group constitute the starting point for manifold transformations as depicted in Fig. 2.2 [107–110]. Since epoxide can be reliably predicted.

2.2.4 *State of the Art: Catalytic Asymmetric Epoxidation of α,β-Unsaturated Ketones*

Since the pioneering report of Wynberg in 1976, the catalytic asymmetric epoxidation of α,β-unsaturated ketones has been the subject of numerous investigations, and a number of useful methodologies have been developed [96–98, 111]. Moreover, the scope of asymmetric Weitz–Scheffer-type epoxidations was extended to various electron-deficient olefins including nitro olefins [112] and α,β-unsaturated sulfones [113]. In the following section, the state of the art in catalytic asymmetric epoxidation of α,β-unsaturated ketones will be outlined.

2.2.4.1 Asymmetric Phase-Transfer Catalysis

Asymmetric phase-transfer catalysis (PTC) is a powerful strategy applicable to a wide range of transformations that proceed through anionic intermediates.

The most commonly used phase-transfer catalysts for the catalytic asymmetric epoxidation of α,β-unsaturated ketones are alkylated Cinchona alkaloids. Based on the seminal studies of Wynberg et al. (cf. Sect. 2.2.2), contributions from several research groups during the late 1990s led to dramatic improvements in terms of scope and level of enantioselection of the enone epoxidation. The groups of Lygo [114, 115] and Corey [116] achieved significant improvements by structural modification of the original Wynberg-type phase-transfer catalysts **19** and **20**

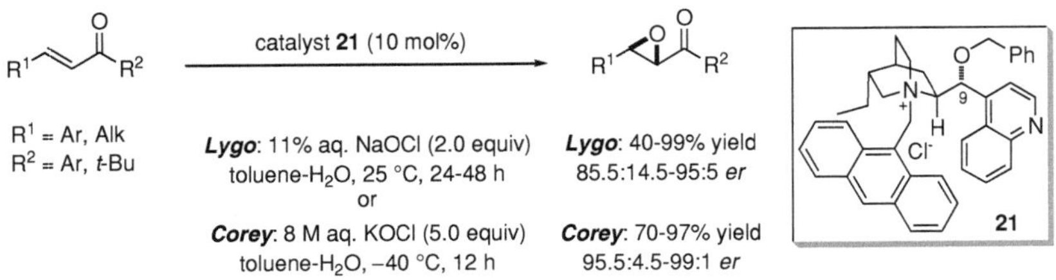

Scheme 2.18 Asymmetric PTC: enone epoxidation according to Lygo and Corey with *N*-9-anthracenylmethyl-*O*-benzyl-modified dihydroquinine-derived phase-transfer catalysts **21**

(Scheme 2.17) [114–117]. In particular, the introduction of a large 9-anthracenylmethyl group in place of the benzyl group at the quinuclidine nitrogen and a benzyl group on the secondary alcohol at C-9 in phase-transfer catalyst **21** was found to have a profound effect on the enantioselectivity of the epoxidation reaction (up to 99:1 *er*) (Scheme 2.18) [114]. Aqueous solutions of hypochlorites emerged as the oxidizing agents of choice giving superior results when compared to hydrogen peroxide, whenever *O*-alkylated Cinchona alkaloid-derived phase-transfer catalysts such as **21** were employed. Corey found that the use of freshly prepared 65% potassium hypochlorite at lower temperature instead of sodium hypochlorite led to further improvement in enantioselectivities [116].

Whereas the epoxidation reaction turned out to be general for a broad range of *trans*-chalcone derivatives, α,β-enones bearing alkyl substituents at the β-position furnished epoxides with reduced enantioselectivity. Moreover, substrates with enolizable alkyl groups adjacent to the ketone (substituent R^2) gave only low conversions under the standard reaction conditions (<10%), and the starting material was recovered in high yield. Competing enolization is the most likely explanation for this observation [118].

Recently, C_2-symmetric axially chiral quaternary *N*-spiroammonium salts such as catalysts **22a** or **b** were introduced by the Maruoka group as powerful chiral phase-transfer catalysts [119]. By means of these catalysts, the high enantioselectivities achieved for chalcone-type substrates were also retained for α,β-enones bearing alkyl substituents at the β-position, and moreover, for β-benzylidene-α-indanone and its tetralone analogue (Scheme 2.19).

2.2.4.2 Polyamino Acid-Mediated Epoxidation (Juliá–Colonna Epoxidation)

The polyamino acid-catalyzed asymmetric epoxidation of α,β-unsaturated ketones was first reported by Juliá and Colonna and co-workers in the early 1980s [120, 121]. The original reaction conditions were triphasic, consisting of the insoluble polyamino acid catalyst [poly-L-alanine (PLA) or poly-L-leucine (PLL)], an aqueous solution of sodium hydroxide, aqueous hydrogen peroxide as the oxidant, and a solution of chalcone in an organic solvent (Scheme 2.20). A range of

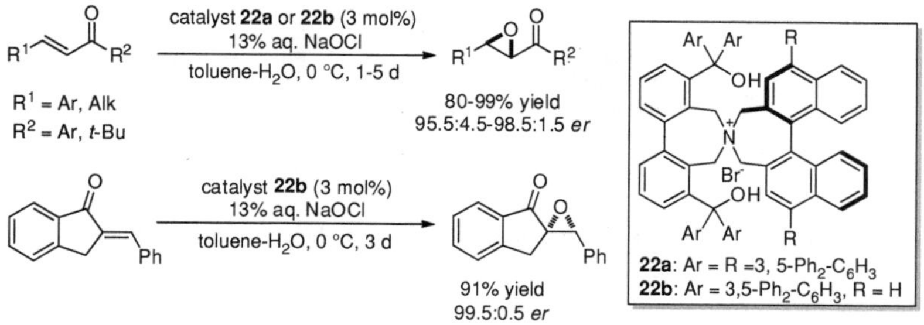

Scheme 2.19 Catalytic asymmetric epoxidation of α,β-unsaturated ketones (including β-benzylidene-α-indanone) with PTCs **22a** or **b** by Maruoka et al

Scheme 2.20 Scope of the Juliá–Colonna epoxidation under triphasic reaction conditions

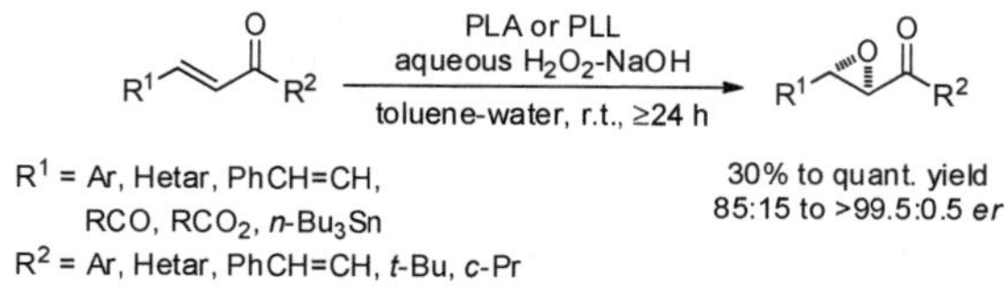

α,β-enones, in particular chalcone and simple analogues, were epoxidized with high enantioselectivities (generally $\geq$95:5 *er*). It was found that the level of enantiocontrol of the epoxidation was dependent on the chain length of the polyamino acid and increased as the average chain length increased from 10 to 30 residues. Somewhat surprisingly, poly-L-valine emerged to be completely ineffective as epoxidation catalyst [122].

Despite the excellent enantioselectivity, a number of problems limited the applicability of this methodology, namely the long reaction times even for relatively reactive substrates such as chalcone and the narrow substrate scope. In particular, enolizable substrates such as benzylidene acetone are found to be epoxidized very slowly or not at all.

Extensive studies by Roberts and co-workers resulted in significant improvements to overcome these limitations. The modified reaction conditions comprised the use of a non-aqueous solvent, an organic base (e. g. 1,8-diazabicyclo[5.4.0]undec-7-ene (DBU)), water-free oxidants such as urea-hydrogen peroxide (UHP) or percarbonate, along with **immobilized** poly-L-leucine as catalyst [123–127]. These non-aqueous, biphasic reaction conditions allowed for greatly enhanced reaction rates (from ca. 24 h to 30 min in the case of chalcone) and lower catalyst loadings (5 instead of 20 mol%). Moreover, substantial expansion of the range of α,β-enones was achieved and previously unreactive enolizable substrates such as benzylidene acetone as well as those bearing alkyl groups at the β-position readily underwent the epoxidation reaction (Scheme 2.21) [125].

The mechanism of the polyleucine-catalyzed epoxidation is still under investigation [102, 128–131]. Kinetic studies indicate that the reaction proceeds via the reversible addition of polyleucine-bound hydroperoxide to the enone [132].

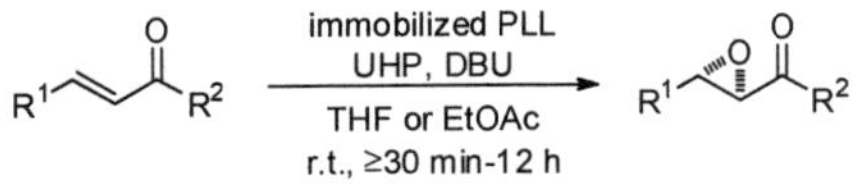

R^1	R^2	yield [%], *er*
Ph	Ph	85, 97.5:2.5
Cy	Ph	91, 94.5:5.5
Ph	Me	70, 90:10
Ph	*i*-Pr	>99, 98.5:1.5
Ph	*t*-Bu	76, 97:3

Scheme 2.21 Expanded substrate scope under Robert's biphasic reaction conditions [125]

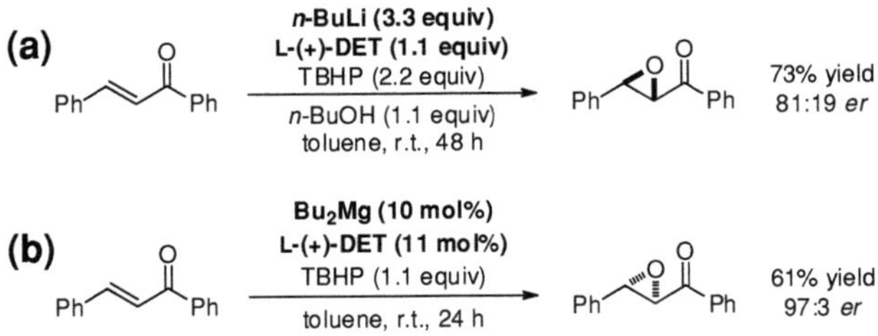

Scheme 2.22 Jackson's asymmetric chalcone epoxidation using either the **a** stoichiometric lithium *tert*-butyl peroxide (+)-DET or the **b** catalytic magnesium *tert*-butyl peroxide (+)-DET system

2.2.4.3 Metal Peroxides in Combination with Chiral Ligands

Several methods for the catalytic asymmetric epoxidation of α,β-unsaturated ketones rely on the use of a chiral ligand coordinated to the metal atom of a metal peroxide, which then effects a Weitz–Scheffer epoxidation. A number of metals and chiral ligands have been used for this purpose and the most successful approaches are discussed below.

Jackson et al. had previously established the use of lithium *tert*-butylperoxide generated *in situ* from *n*-butyl lithium and *tert*-butyl hydroperoxide (TBHP) together with stoichiometric amounts of diethyl tartrate (DET) for the epoxidation of *trans*-chalcone resulting in moderate yield and enantioselectivity (81:19 *er*) of the chalcone epoxide (Scheme 2.22a). However, upon replacement of *n*-butyllithium by dibutylmagnesium, catalytic amounts of the chiral ligand DET and the base were sufficient to provide high levels of asymmetric induction on a range of chalcone-type substrates (Scheme 2.22b) [133].

Intriguingly, the catalytic magnesium system provided epoxides antipodal to those obtained using stoichiometric lithium peroxides, despite employing the identical enantiomer of the chiral ligand (+)-DET. The Jackson group further optimized the magnesium-based catalyst system and identified conditions which not only tolerated chalcone-type substrates but were applicable to a variety of enolizable aliphatic enones without loss of catalyst efficiency (Scheme 2.23) [134, 135].

The addition of 4 Å molecular sieves was crucial to retain as high conversions for aliphatic enones as for chalcone-type substrates. Moreover, the use of tartrate esters derived from secondary or tertiary alcohols instead of DET led to improved

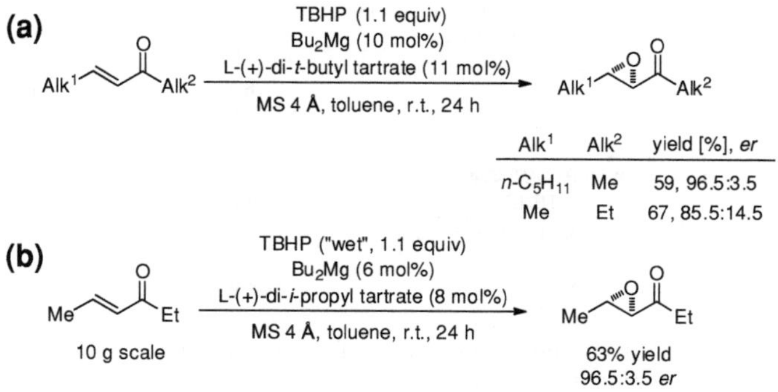

Scheme 2.23 Catalytic asymmetric epoxidation of aliphatic enones with **a** magnesium di-*tert*-butyl tartrate under anhydrous and with **b** magnesium diisopropyl tartrate under "wet" reaction conditions

enantioselectivities with the optimal ligand being di-*tert*-butyl tartrate (Scheme 2.23a) [134]. The use of "wet" *tert*-butyl hydroperoxide in combination with 4 Å molecular sieves at lower catalyst loadings of both dibutylmagnesium and diisopropyl tartrate further enhanced the operational simplicity and practicability of the process which was illustrated by a large scale synthesis of 4,5-epoxy-3-hexanone in good yields along with improved enantioselectivity of 96.5:3.5 *er* (Scheme 2.23b) compared to 85.5:14.5 *er* previously attained (Scheme 2.23a) [135]. Jackson's catalytic asymmetric epoxidation of aliphatic α,β-enones proceeds with the highest enantioselectivities obtained prior to this work for this substrate class.

Probably the most general method for the catalytic asymmetric epoxidation of α,β-unsaturated carbonyl compounds was developed by Shibasaki and co-workers and uses a combination of lanthanoid alkoxides and BINOL or its derivatives with alkyl hydroperoxides as the oxygen source [136, 137]. Two types of catalyst systems exist: (a) one of them is represented by the general structure LnM_3[(*R*)-BINOL] (Ln = lanthanoid, M = alkali metal) and comprises alkali metals whereas (b) the other one is alkali metal-free. In both cases oligomeric complexes between the lanthanoid metal and BINOL are formed.

Alkali metal-free catalyst systems gave superior results in the catalytic asymmetric epoxidation of α,β-unsaturated ketones. The choice of the optimal lanthanoid metal depended on the nature of the enone substrate. The lanthanum-(*R*)-3-hydroxymethyl-BINOL complex (La-(*R*)-**23a**) in combination with cumyl hydroperoxide (CHP) was the ideal choice for the epoxidation of aromatic chalcone-type substrates (Scheme 2.24a), whereas the catalytic activity for aliphatic α,β-unsaturated ketones could be improved by the use of ytterbium complexes along with TBHP as the oxidant (Scheme 2.24b) [138–140].

Remarkably, the alkali metal-free complex ytterbium-(*R*)-3-hydroxymethyl-BINOL (Yb-(*R*)-**23a**) converted ***cis***-enones to the corresponding ***cis***-epoxides in good yields and with high levels of asymmetric induction when using TBHP as the

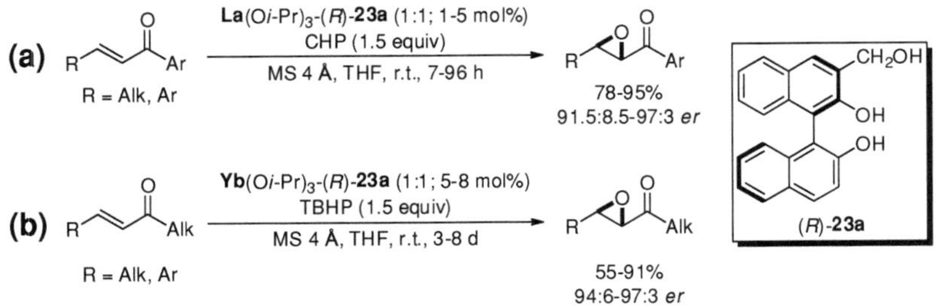

Scheme 2.24 Catalytic asymmetric epoxidation of α,β-unsaturated ketones with lanthanoid-BINOL complexes according to Shibasaki et al. [138]

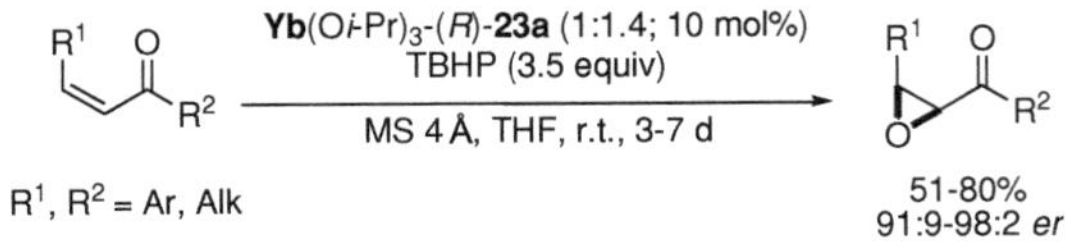

Scheme 2.25 Catalytic asymmetric formation of *cis*-epoxides from *cis*-enones by Shibasaki et al. [141]

oxygen source (Scheme 2.25) [141]. Only for aromatic *cis*-enones (R_2 = Ar) minor amounts (up to 32%) of the corresponding *trans*-epoxy ketones were formed at the same time. This report is remarkable since it constitutes one of the rare examples of the selective formation of ***cis***-epoxides through nucleophilic epoxidation of ***cis***-olefins [118].

Later on, Inanaga et al. [140, 142–144] found that the use of additives such as triphenylphosphine oxide or *tris*(4-flourophenyl)phosphine oxide led to enhanced reaction rates which allowed for significantly shortened reaction times (Scheme 2.26). The optimized conditions applied best to chalcone-type substrates (up to >99.5:0.5 *er*) whereas simple aliphatic enones lagged behind in terms of the level of enantiocontrol (93.5:6.5 *er* with $R^1 = PhCH_2CH_2$, R^2 = Me).

According to the authors, the increased catalyst activity resulted from the deoligomerization of the lanthanoid-BINOL complexes through coordinative saturation of the lanthanoid metal with additional ligands [144].

Shibasaki's lanthanoid-BINOL and related catalyst systems proved to be very general [136], and found widespread application in the catalytic asymmetric epoxidation of α,β-unsaturated esters [145], amides [146, 147], and several ester surrogates such as *N*-acylpyrroles[148–150] or imidazoles [151, 152].

2.2.5 Recent Advances: The Organocatalytic Approach

This chapter focuses on catalytic asymmetric epoxidations of α,β-unsaturated carbonyl compounds mediated by small molecule organocatalysts such as (*S*)-proline and its derivatives. Recent advances within the realm of phase-transfer

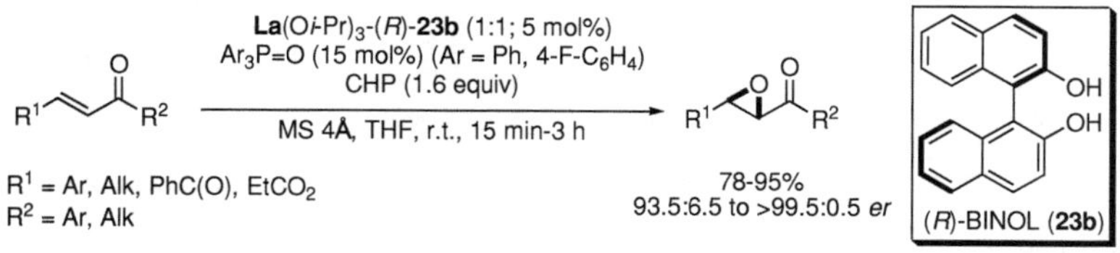

Scheme 2.26 Catalytic asymmetric epoxidation of α,β-unsaturated ketones with a lanthanum-BINOL complex and triarylphosphine oxide as additive according to Inanaga et al. [143]

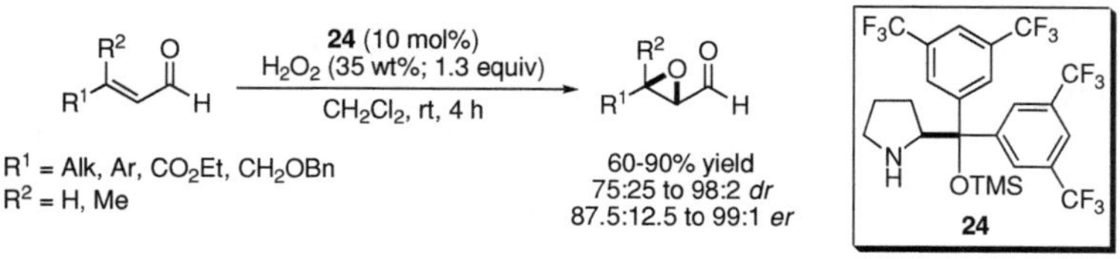

Scheme 2.27 First catalytic asymmetric epoxidation of α,β-unsaturated aldehydes by Jørgensen et al. [153]

catalysis, the Julia-Colonna, and the Shi epoxidation, which can be considered borderline organocatalytic, were covered in the previous chapter (Sect. 2.2.4).

2.2.5.1 Organocatalytic Asymmetric Epoxidation of α,β-Unsaturated Aldehydes

The development of new organocatalytic methods—in particular of those based on iminium ion catalysis—for the asymmetric epoxidation of α,β-unsaturated carbonyl compounds has advanced this research field decisively by extending the substrate scope beyond ester, amide, and chalcone derivatives.

In 2005, Jørgensen and co-workers presented the first direct catalytic asymmetric epoxidation of α,β-unsaturated aldehydes, a reaction that had remained a challenge to chemists [153]. Indeed direct approaches to enantiomerically enriched α,β-epoxy aldehydes were hitherto not available. In the presence of catalytic amounts of the chiral secondary amine α,α-[3,5-bis(trifluoromethyl)phenyl]prolinol trimethylsilyl ether (**24**) and with aqueous hydrogen peroxide as the oxidant, a range of α,β-epoxy aldehydes could be obtained in high yields along with high enantioselectivities from both aromatic and aliphatic enals in a single step (Scheme 2.27).

This seminal work also revealed the compatibility of amine catalysts with various oxidants and dispelled scepticism, which predicted severe difficulties from catalyst degradation via competing catalyst *N*-oxidation.

Alternative protocols for the catalytic asymmetric epoxidation of α,β-unsaturated aldehydes based on iminium ion activation were developed by the MacMillan and List groups.

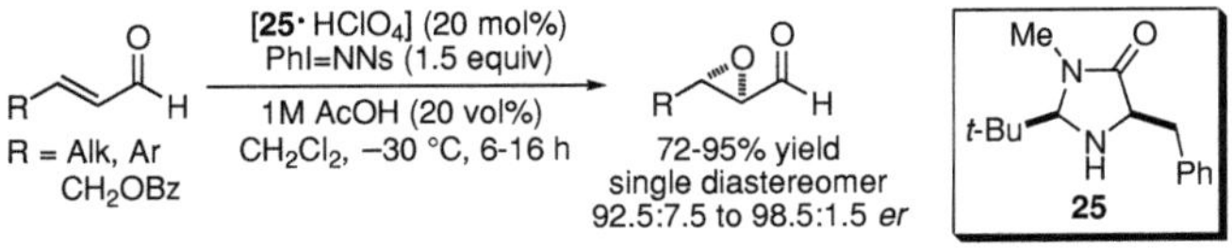

Scheme 2.28 Catalytic asymmetric epoxidation of α,β-unsaturated aldehydes by MacMillan et al.

Scheme 2.29 Catalytic asymmetric epoxidation of α,β-unsaturated aldehydes by List and Wang

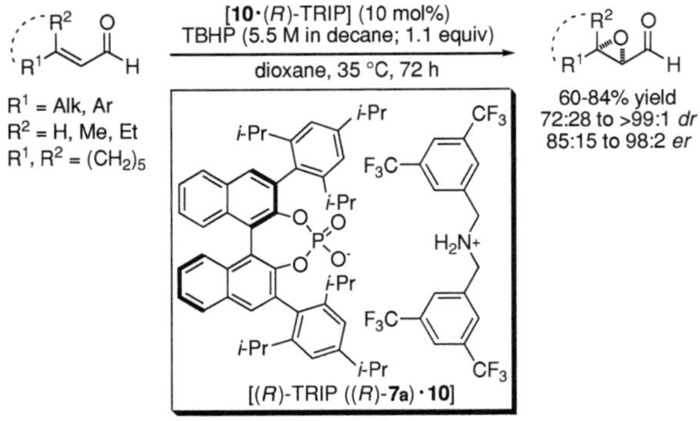

MacMillan and co-workers identified chiral imidazolidinone **25** as its perchlorate salt to effectively mediate the epoxidation reaction with hypervalent iodine reagents as oxidants (which are rarely used in nucleophilic epoxidation reactions) (Scheme 2.28) [154]. A mechanistic study with ^{15}N labelled imidazolidinone catalyst **25** revealed that iodosobenzene indeed brought about slow catalyst degradation by oxidation. Thus, key to achieving high levels of both reaction efficiency and enantioselectivity was the use of [(*N*-Nosylimino)iodo]benzene (PhI = NNs) as an iodosobenzene surrogate which, in the presence of acetic acid, provided a slow release of iodosobenzene over time ('internal syringe pump').

List and Wang successfully applied the ACDC concept to the catalytic asymmetric epoxidation of α,β-unsaturated aldehydes [62]. Among all combinations tested, catalyst salt [**10** · (*R*)-TRIP] comprising an achiral dibenzylamine derivative **10** together with the BINOL phosphate TRIP (**7a**) as the chiral counteranion and only source of chirality, turned out to be the catalyst of choice furnishing the desired α,β-epoxy aldehydes in high yields and enantioselectivities (Scheme 2.29).

Remarkably, symmetrically β,β-disubstituted enals such as senecialdehyde (3-methylbutenal) gave the desired epoxides with excellent enantioselectivities up to 97:3 *er* in the presence of the ACDC catalyst [**10** · (*R*)-TRIP] (Scheme 2.30a). This represents a great advancement compared to the results obtained with the catalyst system described by Jørgensen and co-workers, where this substrate class was converted into the corresponding epoxides with significantly lower enantioselectivities (87.5:12.5 *er*; cf. Scheme 2.30a). Moreover, this observation raises interesting mechanistic questions. Since the initial conjugate addition product is achiral, the stereogenic center is created during the epoxide ring closure.

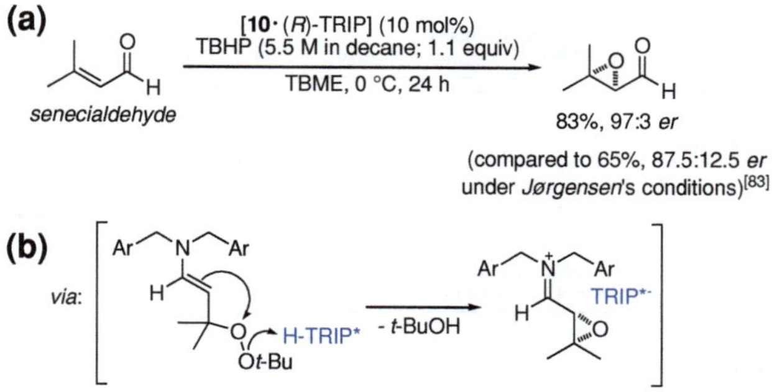

Scheme 2.30 **a** Catalytic asymmetric epoxidation of senecialdehyde according to List and Wang. **b** Proposed reaction pathway

Consequently, the chiral BINOL phosphate (*R*)-TRIP (**7a**) must be involved in this C–O bond-forming event and presumably assists the enantioselective cyclization of the achiral peroxyenamine intermediate (Scheme 2.30b).

2.2.5.2 Organocatalytic Asymmetric Epoxidation of α,β-Unsaturated Ketones

Unfortunately, secondary amine catalysts such as trimethylsilyl diarylprolinol ether **24** and related compounds which efficiently mediated the catalytic asymmetric epoxidation of α,β-unsaturated aldehydes turned out to be less active to completely inactive for the epoxidation of α,β-unsaturated ketones.

Lattanzi and later Zhao and co-workers identified unprotected α,α-diarylprolinol derivatives such as **26a** or **b** and other structurally diverse β-amino alcohols to provide an effective platform for the epoxidation of an array of electron deficient olefins including α,β-unsaturated ketones with *tert*-butylhydroperoxide as the oxidant (Scheme 2.31a and b) [154–164].

Within these reports it was proposed by the authors that α,α-diarylprolinol catalysts may be operating through hydrogen bonding interactions. TBHP may be activated via general base catalysis by the prolinol derivative (Fig. 2.3).

The formation of iminium ions between α,β-unsaturated ketones and catalysts **26a-b** as an alternative activation pathway was ruled out by the authors given the known unreactive nature of enone carbonyls and the detrimental effect of an acid co-catalyst on the outcome of those reactions.[3] Moreover, a strong solvent effect could be detected further supporting the involvement of non-covalent hydrogen bonding interactions. However, whereas this method is well developed for chalcone and derivatives, it was scarcely applied to simple aliphatic enones, and when so, giving inferior results and enantioselectivities of 87.5:12.5 *er* at the most. In addition, the present epoxidation protocol requires the use of TBHP as the

[3] Although smaller in number, reports exist which denote the activation of enone systems as iminium ion in the presence of secondary amine (salt) catalysts.

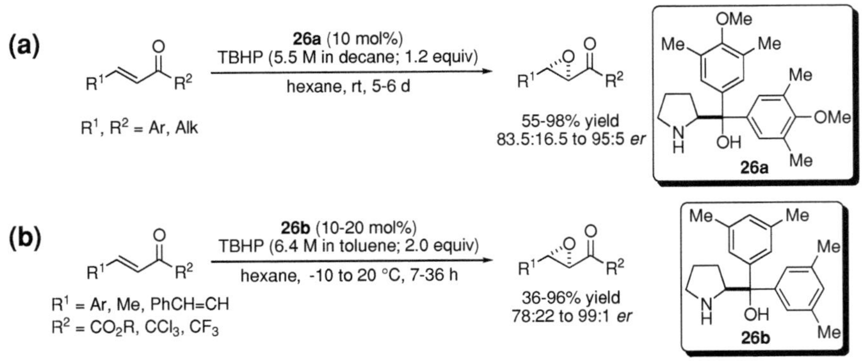

Scheme 2.31 Catalytic asymmetric epoxidation of electron deficient olefins including α,β-unsaturated ketones by **a** Lattanzi [157], and **b** Zhao et al. [164]

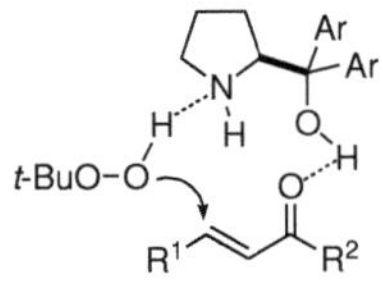

Fig. 2.3 Proposed mode of activation of electron deficient olefins by diarylprolinol derivatives via hydrogen bonding interaction. (R^1 = Ar, Alk; R^2 = Ar, Alk, CO_2R, CCl_3, CF_3)

oxidant. Aqueous hydrogen peroxide which is an inexpensive and environmentally benign alternative cannot be employed [160].

Furthermore, chiral guanidines have been explored as base catalysts in the nucleophilic epoxidation of different types of α,β-enones. Promising, but not yet fully satisfactory levels of stereocontrol have been achieved to date by means of this approach [165–168]. In contrast, chiral guanidinium salts as well as bifunctional guanidinium-urea and guanidinium-hydroxyl organocatalysts delivered α,β-epoxy ketones mostly derived from chalcone-type substrates in high yields and with good to high enantioselectivities with either hydrogen peroxide or TBHP as the oxidant [169–171].

A catalytic asymmetric epoxidation of α,β-unsaturated ketones truly relying on iminium ion activation of the enone carbonyls and a method giving highly enantioenriched α,β-epoxy ketones from simple aliphatic enones had not been put into practice prior to this work.

2.2.6 Asymmetric Epoxidation of Cyclic α,β-Unsaturated Ketones

Cyclic enones constitute a special class of α,β-unsaturated ketones. Up to a ring size with n ≤ 4, they feature a (*Z*)-configured double bond as well as a rigid *s-trans* conformation of the enone moiety in their thermodynamically stable form (Scheme 2.32).

Scheme 2.32 Asymmetric epoxidation of simple cyclic enones

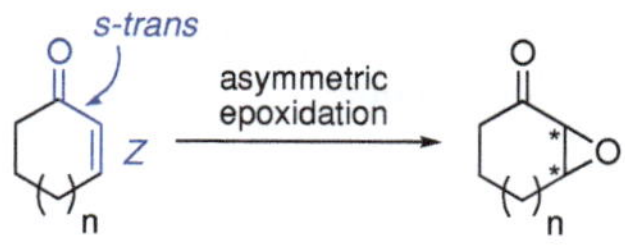

Strained (*E*)-2-cycloheptenone (n = 2), -octenone (n = 3), and—nonenone (n = 4) can only be generated via photoisomerization [172–175]. As the ring size increases, the amount of strain decreases and (*E*)-configured cycloalkenones ($n \geq 5$) are the stable isomers.

Cyclic systems ($n \leq 4$) are restricted to the *s-trans* conformation, which avoids problems arising from *s-cis*/*s-trans* interconversion. By contrast, in open-chain systems the equilibrium between the *s-cis* and *s-trans* conformer has to be taken into account. Based on this, one might suspect that the minimal conformational flexibility inherent in cyclic enones would facilitate their asymmetric epoxidation. Yet, the asymmetric epoxidation of *s-trans*-fixed cyclic enones under Weitz–Scheffer conditions has been little studied to date. Indeed, some asymmetric Weitz–Scheffer-epoxidations rigorously require *s-cis* conformation of acceptors, and are thus not applicable to cyclic enones [176, 177]. Successful implementations of cyclic enones in asymmetric nucleophilic epoxidations had been limited to particular substrate classes such as benzoquinone [178, 179], naphthoquinones [180–184], the respective monoketals [185–188], isoflavones [189], and peri-naphthenone prior to this work [118].

Until 2008, to the best of our knowledge, there was not a single general method available which would allow for the highly enantioselective (catalytic) epoxidation of simple cyclic enones [190]. The literature precedence for the (catalytic) asymmetric epoxidation of 2-cyclohexenone is compiled in Table 2.1.

The highest enantioselectivity of 81.5:18.5 *er* was attained by Baba and co-workers in 1986 by using a quinine-derived bis(phase-transfer catalyst) **27** together with 9-methylflourenyl hydroperoxide (**28**) as the oxidant (Table 2.1) [190, 191]. All other attempts gave inferior results [120, 156, 192–196]. In many cases, 2,3-epoxycyclohexanone was obtained in almost racemic form although the respective catalyst systems efficiently mediated the epoxidation of other types of α,β-unsaturated ketones such as chalcone and derivatives with good to high enantioselectivities. The use of a TADDOL-derived stoichiometric hydroperoxide **29** has recently been studied by Seebach et al. [194]. Whereas an enantiomeric ratio of 91:9 *er* was achieved in the epoxidation of **3-methyl**-2-cyclohexenone, 2,3-epoxycyclohexanone derived from unsubstituted 2-cyclohexenone was obtained in only 55:45 *er*. In contrast, the phase-transfer conditions applied by Wynberg and Marsman were suitable for the asymmetric epoxidation of 2-cyclohexenone, but not so for 3-methyl-2-cyclohexenone since this olefin may be too sterically crowded to be epoxidized with *tert*-butylhydroperoxide [192].

The development of a highly enantioselective and **general** catalytic epoxidation of cyclic enones was desirable in light of the synthetic value of enantiomerically

Table 2.1 Comprehensive data for the (catalytic) asymmetric epoxidation of 2-cyclohexenone

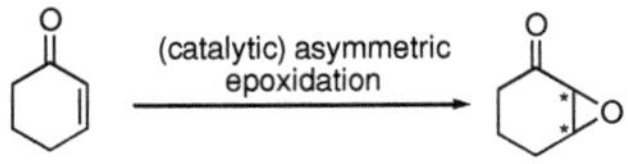

	chiral catalyst	oxidant	*er*
Baba et al.	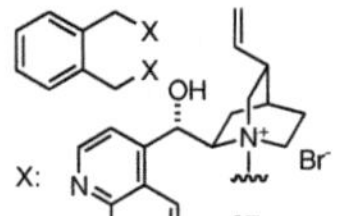 27	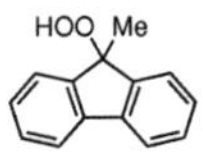 28	81.5:18.5
Strukul et al.	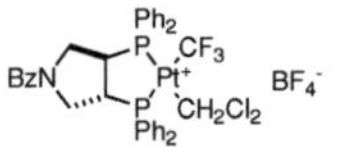	H_2O_2	74.5:25.5
Wynberg and *Marsman*	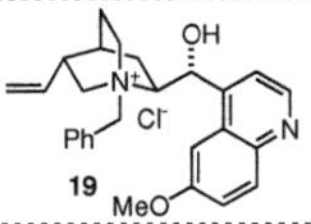 19	*t*-BuOOH	60:40
Lattanzi	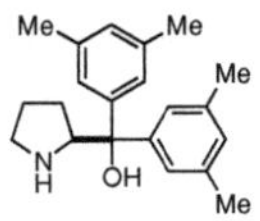	*t*-BuOOH	53:47
Juliá, Colonna et al.	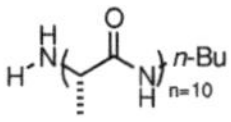	H_2O_2	<52.5:47.5
Armstrong et al.	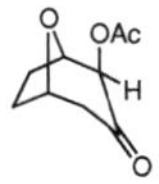	Oxone®	51.5:48.5
Laschat et al.	-----	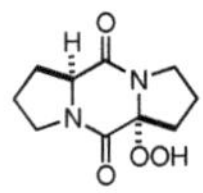	59.5:40.5
Seebach and *Aoki*	-----	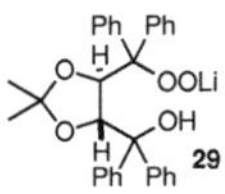 29	55:45

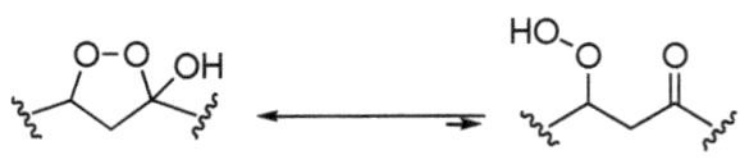

Scheme 2.33 Equilibrium between cyclic peroxyhemiketals and *β*-hydroperoxy carbonyl compounds

pure cyclic *α*,*β*-epoxy ketones [197–202], and even more so due to the absence of such a method prior to this work.[4]

2.3 Synthesis and Relevance of 3-Hydroxy-1,2-dioxolanes and 1,2-Dioxolanes

3-Hydroxy-1,2-dioxolanes (Scheme 2.33) and 1,2-dioxolanes are organic peroxides. Both contain an O–O bond embedded in a five-membered ring. 3-Hydroxy-1,2-dioxolanes are cyclic peroxyhemiketals of *β*-hydroperoxy carbonyl

[4] The asymmetric epoxidation of higher homologues has indeed not been studied at all and for the asymmetric epoxidation of cyclopentenone only one literature report is known to date [190].

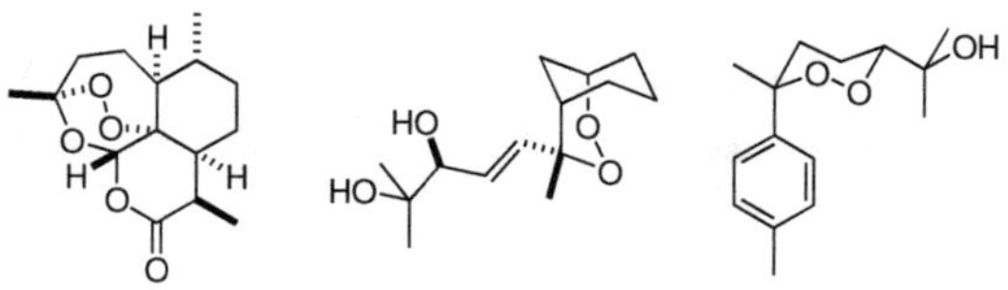

Fig. 2.4 Peroxidic natural products with antimalarial activity

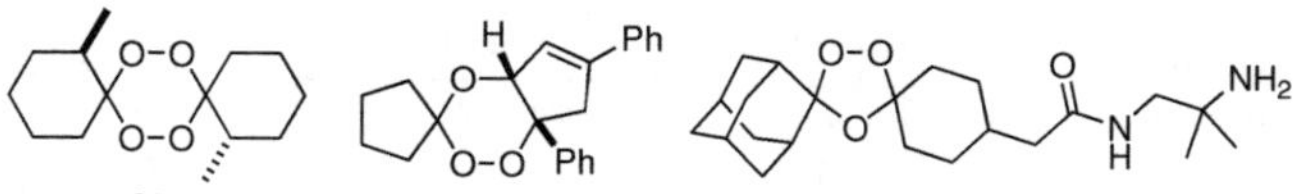

Fig. 2.5 Potent antimalarial drug candidates: tetroxane **33**, trioxane **34**, and trioxolane **35**

compounds with the equilibrium distribution being dependent on the precise structure and substitution pattern of those compounds. However, monocyclic peroxyhemiketals are of remarkable stability and the content of the ring-opened form in the equilibrium mixture is low as detected by ^{13}C NMR (no carbonyl resonances) (*vide infra*).

Peroxide-containing natural products, including several examples of 3-hydroxy-1,2-dioxolanes and related ring systems, occur widely in nature and often possess desirable pharmacological properties as pointed out in the following section [203].

2.3.1 Peroxidic Natural Products

The most prominent example among all peroxide natural products is the active antimalarial agent sesquiterpene 1,2,4-trioxane artemisinin (qinghaosu, **30**) isolated from the common shrub *Artemisia annua* (sweet wormwood) (Fig. 2.4) [204]. Similar antimalarial activities were found for the naturally occurring peroxides yingzhaosu A (**31**) and yingzhaosu C (**32**) containing a 1,2-dioxane core structure [205].

However, the most aggressive parasite, *Plasmodium falciparum*, is showing first resistance effects to artemisinin and its semisynthetic derivatives [206, 207]. Consequently, there is an urgent need for the development of new effective antimalarial remedies. Synthetic cyclic peroxides command increasing attention. They offer a structurally simpler, synthetically readily accessible alternative to artemisinin (**30**) and its analogues. Therefore, much effort is devoted to the development of new strategies for the synthesis of novel cyclic peroxidic compounds of diverse structures to identify a promising new lead in search of efficient antimalarials [208].

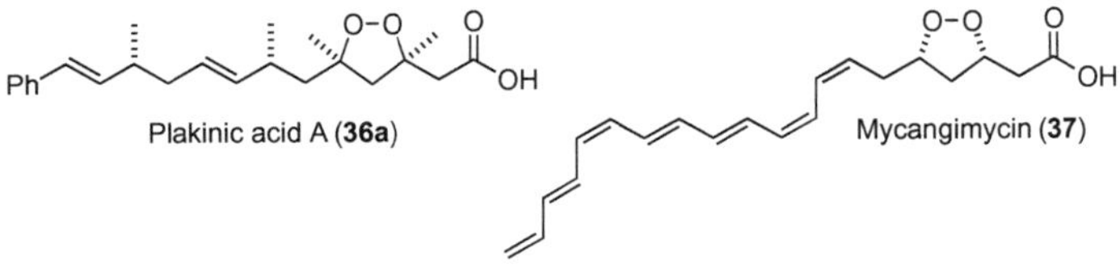

Fig. 2.6 Naturally occurring plakinic acid A (**36a**) and mycangimycin (**37**)

Scheme 2.34 Aerobic oxidation of cyclopropanols **38** to cyclic peroxyhemiketals **41a** [221]

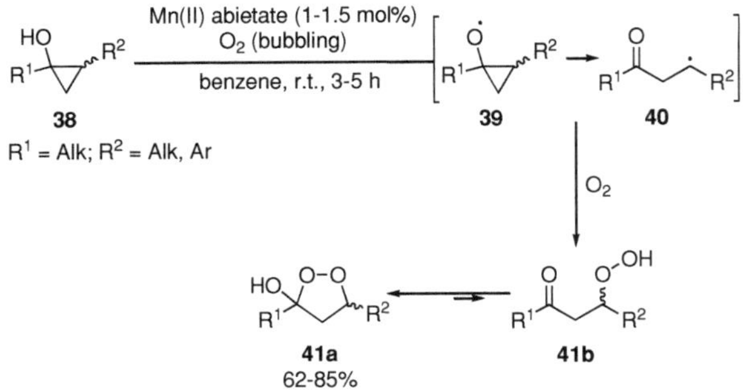

Strikingly, tetroxane **33**, trioxane **34**, or trioxolane **35** show artemisinin-like antimalarial activity although their carbocyclic skeletons bear no resemblance to that of artemisinin (Fig. 2.5) [208, 209].

Accordingly, it is not necessary to simulate the artemisinin framework to secure superior antimalarial potency. The indispensable feature for antimalarial efficacy appears to be the peroxide unit [210].

Naturally occurring five-membered cyclic peroxides containing a 1,2-dioxolane unit such as plakinic acid A **36a** (a member of the plakinic acid natural products family characterized by a common 1,2-dioxolane-3-acetic acid "head" and different aliphatic "tails") [211], or mycangimycin **37**, but also synthetic 1,2-dioxolane-based analogues display anticancer, antifungal, and antiplasmodial activity (Fig. 2.6) [212–214].

2.3.2 General Methods for the Synthesis of 3-Hydroxy-1,2-dioxolanes

The methods to introduce a peroxide functionality into organic compounds are limited. In the synthesis of most peroxides, irrespective of their structural complexity, the peroxide moiety is pre-formed and introduced as either (a) molecular oxygen (through a reaction with singlet oxygen or radical trapping with triplet oxygen), (b) by nucleophilic addition of hydrogen or alkyl peroxides, or (c) by reaction with ozone [215].

Scheme 2.35 Competing reaction pathways under Weitz–Scheffer reaction conditions: epoxide formation versus protonation of *β*-peroxyenolate intermediate **17**. (Base = *i.a.* NaOH, LiOH, *t*-BuNH$_2$)

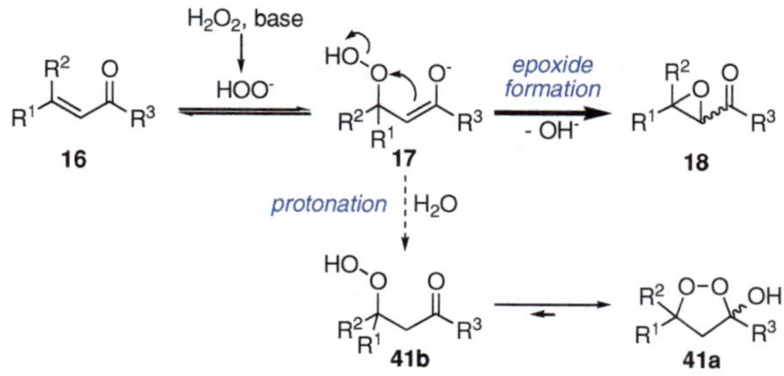

Syntheses of 3-hydroxy-1,2-dioxolanes (cf. Scheme 2.33) have been achieved by various methods [216, 217]. Currently available methodologies mainly rely on the reaction of hydrogen peroxide or singlet oxygen with *α*,*β*-unsaturated carbonyl compounds or exploit the aerobic oxidation of cyclopropanols. These methods will be briefly introduced in the following sections. A direct enantioselective approach to 3-hydroxy-1,2-dioxolanes from simple, readily available starting materials had not been described prior to this work.

2.3.2.1 Aerobic Oxidation of Cyclopropanol Derivatives

Cyclopropanols undergo ring-opening under metal-catalyzed aerobic oxidation; when subjected to single-electron oxidants Fe(acac)$_3$, Cu(acac)$_2$, VO(acac)$_2$, or Mn(II) abietate under an atmosphere of oxygen, they are converted to *β*-hydroperoxo ketones which exist in equilibrium predominantly as the cyclic peroxyhemiketals (*vide infra*, Scheme 2.34) [218–221]. Among the metal salts listed above, Mn(II) abietate and Mn(II) acetylacetonate did not only mediate the aerobic oxidation of bicyclic cyclopropanols but also of simple, readily available cyclopropanols of the general structure **38** [221]. The reaction proceeds via the successive formation of an alkoxy radical **39** and a *β*-carbonyl radical **40**. The oxidation of cyclopropanol derivatives provides expedient access to cyclic peroxyhemiketals **41a**; yet, this transformation holds little promise for the development of an asymmetric version since it proceeds through a radical pathway involving the ablation of preexisting stereocenters.

2.3.2.2 Nucleophilic Addition of Peroxides to *α*,*β*-Unsaturated Ketones

Nucleophilic conjugate addition of alkaline hydrogen peroxide to *α*,*β*-unsaturated ketones **16** generates *β*-peroxyenolate intermediates **17** which undergo ring closure to form *α*,*β*-epoxyketones **18** (cf. Scheme 2.16). Alternatively, intermediates **17** can be intercepted by protonation to afford cyclic peroxyhemiketals **41a** (Scheme 2.35).

Although *β*-peroxyenolates **17** show an overwhelming preference for epoxide formation, cyclic peroxyhemiketals of type **41a** have occasionally been observed as by-products of Weitz–Scheffer-type epoxidations of *α*,*β*-unsaturated ketones **16**.

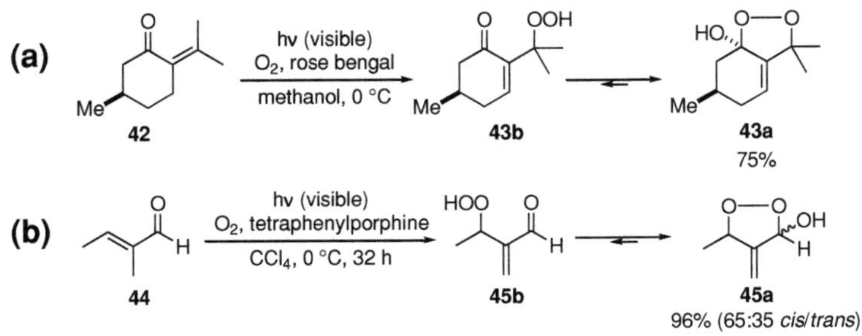

Scheme 2.36 Photooxygenation of α,β-unsaturated carbonyl compounds with singlet oxygen [228, 229]

Their formation was first noticed in 1950 by Nazarov and Akhrem who studied the epoxidation of mesityl oxide [222]. They were the first ones to separate, purify, and characterize a slightly higher boiling by-product, before in 1958, Payne unambiguously elucidated its structure as 3-hydroxy-3,5,5-trimethyl-1,2-dioxolane (**41a** with $R^1 = R^2 = R^3 =$ Me) [113].

3-Hydroxy-l,2-dioxolanes are formed in varying amounts along with the corresponding epoxides depending on the enone structure and the precise reaction conditions [222–226]. Although this method provides in most cases low yields (<20%) of 3-hydroxy-l,2-dioxolanes, it is highly valuable and still used on a preparative scale due to the limited number of operationally simple alternative approaches to this substrate class [227]. Extension of this approach to an asymmetric route to 3-hydroxy-l,2-dioxolanes seems feasible since it has previously been demonstrated in the context of asymmetric Weitz–Scheffer-type epoxidations, that it is possible to render the conjugate addition of hydrogen peroxide to α,β-unsaturated carbonyl compounds enantioselective in the presence of a chiral catalyst.

2.3.2.3 Singlet Photooxygenation

Akin to the reaction with alkaline hydrogen peroxide (*vide supra*), photooxygenation with singlet oxygen ($^1O_2$) provides direct access to 3-hydroxy-1,2-dioxolanes from α,β-unsaturated carbonyl compounds in a single step. As shown in Scheme 2.36a, pulegone (**42**) readily reacts with singlet oxygen to give ene adduct **43b**, which spontaneously cyclizes to the corresponding peroxyhemiketal **43a** [228]. Analogous results were obtained with α,β-unsaturated aldehydes: tiglic aldehyde (**44**) gave peroxyhemiacetal **45** in 96% yield (Scheme 2.36b) [229]. α,β-Unsaturated carbonyl compounds with fixed *s-cis* conformation are more rapidly oxidized by singlet oxygen than conformationally flexible substrates whereas those substrates which are constrained to the *s-trans* conformation did not participate at all in the reaction.

The oxidative thermolysis of cyclic α-azo hydroperoxides developed by Baumstark et al. [230] demonstrates an alternative method for the direct synthesis of 3-hydroxy-1,2-dioxolanes.

Dussault and co-workers have disclosed stereoselective routes to 3-alkoxy-substituted-1,2-dioxolanes. Methoxymethyl (MOM)-protected 3-hydroxy-1,2-dioxolanes were obtained via regio—and diastereoselective photooxygenations of chiral, racemic (*Z*)-allylstannanes by singlet oxygen [231]. Stereospecific cyclization of hydroperoxy acetals onto chiral, enantiomerically enriched oxetanes furnished optically active 3-methoxy-1,2-dioxolanes [232], a strategy which was later exploited in the context of the synthesis of plakinic acid A (**36a**) [211].

References

1. List B, Lerner RA, Barbas CF (2000) J Am Chem Soc 122:2395
2. Ahrendt KA, Borths CJ, MacMillan DWC (2000) J Am Chem Soc 122:4243
3. Pizzarello S, Weber AL (2004) Science 303:1151
4. Liebig Jv (1860) Liebigs Ann Chem 113:246
5. Bredig G, Fiske PS (1913) Biochem Z 46:7
6. Pracejus H (1960) Justus Liebigs Ann Chem 634:9
7. Pracejus H, Maetje H (1964) J Prakt Chem 24:195
8. Hajos ZG, Parrish DR (1974) J Org Chem 39:1615
9. Hajos ZG, Parrish DR (1971) DE 2102623
10. Eder U, Sauer G, Wiechert R (1971) Angew Chem Int Ed Engl 10:496
11. Eder U, Sauer G, Wiechert R (1971) DE 2014757
12. List B (2010) Angew Chem Int Ed 49:1730
13. Dalko PI (ed) (2007) Enantioselective organocatalysis, reactions and experimental procedures. Wiley-VCH, Weinheim
14. Berkessel A, Gröger H (2005) Asymmetric organocatalysis from biomimetic concepts to applications in asymmetric synthesis. Wiley-VCH, Weinheim
15. List B (ed) (2009) Asymmetric organocatalysis Top Curr Chem, vol 291. Springer, Berlin
16. List B (ed) (2007) Chem Rev (Special Issue: Organocatalysis) 107:5413
17. Dondoni A, Massi A (2008) Angew Chem Int Ed 47:4638
18. Seayad J, List B (2005) Org. Biomol Chem 3:719
19. Wurz RP (2007) Chem Rev 107:5570
20. Gaunt MJ, Johansson CCC (2007) Chem Rev 107:5596
21. Knoevenagel E (1896) Ber Dtsch Chem Ges 29:172
22. Knoevenagel E (1898) Ber Dtsch Chem Ges 31:2585
23. Knoevenagel E (1898) Ber Dtsch Chem Ges 31:738
24. Knoevenagel E (1898) Ber Dtsch Chem Ges 31:2596
25. Kuhn R, Hoffer M (1930) Ber Dtsch Chem Ges 63:2164
26. Fischer FG, Marschall A (1931) Ber Dtsch Chem Ges 64:2825
27. Kuhn R, Badstübner W, Grundmann C (1936) Ber Dtsch Chem Ges 69:98
28. Langenbeck W, Borth G (1942) Ber Dtsch Chem Ges 75:951
29. Langenbeck W, Sauerbier R (1937) Ber Dtsch Chem Ges 70:1540
30. Cordes EH, Jencks WP (1962) J Am Chem Soc 84:826
31. Baum JS, Viehe HG (1976) J Org Chem 41:183
32. Jung ME, Vaccaro WD, Buszek KR (1893) Tetrahedron Lett 1989:30
33. Woodward RB, Logusch E, Nambiar KP, Sakan K, Ward DE, Au-Yeung BW, Balaram P, Browne LJ, Card PJ, Chen CH, Chênevert RB, Fliri A, Frobel K, Gais H-J, Garratt DG,

Hayakawa K, Heggie W, Hesson DP, Hoppe D, Hoppe I, Hyatt JA, Ikeda D, Jacobi PA, Kim KS, Kobuke Y, Kojima K, Krowicki K, Lee VJ, Leutert T, Malchenko S, Martens J, Matthews RS, Ong BS, Press JB, Rajan Babu TV, Rousseau G, Sauter HM, Suzuki M, Tatsuta K, Tolbert LM, Truesdale EA, Uchida I, Ueda Y, Uyehara T, Vasella AT, Vladuchick WC, Wade PA, Williams RM, Wong HN-C (1981) J Am Chem Soc 103:3210
34. Kawara A, Taguchi T (1994) Tetrahedron Lett 35:8805
35. Yamaguchi M, Shiraishi T, Hirama M (1996) J Org Chem 61:3520
36. Yamaguchi M, Shiraishi T, Hirama M (1993) Angew Chem Int Ed 32:1176
37. Yamaguchi M, Yokota N, Minami T (1991) J Chem Soc Chem Commun 1088:34
38. Mukherjee S, Yang JW, Hoffmann S, List B (2007) Chem Rev 107:5471
39. Erkkilä A, Majander I, Pihko PM (2007) Chem Rev 107:5416
40. Melchiorre P, Marigo M, Carlone A, Bartoli G (2008) Angew Chem Int Ed 47:6138
41. Bertelsen S, Jorgensen KA (2009) Chem Soc Rev 38:2178
42. Marqués-López E, Herrera RP, Marks T, Jacobs WC, de Könning RM, Figueiredo D, Christmann M (2009) Org Lett 11:4116
43. Bertelsen S, Marigo M, Brandes S, Dinér P, Jørgensen KA (2006) J Am Chem Soc 128:12973
44. de Figueiredo RM, Fröhlich R, Christmann M (2008) Angew Chem Int Ed 47:1450
45. Bertelsen S, Nielsen M, Jørgensen KarlA (2007) Angew Chem Int Ed 46:7356
46. Beeson TD, Mastracchio A, Hong J-B, Ashton K, MacMillan DWC (2007) Science 316:582
47. Sibi MP, Hasegawa M (2007) J Am Chem Soc 129:4124
48. Amatore M, Beeson TD, Brown SP, MacMillan DWC (2009) Angew Chem Int Ed 48:5121
49. List B (2006) Chem Commun 819
50. Enders D, Grondal C, Hüttl MRM (2007) Angew Chem Int Ed 46:1570
51. Grondal C, Jeanty M, Enders D (2010) Nat Chem 2:167
52. Hashimoto T, Maruoka K (2007) Chem Rev 107:5656
53. Llewellyn DB, Arndtsen BA (2005) Tetrahedron Asymmetr 16:1789
54. Llewellyn DB, Adamson D, Arndtsen BA (2000) Org Lett 2:4165
55. Carter C, Fletcher S, Nelson A (1995) Tetrahedron: Asymmetry 2003:14
56. Lacour J, Monchaud D, Marsol C (2002) Tetrahedron Lett 43:8257
57. Smidt SP, Zimmermann N, Studer M, Pfaltz A (2004) Chem Eur J 10:4685
58. Akiyama T, Itoh J, Yokota K, Fuchibe K (2004) Angew Chem Int Ed 43:1566
59. Uraguchi D, Terada M (2004) J Am Chem Soc 126:5356
60. Akiyama T (2007) Chem Rev 107:5744
61. Terada M (2008) Chem Commun 4097
62. Wang X, List B (2008) Angew Chem Int Ed 47:1119
63. Mayer S, List B (2006) Angew Chem Int Ed 45:4193
64. Martin NJA, List B (2006) J Am Chem Soc 128:13368
65. Adair G, Mukherjee S, List B (2008) Aldrichim Acta 41:31
66. García-García P, Lay F, García-García P, Rabalakos C, List B (2009) Angew Chem Int Ed 48:4363
67. Mukherjee S, List B (2007) J Am Chem Soc 129:11336
68. Lacour J, Moraleda D (2009) Chem Commun 7073
69. Hamilton GL, Kang EJ, Mba M, Toste FD (2007) Science 317:496
70. Li C, Wang C, Villa-Marcos B, Xiao J (2008) J Am Chem Soc 130:14450
71. Capon B, Wu ZP (1990) J Org Chem 55:2317
72. Dakin HD (1909) J Biol Chem 7:49
73. Ramasastry SSV, Zhang H, Tanaka F, Barbas CF (2006) J Am Chem Soc 129:288
74. Bassan A, Zou W, Reyes E, Himo F, Córdova A (2005) Angew Chem Int Ed 44:7028
75. Danishefsky S, Cain P (1976) J Am Chem Soc 98:4975
76. Machajewski TD, Wong C-H (2000) Angew Chem Int Ed 39:1352
77. Xu L-W, Luo J, Lu Y (2009) Chem Commun 1807
78. Peng F, Shao Z (2008) J Mol Catal A: Chem 285:1
79. Xu L-W, Lu Y (2008) Org Biomol Chem 6:2047

80. Ishihara K, Nakano K (2007) J Am Chem Soc 129:8930
81. Ishihara K, Nakano K (2005) J Am Chem Soc 127:10504
82. Kano T, Tanaka Y, Osawa K, Yurino T, Maruoka K (2009) Chem Commun 15:1956
83. Song CE (ed) (2009) Cinchona alkaloids in synthesis and catalysis. Wiley-VCH, Weinheim
84. Xie J-W, Chen W, Li R, Zeng M, Du W, Yue L, Chen Y-C, Wu Y, Zhu J, Deng J-G (2007) Angew Chem Int Ed 46:389
85. Chen Y-C (2008) Synlett 13:1919
86. Brunner H, Bügler J, Nuber B (1995) Tetrahedron Asymmetry 6:1699
87. Brunner H, Bügler J (1997) Bull Soc Chim Belg 106:77
88. Bartoli G, Melchiorre P (2008) Synlett 12:1759
89. Galzerano P, Pesciaioli F, Mazzanti A, Bartoli G, Melchiorre P (2009) Angew Chem Int Ed 48:7892
90. Zhang E, Fan C-A, Tu Y-Q, Zhang F-M, Song Y-L (2009) J Am Chem Soc 131:14626
91. Paixão MW, Holub N, Vila C, Nielsen M, Jørgensen KA (2009) Angew Chem Int Ed 48:7338
92. Katsuki T, Sharpless KB (1980) J Am Chem Soc 102:5974
93. Zhang W, Loebach JL, Wilson SR, Jacobsen EN (1990) J Am Chem Soc 112:2801
94. Irie R, Noda K, Ito Y, Matsumoto N, Katsuki T (1990) Tetrahedron Lett 31:7345
95. Wong OA, Shi Y (2008) Chem Rev 108:3958
96. Porter MJ, Skidmore J (2000) Chem Commun 1215
97. Díez D, Núñez MG, Antón AB, García P, Moro RF, Garrido NM, Marcos IS, Urones PBaJG (2008) Curr Org Synth 5:186
98. Lattanzi A (2008) Curr Org Synth 5:117
99. Weitz E, Scheffer A (1921) Chem Ber 54:2344
100. Yang NC, Finnegan RA (1958) J Am Chem Soc 80:5845
101. Bunton CA, Minkoff GJ (1949) J Chem Soc 665
102. Kelly DR, Caroff E, Flood RW, Heal W, Roberts SM (2004) Chem Commun 2016
103. Pluim H, Wynberg H (1980) J Org Chem 45:2498
104. Helder R, Hummelen JC, Laane RWPM, Wiering JS, Wynberg H (1831) Tetrahedron Lett 1976:17
105. Wynberg H, Greijdanus B (1978) J Chem Soc, Chem Commun 427
106. Nourse JG (1975) J Am Chem Soc 97:4594
107. Gorzynski Smith J (1984) Synthesis 8:629
108. Lauret C (2001) Tetrahedron Asymmetry 12:2359
109. Nomura K, Oshima K, Matsubara S (2005) Angew Chem Int Ed 44:5860
110. Ito S, Sato M (1990) Bull Chem Soc Jpn 63:2739
111. Xia QH, Ge HQ, Ye CP, Liu ZM, Su KX (2005) Chem Rev 105:1603
112. Enders D, Kramps L, Zhu J (1998) Tetrahedron Asymmetry 9:3959
113. Lopez-Pedrosa J-M, Pitts MR, Roberts SM, Saminathan S, Whittall J (2004) Tetrahedron Lett 45:5073
114. Lygo B, Wainwright PG (1999) Tetrahedron 55:6289
115. Lygo B, Wainwright PG (1998) Tetrahedron Lett 39:1599
116. Corey EJ, Zhang F-Y (1999) Org Lett 1:1287
117. Arai S, Tsuge H, Shioiri T (1998) Tetrahedron Lett 39:7563
118. Lygo B, Gardiner SD, McLeod MC, To DCM (2007) Org Biomol Chem 5:2283
119. Ooi T, Ohara D, Tamura M, Maruoka K (2004) J Am Chem Soc 126:6844
120. Juliá S, Guixer J, Masana J, Rocas J, Colonna S, Annuziata R, Molinari H (1982) J Chem Soc, Perkin Trans 1:1317
121. Juliá S, Masana J, Vega JC (1980) Angew Chem Int Ed Engl 19:929
122. Colonna S, Molinari H, Banfi S, Juliá S, Masana J, Alvarez A (1983) Tetrahedron 39:1635
123. Bentley PA, Bergeron S, Cappi MW, Hibbs DE, Hursthouse MB, Nugent TC, Pulido R, Roberts SM, Wu LE (1997) Chem Commun 739
124. Porter MJ, Roberts SM, Skidmore J (1999) Bioorg Med Chem 7:2145

125. Adger BM, Barkley JV, Bergeron S, Cappi MW, Flowerdew BE, Jackson MP, McCague R, Nugent TC, Roberts SM (1997) J Chem Soc Perkin Trans 1:3501
126. Allen JV, Bergeron S, Griffiths MJ, Mukherjee S, Roberts SM, Williamson NM, Wu LE (1998) J Chem Soc Perkin Trans 1:3171
127. Geller TP, Roberts SM (1999) J Chem Soc Perkin Trans 1:1397
128. Berkessel A, Gasch N, Glaubitz K, Koch C (2001) Org Lett 3:3839
129. Kelly DR, Roberts SM (2004) Chem Commun 2018
130. Kelly DR, Meek A, Roberts SM (2004) Chem Commun 21:2021
131. Kelly DR, Roberts SM (2006) Biopolymers 84:74
132. Mathew SP, Gunathilagan S, Roberts SM, Blackmond DG (2005) Org Lett 7:4847
133. Elston CL, Jackson RFW, MacDonald SJF, Murray PJ (1997) Angew Chem Int Ed Engl 36:410
134. Jacques O, Richards SJ, Jackson RFW (2001) Chem Commun 2712
135. Hinch M, Jacques O, Drago C, Caggiano L, Jackson RFW, Dexter C, Anson MS, Macdonald SJF (2006) J Mol Catal A: Chem 251:123
136. Shibasaki M, Yoshikawa N (2002) Chem Rev 102:2187
137. Inanaga J, Furuno H, Hayano T (2002) Chem Rev 102:2211
138. Bougauchi M, Watanabe S, Arai T, Sasai H, Shibasaki M (1997) J Am Chem Soc 119:2329
139. Watanabe S, Kobayashi Y, Arai T, Sasai H, Bougauchi M, Shibasaki M (1998) Tetrahedron Lett 39:7353
140. Nemoto T, Ohshima T, Yamaguchi K, Shibasaki M (2001) J Am Chem Soc 123:2725
141. Watanabe S, Arai T, Sasai H, Bougauchi M, Shibasaki M (1998) J Org Chem 63:8090
142. Daikai K, Kamaura M, Inanaga J (1998) Tetrahedron Lett 39:7321
143. Kino R, Daikai K, Kawanami T, Furuno H, Inanaga J (1822) Org Biomol Chem 2004:2
144. Daikai K, Hayano T, Kino R, Furuno H, Kagawa T, Inanaga J (2003) Chirality 15:83
145. Kakei H, Tsuji R, Ohshima T, Shibasaki M (2005) J Am Chem Soc 127:8962
146. Nemoto T, Kakei H, Gnanadesikan V, Tosaki S-y, Ohshima T, Shibasaki M (2002) J Am Chem Soc 124:14544
147. Kakei H, Nemoto T, Ohshima T, Shibasaki M (2004) Angew Chem Int Ed 43:317
148. Matsunaga S, Kinoshita T, Okada S, Harada S, Shibasaki M (2004) J Am Chem Soc 126:7559
149. Kinoshita T, Okada S, Park S-R, Matsunaga S, Shibasaki M (2003) Angew Chem Int Ed 42:4680
150. Matsunaga S, Qin H, Sugita M, Okada S, Kinoshita T, Yamagiwa N, Shibasaki M (2006) Tetrahedron 62:6630
151. Nemoto T, Ohshima T, Shibasaki M (2001) J Am Chem Soc 123:9474
152. Nemoto T, Tosaki S-Y, Ohshima T, Shibasaki M (2003) Chirality 15:306
153. Marigo M, Franzén J, Poulsen TB, Zhuang W, Jørgensen KA (2005) J Am Chem Soc 127:6964
154. Lee S, MacMillan DWC (2006) Tetrahedron 62:11413
155. Russo A, Lattanzi A (2009) Synthesis 9:1551
156. Lattanzi A (2005) Org Lett 7:2579
157. Lattanzi A, Russo A (2006) Tetrahedron 62:12264
158. Lattanzi A (2006) Adv Synth Catal 348:339
159. Russo A, Lattanzi A (2008) Eur J Org Chem 2767
160. Lattanzi A (2009) Chem Commun 3:1452
161. Li Y, Liu X, Yang Y, Zhao G (2006) J Org Chem 72:288
162. Liu X, Li Y, Wang G, Chai Z, Wu Y, Zhao G (2006) Tetrahedron Asymmetry 17:750
163. Cui H, Li Y, Zheng C, Zhao G, Zhu S (2008) J Fluorine Chem 129:45
164. Zheng C, Li Y, Yang Y, Wang H, Cui H, Zhang J, Zhao G (2009) Adv Synth Catal 351:1685
165. McManus JC, Carey JS, Taylor RJK (2003) Synlett 3:365
166. McManus JC, Genski T, Carey JS, Taylor RJK (2003) Synlett 3:369
167. Terada M, Nakano M (1049) Heterocycles 2008:76

168. Ishikawa T, Kumamoto T (2006) Synthesis 5:737
169. Tanaka S, Nagasawa K (2009) Synlett 4:667
170. Shin B, Tanaka S, Kita T, Hashimoto Y, Nagasawa K (2008) Heterocycles 76:801
171. Allingham MT, Howard-Jones A, Murphy PJ, Thomas DA, Caulkett PWR (2003) Tetrahedron Lett 44:8677
172. Corey EJ, Tada M, LaMahieu R, Libit L (1965) J Am Chem Soc 87:2051
173. Eaton PE, Lin K (1965) J Am Chem Soc 87:2052
174. Eaton PE, Lin K (1964) J Am Chem Soc 86:2087
175. Noyori R, Katô M (1974) Bull Chem Soc Jpn 47:1460
176. Adam W, Rao PB, Degen H-G, Saha-Moller CR (2000) J Am Chem Soc 122:5654
177. Adam W, Rao PB, Degen H-G, Saha-Möller CR (2002) Eur J Org Chem 2002:630
178. Dwyer CL, Gill CD, Ichihara O, Taylor RJK (2000) Synlett 5:704
179. Bardhan S, Schmitt DC, Porco JA (2006) Org Lett 8:927
180. Arai S, Tsuge H, Oku M, Miura M, Shioiri T (2002) Tetrahedron 58:1623
181. Arai S, Oku M, Miura M, Shioiri T (1998) Synlett 11:1201
182. Dehmlow EckehardV, Düttmann S, Neumann B, Stammler H-G (2002) Eur J Org Chem 2002:2087
183. Berkessel A, Guixá M, Schmidt F, Neudörfl JM, Lex J (2007) Chem Eur J 13:4483
184. Colonna S, Manfredi A, Annunziata R, Spadoni M (1987) Tetrahedron 43:2157
185. Barrett AGM, Blaney F, Campbell AD, Hamprecht D, Meyer T, White AJP, Witty D, Williams DJ (2002) J Org Chem 67:2735
186. Macdonald G, Alcaraz L, Lewis NJ, Taylor RJK (1998) Tetrahedron Lett 39:5433
187. Alcaraz L, Macdonald G, Ragot JP, Lewis N, Taylor RJK (1998) J Org Chem 63:3526
188. Li C, Pace EA, Liang M-C, Lobkovsky E, Gilmore TD, Porco JA (2001) J Am Chem Soc 123:11308
189. Adam W, Rao PB, Degen H-G, Levai A, Patonay T, Saha-Moller CR (2001) J Org Chem 67:259
190. Baba N, Oda J, Kawaguchi M (1986) Agric Biol Chem 50:3113
191. Baba N, Oda J, Kawahara S, Hamada M (1989) Bull Inst Chem Res, Kyoto Univ 67:121
192. Wynberg H, Marsman B (1980) J Org Chem 45:158
193. Baccin C, Gusso A, Pinna F, Strukul G (1995) Organometallics 14:1161
194. Aoki M, Seebach D (2001) Helv Chim Acta 84:187
195. Kienle M, Argyrakis W, Baro A, Laschat S (1971) Tetrahedron Lett 2008:49
196. Armstrong A, Moss WO, Reeves JR (2001) Tetrahedron Asymmetry 12:2779
197. Sato T, Watanabe M, Honda N, Fujisawa T (1984) Chem Lett
198. Bartel S, Bohlmann F (1989) Tetrahedron Lett 30:685
199. Marino JP, Abe H (1981) J Org Chem 46:5379
200. Guerrab Z, Daou B, Fkih-Tetouani S, Ahmar M, Cazes B (2007) Tetrahedron 63:3367
201. Löcht GVD, Marschall H, Weyerstahl P (1982) Liebigs Ann Chem 1982:1150
202. Ravikumar KS, Chandrasekaran S (1996) Tetrahedron 52:9137
203. Casteel DA (1999) Nat Prod Rep 16:55
204. White NJ (2008) Science 320:330
205. Jefford CW (2007) Drug Discov Today 12:487
206. Woodrow CJ, Krishna S (2006) Cell Mol Life Sci 63:1586
207. Jambou R, Legrand E, Niang M, Khim N, Lim P, Volney B, Ekala MT, Bouchier C, Esterre P, Fandeur T, Mercereau-Puijalon O (1960) Lancet 2005:366
208. Muraleedharan KM, Avery MA (2009) Drug Discov Today 14:793
209. Vennerstrom JL, Arbe-Barnes S, Brun R, Charman SA, Chiu FCK, Chollet J, Dong Y, Dorn A, Hunziker D, Matile H, McIntosh K, Padmanilayam M, Santo Tomas J, Scheurer C, Scorneaux B, Tang Y, Urwyler H, Wittlin S, Charman WN (2004) Nature 430:900
210. Posner GH, O'Neill PM (2004) Acc Chem Res 37:397
211. Dai P, Trullinger TK, Liu X, Dussault PH (2006) J Org Chem 71:2283
212. Oh D-C, Scott JJ, Currie CR, Clardy J (2009) Org Lett 11:633
213. Scott JJ, Oh D-C, Yuceer MC, Klepzig KD, Clardy J, Currie CR (2008) Science 322:63

214. Martyn DC, Beletsky G, Cortese JF, Tyndall E, Liu H, Fitzgerald MM, O'Shea TJ, Liang B, Clardy J (2009) Bioorg Med Chem Lett 19:5657
215. Rappoport Z (ed) (2006) The chemistry of peroxides. Wiley, New York
216. McCullough KJ, Nojimab M (2001) Curr Org Synth 5:601
217. Tsubusaki T, Nishino H (2009) Tetrahedron 65:3745
218. Kirihara M, Kakuda H, Ichinose M, Ochiai Y, Takizawa S, Mokuya A, Okubo K, Hatano A, Shiro M (2005) Tetrahedron 61:4831
219. Barnier J-P, Morisson V, Blanco L (2001) Synth Commun 31:349
220. Morisson V, Barnier J-P, Blanco L (1999) Tetrahedron Lett 40:4045
221. Kulinkovich OG, Astashko DA, Tyvorskii VI, Ilyina NA (2001) Synthesis 10:1453
222. Nazarov IN, Akhrem AA, Obshch Zh (1950) Khim 20:2183
223. Payne G (1958) J Org Chem 23:310
224. Fraile JM, García JI, Mayoral JA, Sebti S, Tahirb R (2001) Green Chem 3:271
225. Cativiela C, Figueras F, Fraile JM, García JI, Mayoral JA (1995) Tetrahedron Lett 36:4125
226. Paquette LA, Carr RVC, Bellamy F (1978) J Am Chem Soc 100:6764
227. Dussault PH, Liu X (1999) Org Lett 1:1391
228. Ensley HE, Carr RVC, Martin RS, Pierce TE (1980) J Am Chem Soc 102:2836
229. Adam W, Griesbeck A (1986) Synthesis 12:1050
230. Baumstark AL, Vasquez PC (1992) J Org Chem 57:393
231. Dussault PH, Zope UR (1995) Tetrahedron Lett 36:2187
232. Dai P, Dussault PH (2005) Org Lett 7:4333

Chapter 3
Objectives of This Ph.D. Work

The vast majority of organic compounds contain oxygenated functionalities. One of the main pathways for the introduction of oxygen into organic molecules is via the epoxidation of carbon–carbon double bonds. Thus, olefin epoxidation constitutes a central transformation in organic chemistry and one of the main routes to access epoxides on both a laboratory and an industrial scale.

Asymmetric olefin epoxidation, in particular catalytic asymmetric olefin epoxidation, has gained increasing importance since optically active epoxides are valuable chiral building blocks and versatile synthetic intermediates [1]. Their enantioselective production is a subject of considerable interest for academia as well as for industry. In view of the industrial applicability, catalytic asymmetric epoxidation methods should employ considerably safe oxidants and produce little waste. Hydrogen peroxide meets these criteria. It is low in cost, safe to handle, readily available, and generates water as the only by-product. Hence, it is probably the best oxidant with respect to environmental and economic considerations [2–4].[1]

Despite the wealth of catalytic asymmetric epoxidations of α,β-unsaturated ketones, particular substrate classes still pose a challenge to synthetic chemists. Indeed, most methods described to date are tailored to chalcone-type substrates but give inferior results with other enone classes. For instance, a highly enantioselective epoxidation of simple cyclic enones (such as cyclohexenone) has not been reported thus far (cf. Sect. 2.2.6). Similarly, there is no general method for the epoxidation of enolizable aliphatic acyclic enones, let alone using hydrogen peroxide as the oxidant (cf. Sect. 2.2.4). The related peroxidation of ketones, which generates the synthetically and medicinally relevant 1,2-dioxolane unit, is equally orphaned of any catalytic asymmetric method that proceeds with high enantioselectivity.

The goal of this Ph.D. work was to develop an efficient and general method for the catalytic asymmetric epoxidation of cyclic as well as acyclic α,β-unsaturated

[1] Indeed, in certain circumstances, hydrogen peroxide is even better than oxygen insofar as oxygen-organic mixtures are prone to spontaneous ignition.

C. Reisinger, *Epoxidations and Hydroperoxidations of α,β-Unsaturated Ketones*,
Springer Theses, DOI: 10.1007/978-3-642-28118-1_3,

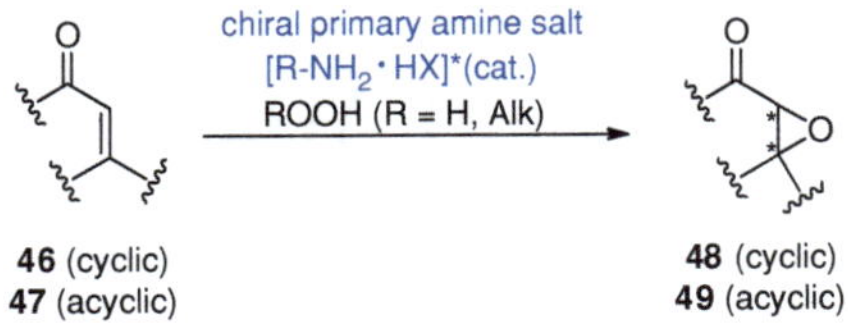

Scheme 3.1 Intended organocatalytic asymmetric epoxidation of (cyclic and acyclic) α,β-unsaturated ketones

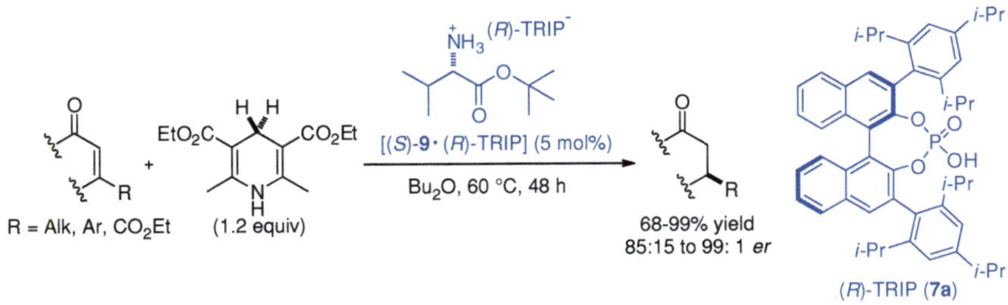

Scheme 3.2 Catalytic asymmetric transfer hydrogenation of α,β-unsaturated ketones catalyzed by chiral primary amine salt [(*S*)-**9** · (*R*)-TRIP (**7a**)]

ketones with hydrogen peroxide as environmentally benign and atom-economic oxidant. In the event, our work also allowed the serendipitous discovery of a highly enantioselective catalytic hydroperoxidation of enones which directly delivers the valuable 3-*hydroxy*-1,2-*dioxolane-moiety*.

Recently, iminium ion catalysis was successfully applied to the enantioselective epoxidation of α,β-unsaturated aldehydes by the groups of *Jørgensen*, *MacMillan*, and our group using catalytic amounts of chiral secondary amines or amine salts (cf. Sect. 2.2.5.1) [5–7]. Inspired by these findings, we reasoned that it should be possible to develop a conceptually similar asymmetric *Weitz-Scheffer*-type epoxidation of α,β-unsaturated ketones with hydroperoxides as potential oxidants (Scheme 3.1) [8].

However, for the activation of α,β-unsaturated ketones as iminium ion, **primary** amine salts have emerged as superior catalysts in many cases due to steric constraints [9]. In particular, the successful application of chiral primary amine salt [(*S*)-**9** · (*R*)-TRIP] in the catalytic asymmetric transfer hydrogenation of cyclic as well as acyclic enones established in our group (Scheme 3.2), [10] encouraged us to accept the challenge of developing a catalytic asymmetric epoxidation process of such substrates based on the same concept.

The common feature of catalysts [**10** · (*R*)-TRIP][2] (cf. Scheme 2.29) and [(*S*)-**9** · (*R*)-TRIP] developed in our group for the enantioselective epoxidation of α,β-unsaturated aldehydes and transfer hydrogenation of enones, respectively,

[2] [bis(3,5-bis(trifluoromethyl)benzyl)amine (**10**) · (*R*)-TRIP (**7a**)].

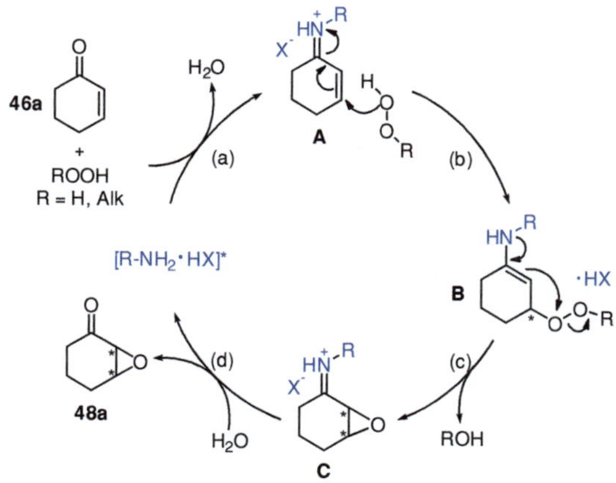

Fig. 3.1 Proposed catalytic cycle for the asymmetric epoxidation of α,β-unsaturated ketones, as exemplified by the reaction of 2-cyclohexenone (**46a**), catalyzed by a chiral primary amine salt [R-NH_2 · HX]*

is the chiral counteranion TRIP. In the enal epoxidation, TRIP is the single source of chiral information in the system, whereas in the enone reduction the matched combination of (*R*)-TRIP with (*S*)-valin-*tert*-butyl ester (**9**) accounts for excellent enantio-selectivi-ties. Thus, both methods illustrate successful implementations of the ACDC-concept in asymmetric catalysis (see Sect. 2.1.3.1).

Our strategy for the catalytic asymmetric epoxidation is outlined in the following section. Mechanistically, the reaction was envisaged to proceed via asymmetric iminium ion catalysis by means of a chiral primary amine salt [R-NH_2 · HX]* (Fig. 3.1). Initial reversible formation of iminium ion **A** may effectively lower the LUMO energy of the substrate and thus activate the enone [e.g. 2-cyclohexenone (**46a**)] for the enantioselective conjugate addition of hydroperoxide ROOH (R = H, Alk). Subsequent intramolecular nucleophilic substitution at the proximal oxygen atom may break the weak peroxide bond of β-peroxy enamine intermediate **B** and close the epoxide ring. Finally, hydrolysis of α,β-epoxy iminium ion **C** releases the product [e.g. 2,3-epoxycyclohexanone (**48a**)] and regenerates the catalyst.

References

1. Xia QH, Ge HQ, Ye CP, Liu ZM, Su KX (2005) Chem Rev 105:1603
2. Grigoropoulou G, Clark JH, Elings JA (2003) Green Chem 5:1
3. Lane BS, Burgess K (2003) Chem Rev 103:2457
4. Campos-Martin JM, Blanco-Brieva G, Fierro JLG (2006) Angew Chem Int Ed 45:6962
5. Marigo M, Franzén J, Poulsen TB, Zhuang W, Jørgensen KA (2005) J Am Chem Soc 127:6964
6. Lee S, MacMillan DWC (2006) Tetrahedron 62:11413
7. Wang X, Reisinger CM, List B (2008) J Am Chem Soc 130:6070
8. Weitz E, Scheffer A (1921) Chem Ber 54:2344
9. Peng F, Shao Z (2008) J Mol Catal A: Chem 285:1
10. Martin NJA, List B (2006) J Am Chem Soc 128:13368

Chapter 4
Results and Discussion

Simple cyclic enones have generally been omitted within the vast majority of studies on the asymmetric epoxidation of α,β-unsaturated ketones. Thus, we decided to focus on these challenging cyclic substrates as a departure point before extending the methodology to different types of α,β-unsaturated ketones.

4.1 Catalytic Asymmetric Epoxidation of Cyclic Enones

4.1.1 Screening Studies

The screening studies were carried out in a collaborative effort with Wang.

4.1.1.1 Development and Optimization of the Catalytic System

Initial experiments were devoted to identifying an effective system for the catalytic asymmetric epoxidation of cyclic enones employing 2-cyclohexenone (**46a**) as the model substrate. The results of our comprehensive screening studies are summarized in Table 4.1. At the outset, we tested dibenzylamine together with (*S*)-TRIP [**50** · (*S*)-TRIP], a combination, which had previously emerged as an efficient catalyst system for the asymmetric epoxidation of α,β-unsaturated aldehydes developed by Wang in our group (cf. Scheme 2.29) [1]. Aqueous hydrogen peroxide was chosen as the oxidant and reactions were conducted in dioxane (0.25 M) at 50 °C. Yet, the activity of [**50** · (*S*)-TRIP] in the asymmetric epoxidation of 2-cyclohexenone (**46a**) was rather low and moreover, 2,3-epoxycyclohexanone (**48a**) was generated with only low enantioselectivity (entry 1). Chiral primary amine salt [(*S*)-**9** · (*R*)-TRIP] had proven to be a powerful catalyst for the asymmetric transfer hydrogenation of cyclic and acyclic ketones (cf. Scheme 3.2) [2].

C. Reisinger, *Epoxidations and Hydroperoxidations of α,β-Unsaturated Ketones*, Springer Theses, DOI: 10.1007/978-3-642-28118-1_4,

Table 4.1 Evaluation of various catalyst systems for the catalytic asymmetric epoxidation of 2-cyclohexenone (**46a**)

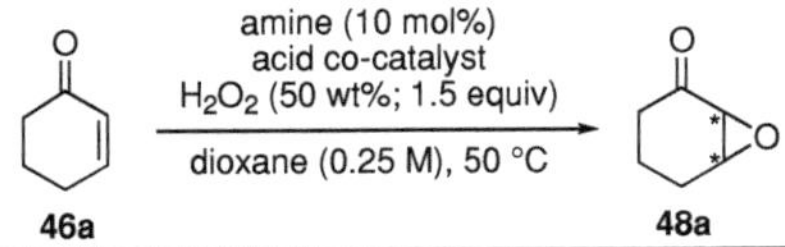

Entry	Amine		Acid co-catalyst	[mol%]	t [h]	Yield [%][a]	*er*[b]
1	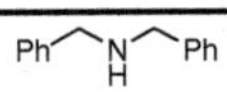	**50**	(*S*)-TRIP	10	72	22	58:42
2	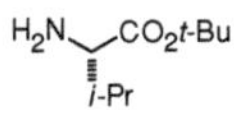	(*S*)-**9**	(*S*)-TRIP	10	24	92	56:44
3			(*R*)-TRIP	10	24	73	62:38
4	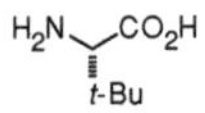	(*S*)-**51**	–	–	72	30	73.5:26.5
5	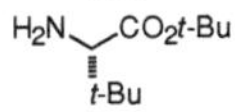	(*S*)-**52**	–	–	72	–	–
6			(*R*)-TRIP	10	72	95	67.5:32.5
7	(*S*)-proline	**1**	–	–	72	90	58.5:41.5
8	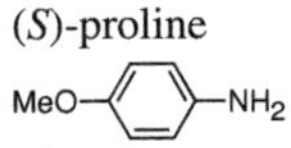	**53**	(*R*)-TRIP	10	48	85	66:34
9	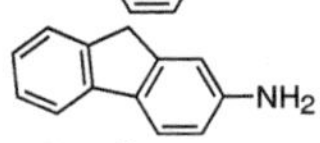	**54**	(*R*)-TRIP	10	48	>99	73.5:26.5
10	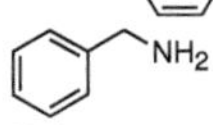	**55**	(*R*)-TRIP	10	24	95	59:41
11	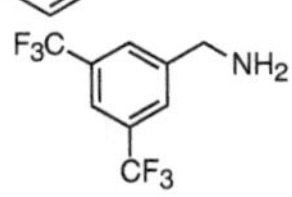	**56**	(*R*)-TRIP	10	24	99	59.5:40.5
12	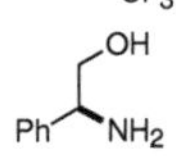	(*S*)-**57**	(*S*)-TRIP	10	24	>99	55.5:44.5
13			(*R*)-TRIP	10	24	>99	56:44
14	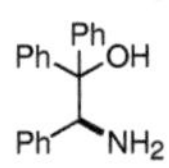	(*S*)-**58**	(*S*)-TRIP	10	24	>99	75:25
15			(*R*)-TRIP	10	24	>99	80:20

[a] Determined by GC
[b] Determined by chiral GC

However, [(*S*)-**9** · (*R*)-TRIP] as well as its diastereomeric salt [(*S*)-**9** · (*S*)-TRIP] performed worse in the asymmetric epoxidation of 2-cyclohexenone (**46a**) and afforded epoxide **48a** with low enantiomeric excess albeit in good yields (entries 2–3). In the presence of the amino acid (*S*)-*tert*-leucine (**51**), the desired product was formed with moderate enantioselectivity of 73.5:26.5 *er*, yet again at very low rates (entry 4). Its *tert*-butyl ester **52**, however, did not mediate the epoxidation reaction (entry 5). In contrast, the combination with (*R*)-TRIP gave epoxide **48a** in high yield and slightly higher optical purity than that attained with the corresponding salt [(*S*)-**9** · (*R*)-TRIP] derived from (*S*)-valine (entry 6 versus entry 3). This result nicely illustrates the crucial effect of an acid co-catalyst (or acidic functionality embedded in the catalyst motif) in the present transformation. Notably, (*S*)-proline (**1**) displayed high catalytic activity but generated epoxide **48a** with low enantioselectivity (entry 7) [3–6].

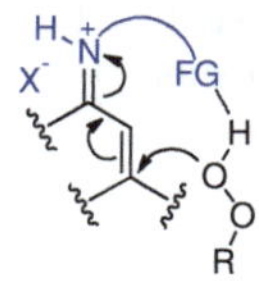

Fig. 4.1 Pre-transition state assembly invoking a directing effect of a bifunctional amine catalyst. (FG = *i. a.* OH, NH_2)

Next, we turned our attention to salts of primary aromatic amines. A broad variety of aniline derivatives were evaluated with *p*-methoxyphenyl amine (**53**) and 2-amino fluorene (**54**) in combination with TRIP giving the best results with good yields and enantiomeric ratios of up to 73.5:26.5 *er* (entries 8–9) [7]. In contrast, benzyl amines **55** and **56** showed superior catalytic activity, yet induced lower levels of enantioselectivity (entries 10–11).

At this point, we speculated that the use of a bifunctional amine catalyst might benefit the catalyst efficiency. A directing group in close proximity to the primary amine functionality could direct the attack of the hydroperoxide nucleophile as suggested in Fig. 4.1.

Based on this hypothesis, we tested various chiral amino alcohols such as (*S*)-2-amino-2-phenylethanol (**57**) and (*S*)-2-amino-1,1,2-triphenylethanol (**58**) with the latter being sterically encumbered and conformationally rigidified due to the geminal diphenyl substitution (entries 12–15). Indeed, amino alcohol (*S*)-**58** in combination (*R*)-TRIP gave full conversion after 24 h along with an increased enantiomeric ratio of 80:20 which was the highest observed at this point of the screening experiments (entry 15).

Encouraged by this result, we went on to study different types of bifunctional primary amines among them (*S*)-BINAM (2,2′-diamino-1,1′-binaphthalene; **59**), (*R*,*R*)-DACH (1,2-diaminocyclohexane; **60**), (*R*,*R*)-DPEN (1,2-diphenylethylenediamine; **12**), (*R*,*R*)-DPEN derivative **61**, (*S*)-2-aminomethyl-pyrrolidine (**62**), as well as primary amines derived from naturally occurring Cinchona alkaloids (Table 4.2). In particular, the TFA salts of (*R*,*R*)-DPEN **12** and 9-amino(9-deoxy)*epi*quinine (9-NH_2-*epi*Q; **13**) emerged as powerful catalysts giving the desired epoxide **48a** with high yield and enantiomeric ratios of 95:5 and 96:4 *er*, respectively (entries 3 and 5). Notably, truncated forms of **13**, the 2-aminomethyl quinuclidines **63** and **64**, proved to be highly active, yet less enantioselective (entries 7–12).

The most promising catalyst candidates (*R*,*R*)-DPEN **12** and 9-amino (9-deoxy)*epi*quinine (9-NH_2-*epi*Q; **13**) were subjected to further screening experiments to refine the catalyst composition with respect to the nature of the acid co-catalyst. In doing so, we took into account the decisive effect of chiral counteranions in catalytic asymmetric transformations as recently revealed in our laboratories [8]. A range of variously 3,3′-disubstituted chiral BINOL-derived phosphoric acids were tested in combination with (*R*,*R*)-DPEN. The first attempts were performed with chiral BINOL phosphate TRIP (**7a**) bearing sterically demanding 2,4,6-triisopropylphenyl substituents at the 3,3′-positions (Table 4.3). TRIP had already surfaced as the chiral counteranion of choice in several catalytic asymmetric transformations developed in our group. Indeed, also in the epoxidation of 2-cyclohexenone (**46a**), none of the chiral BINOL phosphates tested could

Table 4.2 Evaluation of various bifunctional amine salts for the catalytic asymmetric epoxidation of 2-cyclohexenone (**46a**)

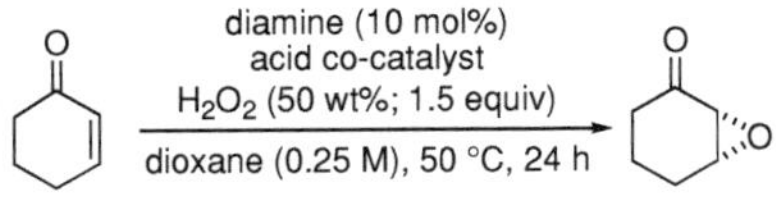

46a (*R*,*R*)-**48a**

Entry	Amine		Acid co-catalyst	[mol%]	Yield [%][a]	*er*[b]
1	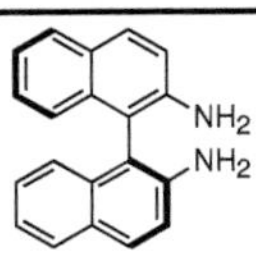	(*S*)-**59**	TFA	10	71	38:62
2	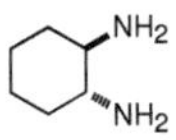	(*R*,*R*)-**60**	TFA	10	98	82:18
3	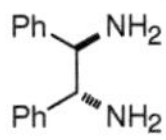 (*R*,*R*)-DPEN	(*R*,*R*)-**12**	TFA	10	95	95:5
4	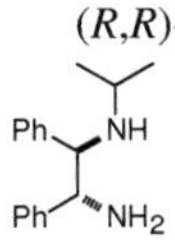	(*R*,*R*)-**61**	TFA	10	77	93:7
5	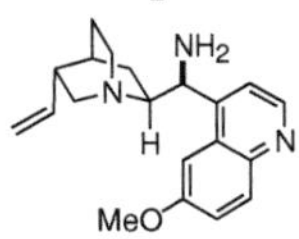 9-NH_2-*epiQ*	**13**	TFA	20	76	4:96
6	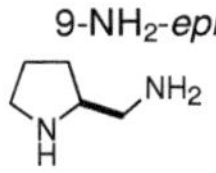	(*S*)-**62**	(*S*)-TRIP	10	84	94:6
7[c]	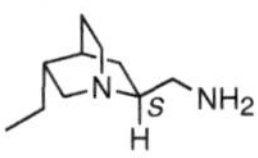	(8*S*)-**63**	TFA	10	>99	56:44
8[c]		(8*S*)-**63**	TFA	20	>99	58:42
9[c]		(8*S*)-**63**	(*S*)-TRIP	10	>99	63:37
10[c]	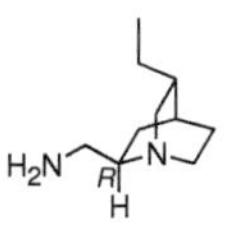	(8*R*)-**64**	TFA	10	96	38:62
11[c]		(8*R*)-**64**	TFA	20	>99	31.5:68.5
12[c]		(8*R*)-**64**	(*S*)-TRIP	10	>99	18:82

[a] Determined by GC
[b] Determined by chiral GC
[c] 12 h

compete with TRIP in terms of both activity and enantioselectivity. Pairing of (*R*,*R*)-DPEN **12** with (*S*)-TRIP in a 1:1 ratio improved both the yield and the enantioselectivity compared to the reaction catalyzed by [(*R*,*R*)-DPEN · TFA] (entry 3). In contrast, the results obtained with [(*R*,*R*)-DPEN · (*R*)-TRIP] were inferior and denoted a significant matched/mismatched scenario (entry 5). Moreover, we evaluated other chiral acid motifs. In the presence of chiral BINOL-derived *N*-triflyl phosphoramide **65**, (*R*,*R*)-2,3-epoxy-cyclo-hexanone (**48a**) was formed in only moderate yield and reduced enantioselectivity (86:14 *er*) (entry 6).

Table 4.3 Evaluation of different acid co-catalysts in combination with (*R*,*R*)-DPEN [(*R*,*R*)-**12**]

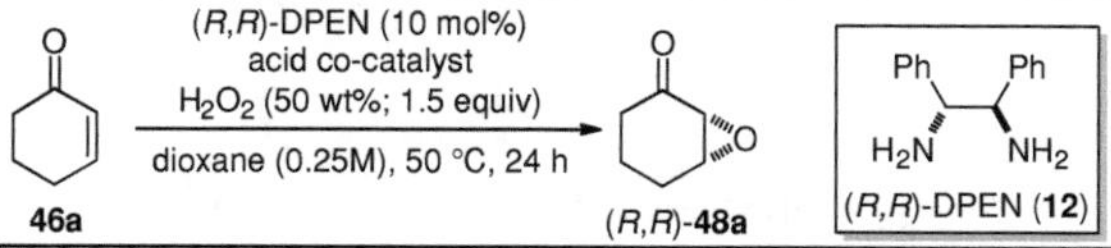

Entry	Acid co-catalyst	[mol%]	Yield [%][a]	*er*[b]
1	TFA	10	95	95:5
2	TFA	20	88[c]	94:6
3	(*S*)-TRIP	10	99	96:4
4	(*S*)-TRIP	20	91	92:8
5	(*R*)-TRIP	10	49	90:10
6	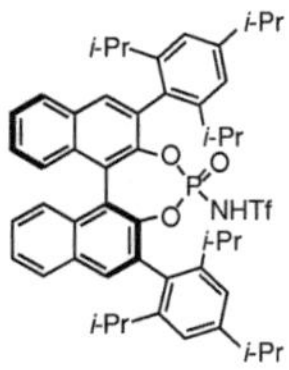 (*S*)-**65**	10	38	86:14
7	$(PhO)_2PO(OH)$ (**66**)	10	77[c]	96:4
8	HCl	10	60	90.5:9.5
9	*p*-TsOH · H_2O	10	92	89.5:10.5
10	TfOH	10	30	73:27
11[d]	(*R*)-TRIP	10	95	42:58

[a] Determined by GC
[b] Determined by chiral GC
[c] By-product formation was observed
[d] *meso*-DPEN (*meso*-**12**) was used

Notably, the use of achiral diphenyl phosphate salt [(*R*,*R*)-DPEN · **66**] resulted in excellent enantioselectivity of 96:4 *er*; nevertheless increased by-product formation was observed (entry 7). When TRIP was used in combination with *meso*-DPEN (*meso*-**12**), and thus was the only source of chirality in the system, the enantioselectivity remained low (entry 11).

In addition, we evaluated a variety of DPEN derivatives bearing either electron-donating or electron-withdrawing substituents at the phenyl rings in combination with TRIP. Yet, none of the derivatives studied was able to enhance the stereocontrol obtained with unsubstituted (*R*,*R*)-DPEN ((*R*,*R*)-**12**).

In parallel, we investigated the use of Cinchona alkaloid-derived primary amines in the epoxidation of 2-cyclohexenone (**46a**) (Table 4.4). In particular, 9-amino (9-deoxy)*epi*quinine (**13**) generated epoxide **48a** in high yield along with an excellent enantioselectivity of 96:4 *er* (entry 1). The results obtained with analogous primary amines **67–70** derived from quinidine, cupreidine, cinchonidine, or cinchonine, respectively, were slightly inferior (entries 4–7). In contrast, amine **71** derived from 6′-isopropoxycinchonidine and 9-amino(9-deoxy)*epi*dihydro-quinine (**72**) both furnished epoxide **48a** with comparable efficiencies (entries 8–9); the former one even afforded a slightly higher yield and enantioselectivity of 96.5:3.5 *er* than 9-amino(9-deoxy)*epi*quinine (**13**) under otherwise identical conditions.

Table 4.4 Screening of various Cinchona alkaloid-derived primary amines

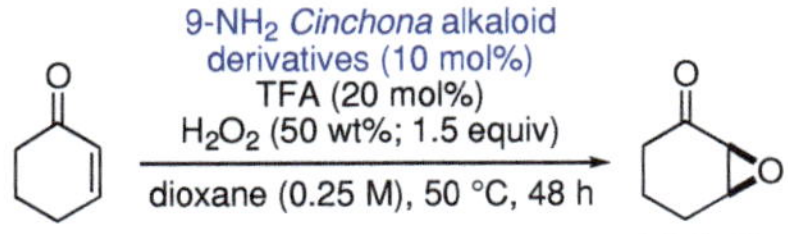

46a (*S*,*S*)-**48a**

Entry	9-NH_2 Cinchona alkaloid derivative	Conv. [%][a]	Yield [%][a]	*er*[b]
1	**13** (9-NH_2-*epi*Q)	97	76	96:4
2[c]	**13** (9-NH_2-*epi*Q)	95	82	96:4
3[d]	**13** (9-NH_2-*epi*Q)	97	93	97:3
4	**67** (9-NH_2-*epi*QD)	94	87	6:94
5[d]	**68**	98	89	91:9
6	**69**	full	77	94.5:5.5
7	**70**	full	78	7:93
8	**71**	96	83	96.5:3.5
9	**72**	96	75	95.5:4.5
10[e]	**73** (9-NH_2-Q)	82	80	21:79

[a] Determined by GC
[b] Determined by chiral GC
[c] 24 h
[d] At 30 °C
[e] 72 h

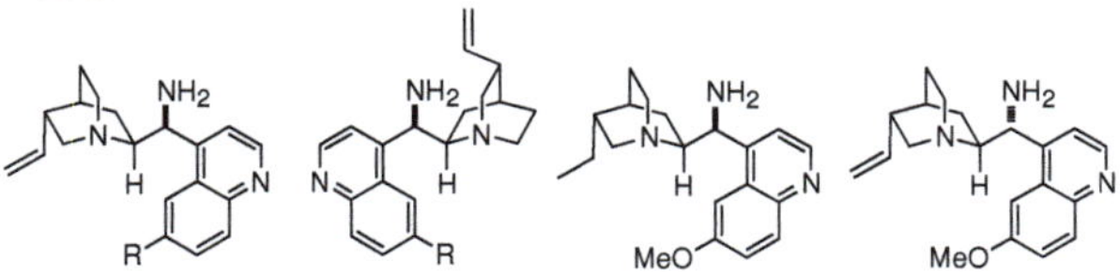

R = OMe: **13**
H: **69**
OH: **68**
O*i*-Pr: **71**

R = OMe: **67**
H: **70**

72

73

However, we have selected amine **13** for further investigations since it is readily accessible from quinine via a two step-one pot sequence whereas the synthesis of the 6′-*iso*propoxy derivative **71** requires a lengthy five-step procedure (cf. Sect. 4.7.1). Intriguingly, 9-amino(9-deoxy)quinine (9-NH_2-Q; **73**) featuring the natural configuration of quinine at C-9 was less active and significantly less enantioselective compared to amine **13**. Indeed, the opposite enantiomer (*R*,*R*)-**48a** was generated with only 79:21 *er* (entry 10). In general, pseudoenantiomeric amine catalysts such as quinine-derived **13** and quinidine-derived **67** provide antipodal 2,3-epoxycyclohexanone (**48a**) with similar levels of enantiocontrol (entries 1 and 4).

In the course of our screening studies, we noted that prolonged reaction times had a detrimental effect on the yield. Whereas a yield of 82%[1] of epoxide **48a** was detected after 24 h (entry 2), only 76% (see footnote 1) were present in the reaction mixture after continued heating for 48 h (entry 1). In addition, increased side product

[1] Yield determined by GC with an internal standard method.

Table 4.5 Screening of acid co-catalysts for the epoxidation of 2-cyclohexenone (**46a**) with 9-amino(9-deoxy)*epi*quinine (**13**)

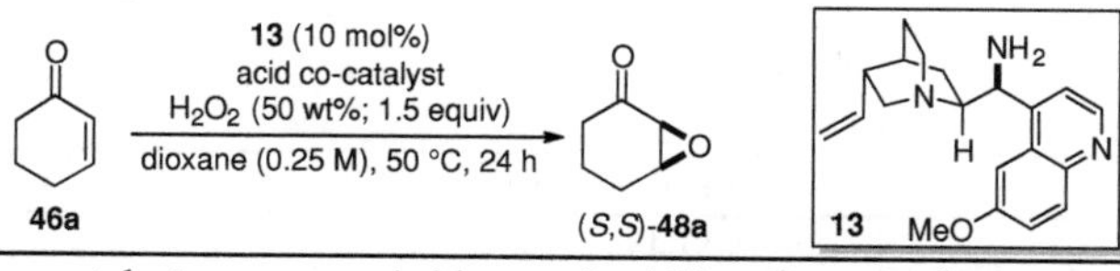

Entry	Acid co-catalyst		Acid conc. [mol%]	Conv. [%][a]	Yield [%][a]	*er*[b]
1[c]	TFA		20	97	93	97:3
2[c]			10	90	89	96.5:3.5
3[c]			30	97	90	96.5:3.5
4	HCl		10	79	70	87:13
5			20	85	69	92:8
6	DNBA		10	92	92	93:7
7		**74**	10	85	85	93.5:6.5
8			20	88	88	93.5:6.5
9	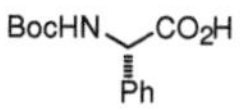	(*S*)-**75**	10	88	88	93.5:6.5
10			20	91	91	94:6
11	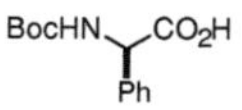	(*R*)-**75**	10	89	89	94.5:5.5
12			20	93	89	95:5
13	$(PhO)_2PO(OH)$	**66**	10	94	94	96:4
14	(*R*)-TRIP	**7a**	10	56	50	97:3
15	(*S*)-TRIP	**7a**	10	56	50	87:13
16[d]	(*R*)-TRIP	**7a**	10	71	69	15:85
17[d]	(*S*)-TRIP	**7a**	10	65	65	4:96

[a] Determined by GC
[b] Determined by chiral GC
[c] At 30 °C for 20 h
[d] 9-Amino(9-deoxy)*epi*quinidine (**67**; 10 mol%) was used

formation presumably resulting from product decomposition was noticable under these conditions. Similarly, decreasing the temperature from 50–30 °C proved beneficial to the reaction (entry 3). Taking these observations into account, (*S*,*S*)-2, 3-epoxycyclohexanone (**48a**) was obtained in 93% yield (see footnote 1) and with excellent enantioselectivity of 97:3 *er* in the presence of 10 mol% of [**13** · 2 TFA].

Attempts to refine the catalyst composition with regard to the acid co-catalyst (Table 4.5), confirmed [**13** · 2 TFA] as the optimum choice. A ratio of 1:1 amine **13**/TFA resulted in somewhat lower catalytic activity (entry 2). However, both increasing and decreasing the loading of TFA only marginally influenced the reaction outcome in terms of yields and enantioselectivities (entries 2–3). With diphenyl phosphoric acid (**66**) as co-catalyst, 2,3-epoxycyclohexanone was obtained in 94% yield and 96:4 *er* (entry 13). This value could be further enhanced to 97:3 *er* by pairing 9-amino(9-deoxy)*epi*quinine (**13**) with (*R*)-TRIP (entry 14). However, [**13** · (*R*)-TRIP] exhibited significantly decreased catalytic activity. Notably, combining (*S*)-TRIP with 9-amino(9-deoxy)*epi***quinidine** (**67**) furnished the (*R*,*R*)-isomer of **48a** with 96:4 *er* (entry 17) whereas catalyst [**67** · 2 TFA]

Table 4.6 Catalytic asymmetric epoxidation of 2-cyclohexenone (**46a**)

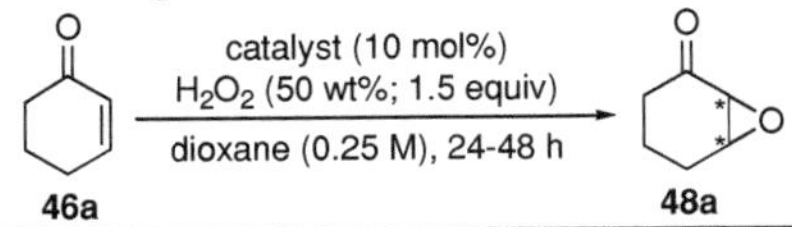

Entry	Catalyst	Product	Temp. [°C]	Conv. [%][a]	Yield [%][b]	*er*[c]
1	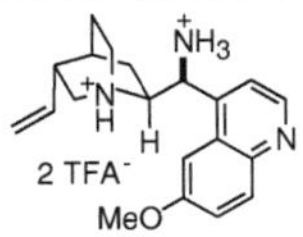 [9-NH$_2$-*epi*Q (**13**) • 2 TFA]	(*S*,*S*)-**48a**	30	97	58 (93)	97:3
2	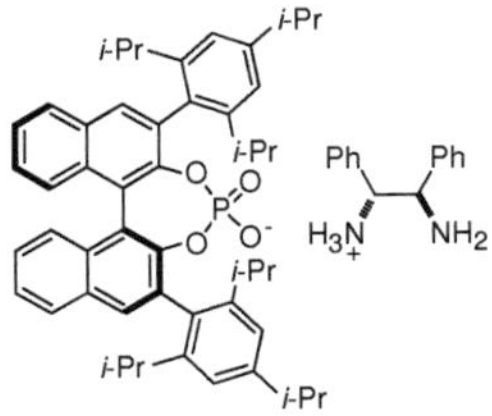 [(*R*,*R*)-DPEN • (*S*)-TRIP]	(*R*,*R*)-**48a**	50	full	68 (99)	96:4

[a] Determined by GC
[b] Yields of pure, isolated products. Values in parenthesis correspond to yields determined by GC with an internal standard method
[c] Determined by chiral GC

generated epoxide (*R*,*R*)-**48a** with a reduced 94:6 *er* (cf. Table 4.4, entry 5). Subsequent optimization experiments were carried out with [**13** · 2 TFA].

4.1.1.2 Optimization of the Reaction Conditions

After having identified two efficient catalytic systems—[9-NH$_2$-*epi*Q (**13**) · 2 TFA] and [(*R*,*R*)-DPEN · (*S*)-TRIP] (Table 4.6), our objective was to further optimize the reaction conditions prior to investigating the scope of the asymmetric epoxidation reaction.

With both catalyst systems, the crude reaction mixtures were extremely clean as determined by GC and ^{1}H NMR and cyclohexenone epoxide **48a** was the only product detected; only its pronounced volatility precluded the obtention of high isolated yields (Table 4.6). The absolute configuration of 2,3-epoxycyclohexanone (**48a**) was established by comparing the optical rotation with literature values [9].

In the context of the reaction optimization, we also wanted to evaluate the applicability of different oxygen sources. Not only is aqueous hydrogen peroxide unquestionably the most favourable oxidant besides molecular oxygen, but it also proved superior to all the other oxygen sources tested (Table 4.7). In particular, 50 wt% aqueous hydrogen peroxide afforded cyclohexenone epoxide **48a** with slightly higher enantioselectivity and higher yield than the less concentrated

Table 4.7 Evaluation of different oxidants for the epoxidation of 2-cyclohexenone (**46a**) catalyzed by [(*R*,*R*)-DPEN · (*S*)-TRIP]

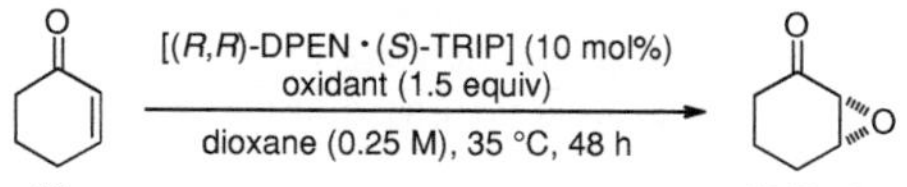

Entry	Oxidant	Yield [%][a]	*er*[b]
1	H_2O_2 (50 wt%)	99	96:4
2	H_2O_2 (30 wt%)	96	95:5
3	UHP	69	87:13
4	$(H_2O_2)_3(Na_2CO_3)_2$	10	52.5:47.5
5[c]	TBHP (5.5 M in decane)	71	74:26
6[c]	CMHP (~80% in cumene)	10	71:29
7[c]	PhI = O	<5	n.d.
8[c]	$KHSO_5$	20	79.5:20.5
9[c]	*m*-CBPA	18	n.d.
10[c]	$NaBO_3 \cdot H_2O$	12	67.5:32.5

[a] Determined by GC
[b] Determined by chiral GC
[c] 72 h

reagent (entries 1–2). The use of anhydrous H_2O_2 sources [urea-hydrogen peroxide (UHP) or sodium percarbonate] gave inferior results in terms of both reactivity and selectivity (entries 3–4). Alkylhydroperoxides were also tested (entries 5–6). Among them, *tert*-butylhydroperoxide furnished the desired product in good yield but with moderate optical purity. As expected, electrophilic oxidants were generally less effective in epoxidizing the electron-deficient double bond of 2-cyclohexenone (**46a**) (entries 7–10).

Both catalyst components of [(*R*,*R*)-DPEN · (*S*)-TRIP] are commercially available. Yet, due to the high molecular weight of the BINOL-derived phosphoric acid (*S*)-TRIP, a reduction in catalyst loading seemed desirable. However, this led to a drop in activity while the enantioselectivity remained essentially unaffected (Table 4.8, entries 2–4). We hypothesized that it might be possible to counter the loss of acitivity by keeping the concentration of the amine component (*R*,*R*)-DPEN constant while lowering the loading of (*S*)-TRIP. Previously, this strategy had proven successful in the catalytic asymmetric transfer hydrogenation of α,β-unsaturated ketones [10]. However, the results depicted in Table 4.8 indicate that a 1:1 (*R*,*R*)-DPEN/(*S*)-TRIP ratio is crucial to obtain high enantioselectivity. Lowering the loading of (*S*)-TRIP while keeping the concentration of (*R*,*R*)-DPEN constant had a detrimental effect on both the activity and selectivity of the catalyst system (entries 5–8).

Screening of different solvents revealed that ethereal solvents are vital for high catalytic activity (Table 4.9). Among all ethereal solvents tested, 1,4-dioxane afforded the best results (entries 1 and 10). With catalyst [(*R*,*R*)-DPEN · (*S*)-TRIP], excellent results were also obtained in dimethoxyethane (DME) (entry

Table 4.8 Optimization of the loading of the (*R*,*R*)-DPEN/(*S*)-TRIP catalyst system

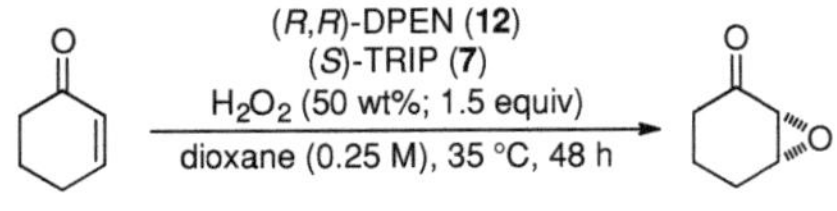

46a (*R*,*R*)-**48a**

Entry	(*S*)-TRIP [mol%]	(*R*,*R*)-DPEN [mol%]	Yield [%][a]	*er*[b]
1	20	20	91	96:4
2	10	10	99	96:4
3	5	5	78	95:5
4[c]	1	1	38	97:3
5	5	10	80	95:5
6	3	10	65	93.5:6.5
7	2	10	49	90.5:9.5
8	1	10	30	85:15

[a] Determined by GC
[b] Determined by chiral GC
[c] 72 h

Table 4.9 Solvent screening

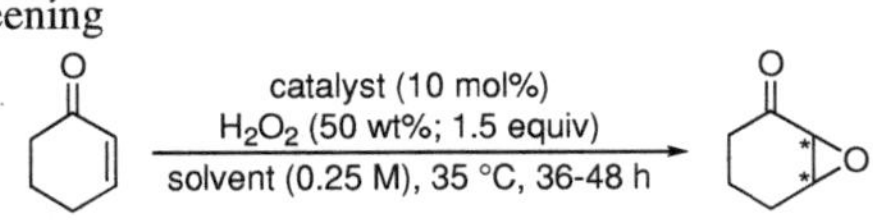

46a **48a**

Catalyst	[**13** · 2 TFA]			[(*R*,*R*)-DPEN · (*S*)-TRIP]		
Solvent	Entry	Yield [%][a]	*er* (*S*,*S*)[b]	Entry	Yield [%][a]	*er* (*R*,*R*)[b]
1,4-dioxane	1	93	96.5:3.5	10	99	96:4
MTBE	2	89	84:16	12	99	95:5
DME	3	90	91:9	13	96	96:4
Et_2O	4	86	73.5:26.5	14	97	94.5:5.5
n-Bu_2O	5	69	57:43	15	88	90.5:9.5
THF	6	n.d.	n.d.	16	99	94:6
hexane	7	69	59.5:40.5	17	83	84:16
toluene	8	50	58.5:41.5	18	n.d.	n.d.
CH_2Cl_2	9	41	74:26	19	n.d.	n.d.

[a] Determined by GC
[b] Determined by chiral GC

13). In chlorinated solvents and non-polar solvents such as hexane and toluene the catalytic efficiency was greatly reduced (entries 7–9, 17). Polar solvents including methanol, ethanol, DMSO, and DMF had been tested previously under slightly modified conditions, and overall gave inferior results than all the ethereal solvents.

Before investigating the scope of the reaction, we sought the possibility of increasing the substrate concentration (thus minimizing the amount of solvent required). However, a substrate concentration of 0.5 M brought about reduced enantioselectivity and increased the formation of by-products in the epoxidation of 2-cyclohexenone (**46a**). We were similarly interested in decreasing the loading of hydrogen peroxide. However, a slight reduction to 1.25 equivalents of hydrogen peroxide resulted in a pronounced drop in the reaction rate as such that the reaction did not reach full conversion within 48 h of reaction time.

4.1.2 Reaction Scope and Discussion

After having established optimal conditions for the highly enantioselective epoxidation of 2-cyclohexenone (**46a**), we became interested in exploring the scope of this new catalytic asymmetric transformation to cyclic α,β-unsaturated ketones of different ring sizes and with different substitution patterns. The following protocol was applied: catalyst salts [(*R*,*R*)-DPEN · (*S*)-TRIP] or [9-NH_2-*epi*Q (**13**) · 2 TFA] were prepared *in situ* by mixing chiral primary amines ((*R*,*R*)-DPEN or 9-NH_2-*epi*Q (**13**); 10 mol%) with the respective acids ((*S*)-TRIP: 10 mol%; TFA: 20 mol%) in dioxane for 10 min at room temperature. Then, cyclic enone **46** (0.25 M) was added followed by the addition of aqueous hydrogen peroxide (50 wt%; 1.5 equiv) and the reaction mixture was stirred at 30–50 °C for 24–48 h.

4.1.2.1 Scope of Substituted Cyclohexenones

At the outset, we focused on 2-cyclohexenones bearing substituents at the 4-, 5-, and 6-positions (Table 4.10). Both catalyst systems [(*R*,*R*)-DPEN · (*S*)-TRIP] and [**13** · 2 TFA] efficiently mediated the epoxidation of 4,4-dimethyl-2-cyclohexenone (**46b**) to afford the corresponding epoxides (*R*,*R*)- and (*S*,*S*)-**48b**, respectively, in high yields along with high enantiomeric ratio of 97:3 in both cases (entries 3–4). Substitution at the 5-position was well tolerated by the catalyst system [(*R*,*R*)-DPEN · (*S*)-TRIP], whereas [**13** · 2 TFA] provided 5,5-dimethyl-substituted 2,3-epoxycyclohexanones **48c** and **d** with slightly lower enantioselectivity, albeit in a comparable yield (entries 5–8).

To evaluate the effect of substituents at the 6-position, we studied the epoxidation of racemic 6-methyl-2-cyclohexenone (*rac*-**46e**). This substrate exhibited with both catalyst systems a significantly reduced reactivity compared to 4- and 5-substituted cyclohexenone: With [**13** · 2 TFA] (10 mol%) after 48 h at 50 °C, a mixture of *cis*- and *trans*-epoxides **48e** were formed in 47% yield; and in even lower yield of 34% in the presence of the catalyst [(*R*,*R*)-DPEN · (*S*)-TRIP] (10 mol%) under otherwise identical conditions (entries 9–10). Presumably, steric constraints may hinder iminum ion formation and may account for the low reactivity of 6-methyl-2-cyclohexenone (**46e**).

Next, we investigated the epoxidation of 3-methyl-2-cyclohexenone (**46f**). Both catalyst systems were tested and to our delight, [**13** · 2 TFA] gave the desired

Table 4.10 Substrate scope I: 4-, 5-, and 6-substituted 2-cyclohexenones

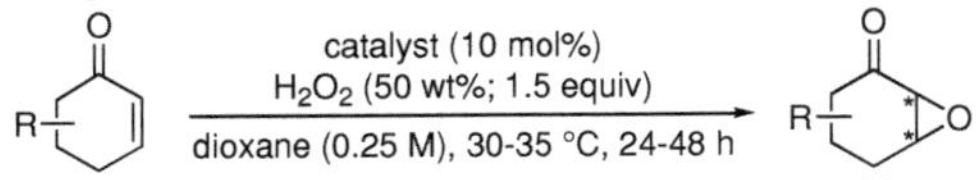

46 R = H, Me → **48**

Entry	Catalyst	Enone	Product		Yield [%][a]	*er*[b]
1	[**13** · 2 TFA]	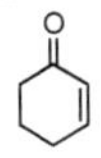 **46a**	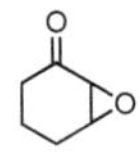 **48a**	(*S*,*S*)-**48a**	58 (93)	97:3
2	[(*R*,*R*)-**12** · (*S*)-TRIP]			(*R*,*R*)-**48a**	68 (99)	96:4
3	[**13** · 2 TFA]	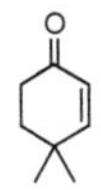 **46b**	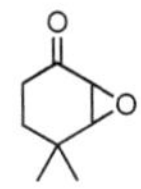 **48b**	(*S*,*S*)-**48b**	84	97:3
4	[(*R*,*R*)-**12** · (*S*)-TRIP]			(*R*,*R*)-**48b**	80	97:3
5	[**13** · 2 TFA]	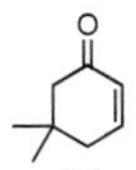 **46c**	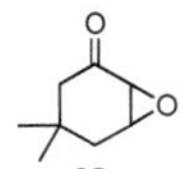 **48c**	(*S*,*S*)-**48c**	72	96:4
6	[(*R*,*R*)-**12** · (*S*)-TRIP]			(*R*,*R*)-**48c**	76	98:2
7	[**13** · 2 TFA]	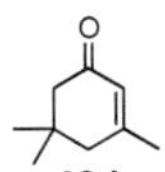 **46d**	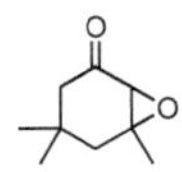 **48d**	(*S*,*S*)-**48d**	52	95.5:4.5
8	[(*R*,*R*)-**12** · (*S*)-TRIP]			(*R*,*R*)-**48d**	63	96.5:3.5
9[c]	[**13** · 2 TFA]	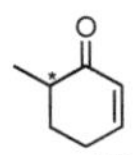 *rac*-**46e**	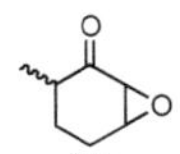 **48e**	(2*S*,3*S*)-**48e**	47	n.d.
10[c]	[(*R*,*R*)-**12** · (*S*)-TRIP]			(2*R*,3*R*)-**48e**	34	n.d.

Results obtained in collaboration with Wang

[a] Yields of pure, isolated products. Values in parenthesis correspond to yields determined by GC

[b] Determined by chiral GC

[c] At 50 °C

epoxide **48f** as the (*S*,*S*)-isomer in good yield along with high enantioselectivity (70%, 98:2 *er*) (Table 4.11, entry 1), while [(*R*,*R*)-DPEN · (*S*)-TRIP] was both considerably less active and less enantioselective providing (*R*,*R*)-**48f** in only 90.5:9.5 *er* (entry 3). In combination with TFA as acid co-catalyst, (*R*,*R*)-DPEN displayed slightly higher catalytic efficiency (entries 4–5). By using catalytic amounts of [(*R*,*R*)-DPEN · TFA], (*R*,*R*)-3-methyl-2,3-epoxycyclohexanone (**48f**) was formed in 71% yield along with 92.5:7.5 *er* (entry 4).

While studying the scope of 3-substituted cyclohexenones, [9-NH_2-*epi*Q (**13**) · 2 TFA] emerged as highly efficient and general catalyst giving a broad range of 3-substituted 2,3-epoxycyclohexanones **48f-m** in good to high yields and with excellent enantiomeric ratios of up to 99.5:0.5 (Table 4.12, entries 1–9).[2] A wide

[2] Catalyst system [(*R*,*R*)-**DPEN** · (*S*)-**TRIP**] exhibited low catalytic efficiency also with other 3-substituted 2-cyclohexenones; *i.a.* (2*R*,3*S*)-3-benzyl-2,3-epoxycyclohexanone (**48m**) was obtained in 35% yield and 86:14 *er* and (2*R*,3*S*)-3-isopropyl-2,3-epoxycyclohexanone (**48i**) in 36% yield and 89.5:10.5 *er*, respectively.

Table 4.11 Catalytic asymmetric epoxidation of 3-methyl-2-cyclohexenone (**46f**)

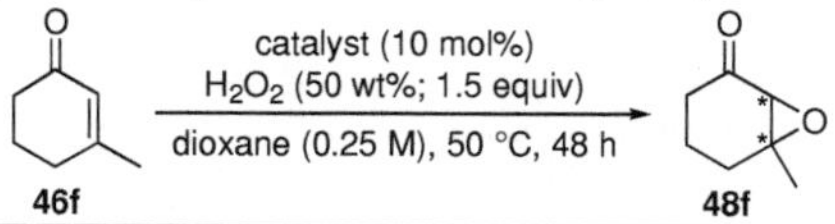

46f **48f**

Entry	Catalyst	Product	Conv. [%][a]	Yield [%][b]	*er*[c]
1[d]	[**13** · 2 TFA]	(*S*,*S*)-**48f**	95	70 (91)	98:2
2	[**13** · 2 TFA]	(*S*,*S*)-**48f**	93	(79)	97:3
3	[(*R*,*R*)-**12** · (*S*)-TRIP]	(*R*,*R*)-**48f**	66	(47)	90.5:9.5
4	[(*R*,*R*)-**12** · TFA]	(*R*,*R*)-**48f**	81	(71)	92.5:7.5
5	[(*R*,*R*)-**12** · 2 TFA]	(*R*,*R*)-**48f**	90	(58)	90:10

[a] Determined by GC
[b] Yields of pure, isolated products. Values in parenthesis correspond to yields determined by GC
[c] Determined by chiral GC
[d] At 30 °C for 24 h

variety of linear and branched alkyl substituents including ethyl, isopropyl, isobutyl, benzyl, and phenethyl groups were well tolerated and the corresponding epoxides **48** were obtained with consistently high enantioselectivity. Even cyclohexenone **46j** bearing a sterically demanding *tert*-butyl group at the 3-position was readily converted into the corresponding epoxide **48j** with essentially perfect stereocontrol (99.5:0.5 *er*) (entry 5). Pleasingly, for 3-substituted cyclohexenones, the pseudo-enantiomeric catalyst [**67** · 2 TFA] (with amine **67** derived from quinidine, cf. Table 4.4) provided access to antipodal epoxides with equally high enantioselectivities as demonstrated for 3-benzyl-2-cyclohexenone (**46m**) (entry 9).

In the case of 3-allyl-2-cyclohexenone (**46k**), partial isomerization of the terminal double bond into conjugation with the α,β-enone moiety accounts for the low yield of 23% of 3-allyl-2,3-epoxycyclohexanone (**48k**) (entry 6). This facile isomerization is probably enhanced upon formation of the iminium ion **A**, which should lead to increased acidity in the γ'-position (Scheme 4.1).

3-(Prop-1-en-1-yl)-2-cyclohexenone (**76**) is not susceptible to epoxidation under our reaction conditions neither at the endocyclic double bond (1,4-addition of hydrogen peroxide) nor at the acyclic double bond (1,6-addition). Accordingly, both 3-vinyl- and 3-ethinyl-2-cyclohexenones **46o** and **p** as further examples of α,β-γ,δ-unsaturated cyclic ketones could not be converted into the desired epoxides (Table 4.12, entries 11 and 12). The related 3-phenyl-2-cyclohexenone **46n**, as a representative of 3-aryl substituted cyclohexenones, did not productively engage in the reaction either (entry 10). Extended conjugation either through an aromatic or olefinic substituent at the 3-position seems to affect the electronic properties of the system to the point that conjugate addition of hydrogen peroxide does not proceed.

Ketoisophorone (**46q**) gave the corresponding epoxide **48q** in a reasonable yield along with high enantioselectivity of 96:4 *er* (Scheme 4.2). However, this

Table 4.12 Substrate scope II: 3-substituted 2-cyclohexenones

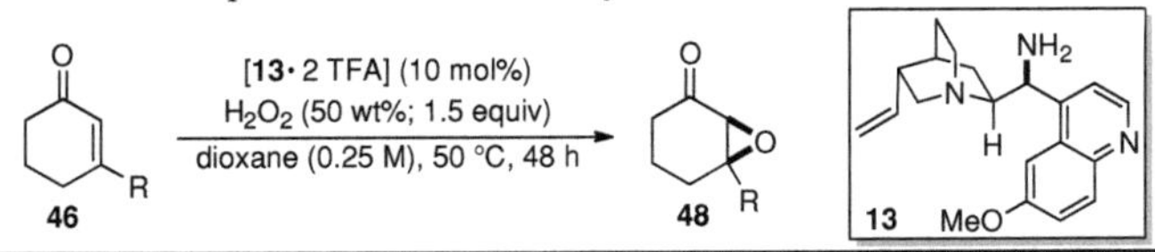

Entry	R	Enone	Product	Yield [%][a]	*er*[b]
1[c]	Me	**46f**	(2*S*,3*S*)-**48f**	70	98:2
2[d]	Et	**46g**	(2*S*,3*S*)-**48g**	73	98.5:1.5
3	*i*-Bu	**46h**	(2*S*,3*R*)-**48h**	73	98:2
4	*i*-Pr	**46i**	(2*S*,3*R*)-**48i**	79	99:1
5	*t*-Bu	**46j**	(2*S*,3*R*)-**48j**	68	99.5:0.5
6	allyl	**46k**	(2*S*,3*R*)-**48k**	23	97.5:2.5
7	CH_2CH_2Ph	**46l**	(2*S*,3*R*)-**48l**	84	98.5:1.5
8	CH_2Ph	**46m**	(2*S*,3*R*)-**48m**	78	99:1
9[e]	CH_2Ph	**46m**	(2*R*,3*S*)-**48m**	77	1.5:98.5
10	Ph	**46n**	(2*S*,3*R*)-**48n**	0	–
11	vinyl	**46o**	(2*S*,3*R*)-**48o**	0	–
12	ethinyl	**46p**	(2*S*,3*R*)-**48p**	0	–

[a] Yields of pure, isolated products
[b] Determined by chiral GC
[c] At 30 °C for 24 h
[d] 30 h
[e] [**67** · 2 TFA] (10 mol%) was used as the catalyst

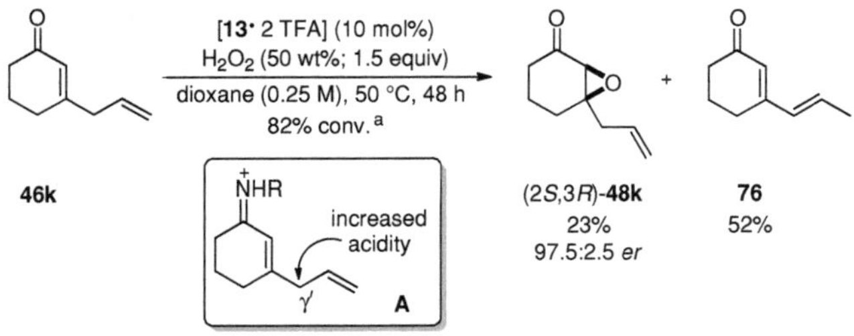

Scheme 4.1 Catalytic asymmetric epoxidation of 3-allyl-2-cyclohexenone (**46k**). ([a]Determined by GC)

enedione substrate turned out to be less reactive, and therefore 20 mol% [**13** · 2 TFA] were employed in the reaction.[3]

Unfortunately, physiologically relevant vitamin K_3 (menadion, **77**) did not participate in the epoxidation reaction under the present reaction conditions, neither did benzoquinone dimethylmonoacetal (**78**) (Fig. 4.2).

[3] [(*R*,*R*)-DPEN · TFA] and [(*R*,*R*)-DPEN · (*S*)-TRIP] were tested, but gave inferior results (~40% conversion, 80:20 and 93:7 *er*, respectively).

Scheme 4.2 Catalytic asymmetric epoxidation of ketoisophorone (46q). ([a]Determined by GC)

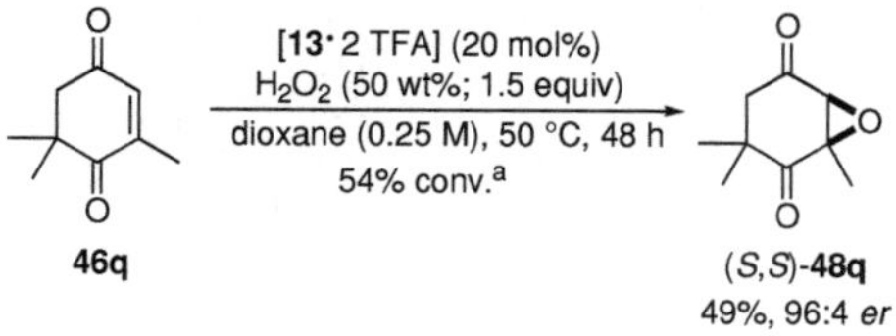

Fig. 4.2 Current limitations of the catalytic asymmetric enone epoxidation

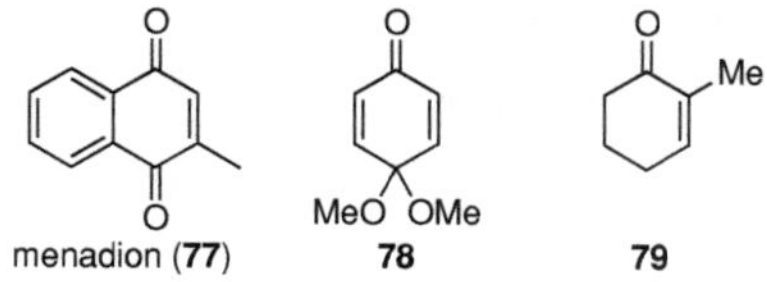

In addition, α-substituted cyclic enones did not respond productively to iminium ion catalysis as exemplified by the attempted epoxidation of 2-methyl-2-cyclohexenone (**79**). Even after prolonged reaction times up to five days at 50 °C in the presence of either catalyst [9-NH_2-*epi*Q (**13**) · 2 TFA] or [(*R*,*R*)-DPEN · (*S*)-TRIP], only a trace amount of product could be detected by GC-MS.

The inherent difficulties in generating congested iminium ion intermediates from ketone substrates have been partly overcome with the rising use of primary amine catalysts. Owing to reduced steric constraints, primary amines are (in contrast to most secondary amines) able to activate sterically demanding ketone substrates. Nevertheless, the activation of α-**substituted** α,β-unsaturated ketones remains challenging in this context and has, to the best of our knowledge, not been implemented to date (cf. Scheme 2.11).

4.1.2.2 Scope of Cyclic Enones of Different Ring Sizes

Cyclic α,β-unsaturated ketones of different ring sizes were subjected to our catalytic asymmetric epoxidation protocol in the presence of catalytic amounts of [9-NH_2-*epi*Q (**13**) · 2 TFA] or [(*R*,*R*)-DPEN · (*S*)-TRIP]. Except for 2-cyclopentenone (**46r**), the corresponding epoxides **48a, 48s–x** were generated in high yields and with high optical purity by using either of the two catalyst systems (Table 4.13).

Cycloheptenone oxide **48s** was provided by both catalyst systems with excellent enantiomeric ratios of 99:1 with [(*R*,*R*)-DPEN · (*S*)-TRIP] and > 99.5:0.5 with [**13** · 2 TFA] (entries 5–6). Moreover, catalyst [**13** · 2 TFA] afforded 3-substituted cycloheptenone oxides **48t-u** in high yields along with outstanding enantioselectivities (entries 7–8). The corresponding eight-membered ring epoxide **48v** was obtained from 2-cyclooctenone (**46v**) in good yield and high enantioselectivity (55%, 98:2 *er*) (entry 9). In addition, we studied macrocyclic enones in the epoxidation reaction. These proved to be somewhat more reactive than the seven- and eight-membered ring analogues, just as cyclohexenone and its derivatives. As the ring size increased, the 9-amino(9-deoxy)*epi*quinine TFA salt [**13** · 2 TFA]

Table 4.13 Substrate scope III: cyclic enones of different ring sizes

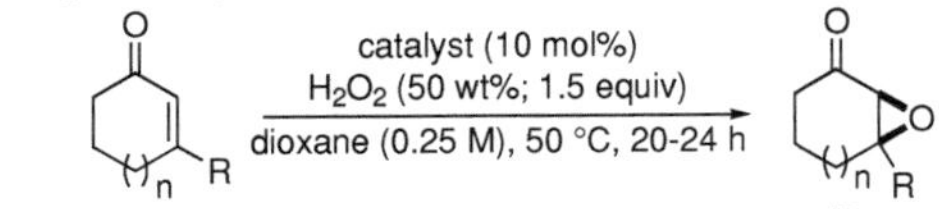

46
n = 0-10
R = H, Et, CH_2Ph

48

Entry	Catalyst	Enone	Product		Yield [%][a]	*er*[b]
1[c]	[**13** · 2 TFA]	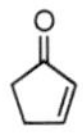 **46r**	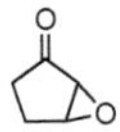 **48r**	(*S*,*S*)-**48r**	(33)	87:13
2[c]	[(*R*,*R*)-DPEN · (*S*)-TRIP]			(*R*,*R*)-**48r**	(33)	90:10
3[d]	[**13** · 2 TFA]	 **46a**	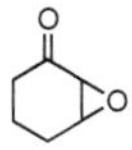 **48a**	(*S*,*S*)-**48a**	58 (93)	97:3
4[d]	[(*R*,*R*)-DPEN · (*S*)-TRIP]			(*R*,*R*)-**48a**	68 (> 99)	96:4
5	[**13** · 2 TFA]	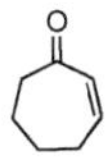 **46s**	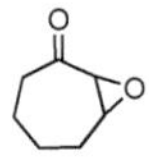 **48s**	(*S*,*S*)-**48s**	62	>99.5:0.5
6	[(*R*,*R*)-DPEN · (*S*)-TRIP]			(*R*,*R*)-**48s**	65[e]	99:1
		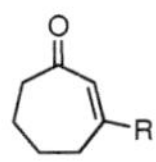	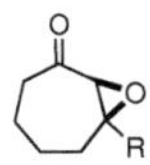			
7	[**13** · 2 TFA]	R = Et: **46t**	R = Et: **48t**	(*S*,*S*)-**48t**	82	>99.5:0.5
8	[**13** · 2 TFA]	R = CH_2Ph: **46u**	R = CH_2Ph: **48u**	(*S*,*R*)-**48u**	85	>99.5:0.5
9[f]	[**13** · 2 TFA]	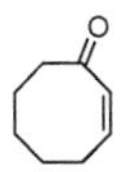 **46v**	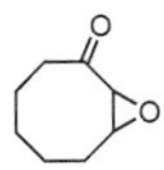 **48v**	(*S*,*S*)-**48v**	55	98:2
10	[**13** · 2 TFA]	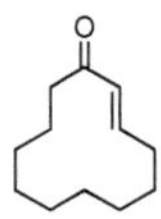 **46w**	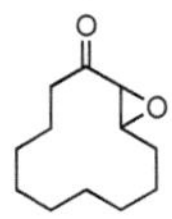 **48w**	(2*S*,3*R*)-**48w**	92	99.5:0.5
11[g]	[(*R*,*R*)-DPEN · (*S*)-TRIP]			(2*R*,3*S*)-**48w**	92	97:3
12	[**13** · 2 TFA]	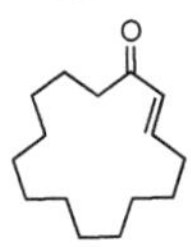 **46x**	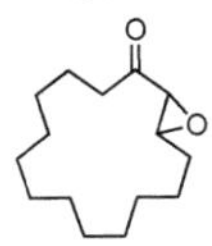 **48x**	(2*S*,3*R*)-**48x**	87	99.5:0.5
13	[**67** · 2 TFA]			(2*R*,3*S*)-**48x**	86	99.5:0.5
14[g]	[(*R*,*R*)-DPEN · (*S*)-TRIP]			(2*R*,3*S*)-**48x**	52	95.5:4.5

[a] Yields of pure, isolated products. Values in parenthesis correspond to yields determined by GC
[b] Determined by chiral GC
[c] 48 h
[d] At 30/35 °C
[e] Determined by ^1H NMR analysis of the crude mixture
[f] Crude product was treated with 1 N NaOH (1 equiv) in THF
[g] At 35 °C for 48 h

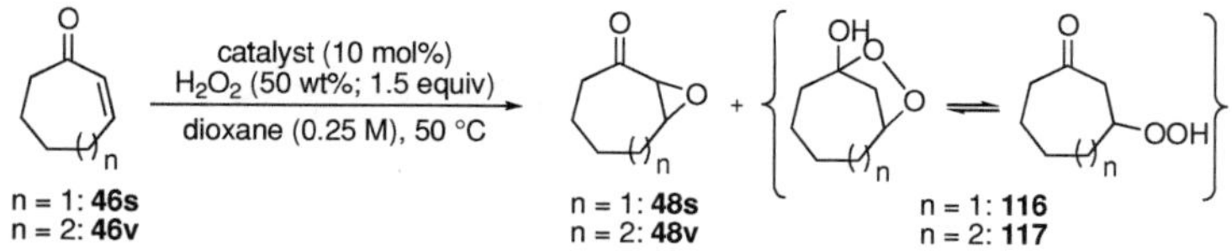

Scheme 4.3 Primary amine salt-catalyzed reaction of seven- and eight-membered cyclic enones **46s** and **v** with hydrogen peroxide

gave superior results compared to the catalyst system [(*R*,*R*)-DPEN · (*S*)-TRIP]. The two examples of macrocyclic enones nicely illustrate this trend: whereas epoxycyclododecanone **48w** was still obtained in an enantiomeric ratio of 97:3 *er* with [(*R*,*R*)-DPEN · (*S*)-TRIP] (entry 11), epoxycyclopentadecanone **48x** was formed in a slightly reduced 95.5:4.5 *er* (entry 14). Since both rings are large enough to allow *s-cis* and *s-trans* conformational interconversion, they resemble acyclic enones in their conformational flexibility. Indeed, we had found, during our studies with acyclic enones as substrates, that catalyst [(*R*,*R*)-DPEN · (*S*)-TRIP] gave inferior results compared to [**13** · 2 TFA] with this substrate class (cf. Table 4.21).

Yet, the quinine-derived catalyst system [**13** · 2 TFA] proved highly efficient and gave macrocyclic epoxides **48w-x** in high yields along with outstanding enantioselectivities of 99.5:0.5 *er* in both cases (entries 10 and 12). Pleasingly, the pseudo-enantio-meric catalyst salt [**67** · 2 TFA] derived from quinidine gave epoxide **48x** with the opposite sense and identical magnitude of stereoinduction (0.5:99.5 *er*) (entry 13).

Cyclopentenone epoxide **48r** was obtained in moderate yield along with a good enantioselectivity of 90:10 *er* with catalyst [(*R*,*R*)-DPEN · (*S*)-TRIP] and still with 87:13 *er* in the presence of [9-NH_2-*epi*Q (**13**) · 2 TFA] (entries 1–2), which are the highest enantio-selec-tivities reported to date for the catalytic asymmetric epoxidation of 2-cyclopentenone (**46r**) [11]. Nevertheless, the results obtained with 2-cyclopentenone are only moderate if compared with those for cyclohexenone and higher homologues, which is a general phenomenon observed in catalytic asymmetric conjugate additions to cyclic enones [12, 13]. Likewise, cyclopentenone is a special case in our epoxidation reaction and requires an individual treatment. Thus, a separate catalyst screening was conducted dedicated to the epoxidation of 2-cyclopentenone (**46r**) and the results are summarized and discussed in Sect. 4.1.2.5.

Whereas the epoxides were essentially the only products observed in the reaction of 2-cyclopentenone, 2-cyclohexenone, as well as the macrocyclic enones, 2-cycloheptenone (**46s**) and 2-cyclooctenone (**46v**) afforded a second product along with the expected epoxide. This product was identified as the conjugate addition product of hydrogen peroxide. Such β-hydroperoxy-cycloalkanones exist in equilibrium with the bicyclic peroxyhemiketals as depicted in Scheme 4.3.

A detailed analysis of this observation, as well as a study of the influence of different catalysts and reaction conditions on the product distribution, will be presented in Sect. 4.2.2.6.

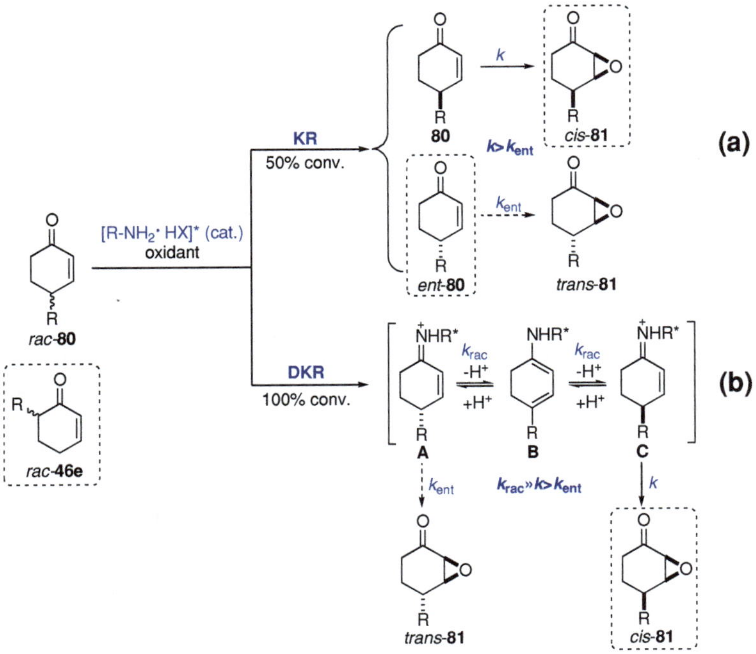

Scheme 4.4 General scheme for the **a** kinetic resolution (KR), or **b** dynamic kinetic resolution (DKR) of 4- (or 6-) substituted cyclohexenones via asymmetric epoxidation

4.1.2.3 Toward (Dynamic) Kinetic Resolutions of 4-Substituted Cyclohexenones

At this juncture, we were intrigued by the asymmetric epoxidation of racemic 4-substituted cyclohexenones of the general structure *rac*-**80**. We envisaged that one enantiomer of the starting material *rac*-**80** might react preferentially in the sense of a kinetic resolution (KR) of cyclohexenones via asymmetric epoxidation as depicted in Scheme 4.4a. Thus, synthetically valuable enantioenriched 4-substituted cyclohexenones **80** and enantioenriched epoxy cyclohexanones **81** would be concomitantly obtained. We might even encounter a case of dynamic kinetic resolution (DKR) if interconversion of iminium ions **A** and **C** via a dienamine intermediate **B** would proceed at rates greater than those of the epoxidation pathways ($k_{rac} \gg k > k_{ent}$; Scheme 4.4b). An analogous scenario may be anticipated for the chiral amine salt-catalyzed asymmetric epoxidation of 6-substituted cyclohexenones *rac*-**46e** (Scheme 4.4). Racemization of the corresponding 6-substituted α,β-unsaturated iminium ions (equivalent to the iminium ions **A**) via a cross-conjugated dienamine intermediate may occur more readily due to the increased acidity of the α'-position of cyclohexenone compared to the vinylogous γ-position [14]. However, since the iminium ion formation with substrate *rac*-**46e** seems to be hindered by steric constraints caused by the substituent

Table 4.14 Asymmetric epoxidation of racemic 4-methyl-2-cyclohexenone (*rac*-**80a**)

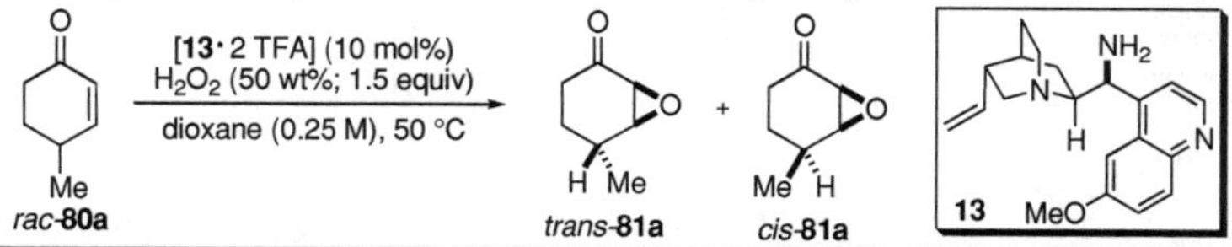

Entry	t [h]	conv. [%][a]	*dr* (*trans*/*cis*)[a]	*er* (*trans*-**81a**)[b]	*er* (*cis*-**81a**)[b]	*er* (**80a**)[b]
1	4	55	54:46	n.d.	n.d	n.d
2	26	94	54:46	93.5:6.5	99:1	56:44

[a] Determined by GC
[b] Determined by chiral GC

at the 6-position (Table 4.10, entries 9–10), we decided to start our investigations with 4-substituted racemic substrates *rac*-**80**.

To test our hypothesis, 4-methyl-2-cyclohexenone (*rac*-**80a**) was subjected to the standard reaction conditions and samples were taken at different time intervals (Table 4.14). GC analysis of the samples revealed that both diastereomeric epoxides *cis*- and *trans*-**81a** were formed with high enantioselectivities at similar rates [54:46 *dr*(*trans/cis*)]. In stark contrast, the racemic synthesis with alkaline hydrogen peroxide in methanol proceeded with a diastereoselectivity of 74:26 *dr* in favour of the *trans*-epoxide **81a** [15–17].

This result represents a manifest case of strong catalyst control. The configuration of the substrate appears to play only a minor role and, therefore, (dynamic) kinetic resolution did not take place—at least not to a synthetically useful extent. At 55% conversion the starting material exhibited an enantiomeric ratio of only 56:44.

Interestingly, the *cis*- and *trans*-epoxides **81a** were obtained with in both cases high, yet noticeably different levels of enantiocontrol. Whereas *cis*-**81a** was formed in an excellent enantiomeric ratio of 99:1 *er*, corresponding *trans*-**81a** exhibited a con-siderably lower optical purity (93.5:6.5 *er*). This can be rationalized through analysis of the putative pre-transition state assemblies depicted in Scheme 4.5.

Singleton and co-workers have found that the lowest-energy transition state structure for the conjugate addition of *tert*-butyl hydroperoxide to 4-methyl-2-cyclohexenone is the one which combines the prerequisite axial attack of the nucleophile with the methyl group adopting a pseudo-axial rather than a pseudo-equatorial orientation [18]. Despite the concomitant strain caused by the latter orientation, this transition state structure is favored since an equatorial substituent at C-4 would suffer from unfavorable interactions with the incoming *tert*-butyl peroxy anion.

Accordingly, Scheme 4.5 shows the possible pre-transition state assemblies **A-D** for the conjugate addition of hydrogen peroxide to the α,β-unsaturated iminium ions generated from (*S*)- and (*R*)-4-methyl-2-cyclohexenone (**A/B** and **C/D**, respectively) and 9-amino(9-deoxy)*epi*-quinine (**13**) (simplified as R-NH_2). *Trans*-epoxide *trans*-**81a** arises from a reaction proceeding through half-chair **C**. Both catalyst and substrate direct the hydrogen peroxide attack to the *si*-face of the double bond. The addition of hydrogen peroxide through half-chair **A** to

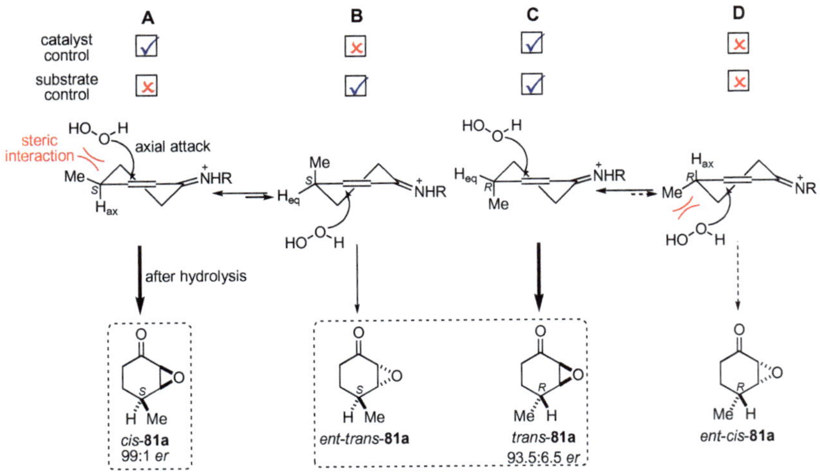

Scheme 4.5 Pre-transition state assemblies **A-D** for the conjugate addition of hydrogen peroxide to 4-methyl-2-cyclohexenone-derived iminium ions

generate *cis*-epoxide *cis*-**81a** occurs as well in accordance with catalyst control on the *si*-face of the molecule. However, in this setting, the pseudo-equatorial methyl group at C-4 experiences steric interactions with the incoming nucleophile. Thus, half-chair **A** may have an inherent tendency to undergo ring inversion and place the methyl group in pseudo-axial orientation (cf. half-chair **B**). Subsequent axial attack of hydrogen peroxide to half-chair **B** would afford the opposite enantiomer of *trans*-**81a** accounting for its lower experimentally obtained enantioselectivity compared to the corresponding *cis*-**81a**.

We tried to improve the efficiency of the envisaged kinetic resolution by increasing the size of the substituent R in *rac*-**80**.

4-*tert*-Butyl-2-cyclohexenone **80b** gave *trans*-epoxide **81b**[4] in 48% yield, after 4 d at 50 °C in the presence of catalyst salt [**13** · 2 TFA], with a diastereomeric ratio of 92:8 (*trans*/*cis*).[5] Although the remaining starting material (44%) was significantly enantioenriched with 86:14 *er*, the enantiomeric ratio of *trans*-epoxide **81b** remained relatively low (79.5:20.5 *er*) (Scheme 4.6).

We rationalized this observation in an analogous manner to the reaction with 4-methyl-2-cyclohexenone (**80a**), by analyzing the corresponding plausible pre-transition state assemblies (cf. Scheme 4.5). In addition, we take into account the increased bulkiness of the *tert*-butyl group at C-4 and its resultant reluctance towards adopting a pseudo-axial orientation. The large size of this substituent may further multiply steric interactions arising from *cis*-selective axial attack of hydrogen peroxide. Thus, we invoke pre-transition state assemblies **A′** and **D′**, with

---

[4] The relative configuration was assigned by NOE analysis.

[5] Racemate was obtained with alkaline hydrogen peroxide in methanol with 98:2 *dr*.

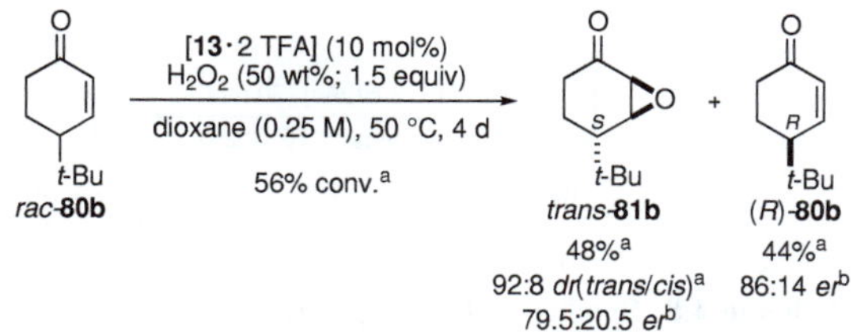

Scheme 4.6 Asymmetric epoxidation of racemic 4-*tert*-butyl-2-cyclohexenone (*rac*-**80b**). ([a]Determined by GC; [b]Determined by chiral GC.)

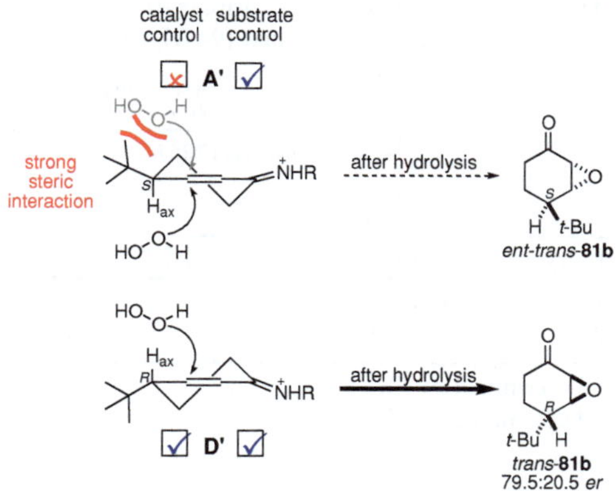

Scheme 4.7 Pre-transition state assemblies **A′** and **D′** for the conjugate addition of hydrogen peroxide to 4-*tert*-butyl-2-cyclohexenone-derived iminium ions

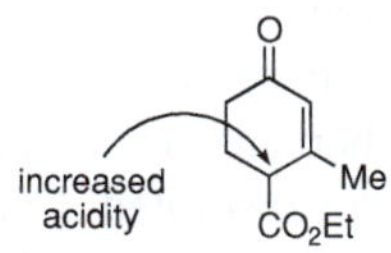

Fig. 4.3 Hagemann's ester

the *tert*-butyl group in pseudo-equatorial orientation, and assume an equatorial attack of hydrogen peroxide, which is in line with *trans*-epoxide **81b** being formed as essentially a single diastereomer (Scheme 4.7). The observed enantioselectivity, however, remains only moderate (79.5:20.5 *er*) as a consequence of this.

Further attempts to achieve higher efficiency in the kinetic resolution of racemic 4-substituted cyclohexenones *rac*-**80** via asymmetric epoxidation should strive to assist the interconversion of the enantiomers of the starting material. For instance, if Hagemann's ester (Fig. 4.3) is employed, the expected increased acidity at the γ-position may render dynamic kinetic resolutions via dienamine formation more feasible (cf. Scheme 4.4) [19, 20].

4.1.2.4 Epoxidation of Enantioenriched 5-Substituted Cyclohexenones

We were further interested how our catalyst system would deal with pre-existing stereocenters embedded in the ring scaffold which cannot be equilibrated in the

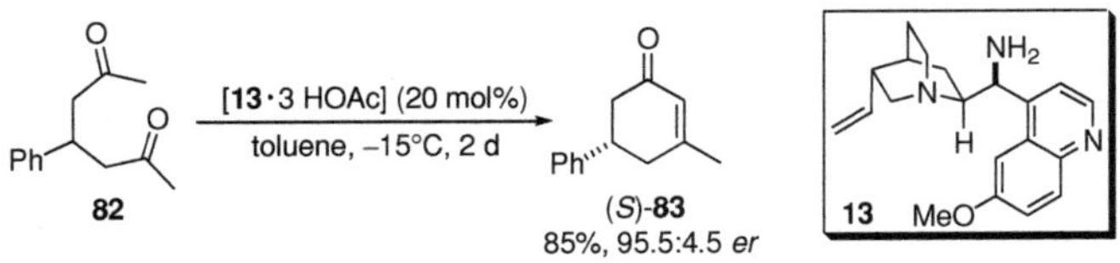

Scheme 4.8 Synthesis of (*S*)-5-phenyl-3-methyl-2-cyclohexenone [(*S*)-**83**] [7]

Table 4.15 Catalytic asymmetric epoxidation of (*S*)-**83** with pseudoenantiomeric catalysts [9-NH_2-*epi*Q (**13**) · 2 TFA] and [9-NH_2-*epi*QD (**67**) · 2 TFA]

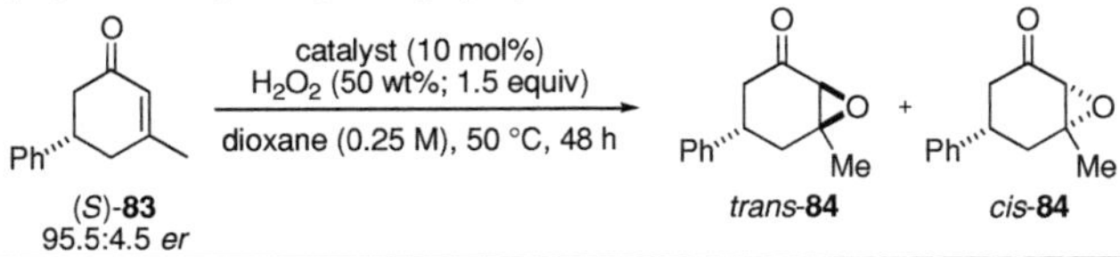

Entry	Catalyst	Conv. [%][a]	Yield [%][a]	*dr* (*trans*/*cis*)[a]	*er* (*trans*-**84**)[b]
1	[**13** · 2 TFA]	83	75	97:3	98.5:1.5
2	[**67** · 2 TFA]	79	74	63:37	92.5:7.5

[a] Determined by GC
[b] Determined by chiral GC

course of the reaction. To this end, we prepared chiral enantioenriched 5-phenyl-3-methyl-2-cyclohexenone [(*S*)-**83**] according to a method previously developed in our group (Scheme 4.8) [7]. The aldol cyclodehydration of 4-substituted 2,6-heptandiones such as **82** in presence of the chiral catalyst [**13** · 3 HOAc] afforded 5-substituted-3-methyl-2-cyclohexenones in high yields and with enantioselectivities greater than 95:5 *er*.

We then subjected enantiomerically enriched (*S*)-**83** to our epoxidation reaction. The results are depicted in Table 4.15.[6]

Both catalysts tested—[**13** · 2 TFA] and its pseudoenantiomeric form [**67** · 2 TFA]—mediated the epoxidation of cyclohexenone (*S*)-**83** with similar rates. However, whereas [**13** · 2 TFA] gave epoxide **84** with a diastereomeric ratio of 97:3 (*trans*/*cis*) and an enantiomeric ratio of 98.5:1.5 (*trans*), the latter gave epoxide **84** as a 63:37 (*trans*/*cis*) mixture of diastereomers along with inferior enantioselectivity of 92.5:7.5 *er* (major).[7]

In the presence of catalyst [**13**· 2 TFA], the epoxidation of the major enantiomer (*S*)-**83** is directed by both the catalyst and the substrate to the *si*-face of the enone (Scheme 4.9**A**). On the contrary, the reaction of the minor enantiomer (*R*)-**83**

[6] One-pot syntheses of epoxide **84** starting from diketone **82** are possible in the presence of catalyst [**67** · 2 TCA] but have not been optimized.

[7] Racemate was obtained with alkaline hydrogen peroxide in methanol with 94:6 *dr* from racemic enone **83**.

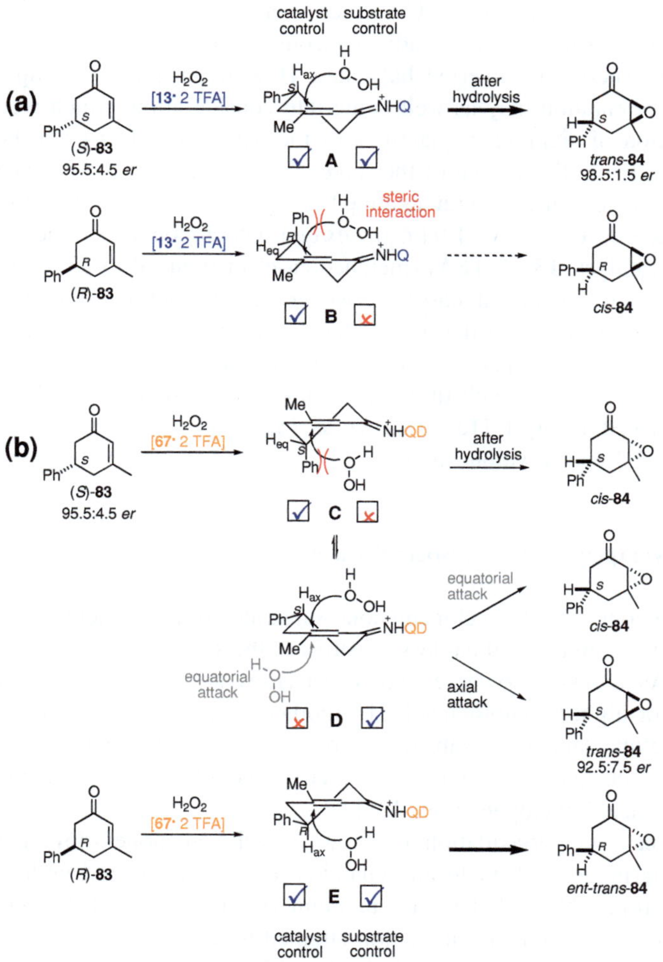

Scheme 4.9 Pre-transition state assemblies **A**-**E** for the conjugate addition of hydrogen peroxide to (*S*)-**83**- and (*R*)-**83**-derived iminium ions with catalysts **a** [**13** · 2 TFA], and **b** [**67** · 2 TFA]. (*Q* quinine, *QD* quinidine)

(via **B**) is kinetically disfavoured and thus, we observed an enantiomeric enrichment in *trans*-epoxide **84**.

In general, pseudoenantiomeric catalyst [**67** · 2 TFA] derived from quinidine preferentially delivers hydrogen peroxide from the *re*-face. However, the asymmetric epoxidation of cyclohexenone (*S*)-**83** in the presence of [**67** · 2 TFA] furnished a mixture of diastereomers (63:37 *dr*) in favour of the *trans*-epoxide **84** which was obtained in reduced optical purity of only 92.5:7.5 *er*. Putative pre-transition state assemblies **C**-**E** imply that the reaction of the minor enantiomer (*R*)-**83** is kinetically favoured in the presence of catalyst [**67** · 2 TFA]. Moreover, *trans*-epoxide **84** is obtained as the major product; this presumably arises since pre-transition state assembly **C**, which may account for the formation of *cis*-epoxide **84**, suffers from disadvantageous steric interactions with the phenyl substituent. Therefore, inverted half-chair **D** with the phenyl group in pseudo-equatorial orientation may as well come into consideration, which would explain the formation of significant quantities of the *trans*-epoxide **84** via the preferred axial attack (from the *si*-face of the molecule regardless of the catalyst's bias).

The racemate synthesis with alkaline hydrogen peroxide in methanol proceeded with a diastereoselectivity of 94:6 (*trans*/*cis*), which could be either enhanced to 97:3 with catalyst [**13** · 2 TFA] (matched case) or reduced to 63:37 with catalyst [**67** · 2 TFA] (mismatched case). However, it was not possible to invert the diastereoselectivity within the bias of the cyclic system. In addition, this experiment lends further experimental support to the assignment of the absolute stereochemistry of 3-substituted-2,3-epoxycyclohexanones generated with quinine-derived catalyst [**13** · 2 TFA] as (*S*,*S*) (as we have also deduced by analogy to 2,3-epoxycyclohexanone).

4.1.2.5 Cyclopentenone: A Special Case

A problem connected to cyclopentenone substrates is a considerable "flatness" of the molecules, rendering them less sensitive to the steric requirements of the chiral catalyst. As a result, lower enantioselectivities are obtained than with cyclohexenone or higher homologues [12]. Another common issue associated with the use of cyclopentenone as substrate pertains to its high reactivity. The enolate generated after conjugate addition is reactive enough to undergo conjugate addition to unreacted cyclopentenone [21].[8]

In the catalytic epoxidation reaction, 2-cyclopentenone (**46r**) gave inferior results in terms of both yield and enantioselectivity compared with other cyclic enone substrates. Nevertheless, an enantiomeric ratio of 90:10 still represents the highest level of asymmetric induction attained to date for the epoxidation of this particular substrate (Table 4.16, cf. Table 4.13).

Indeed, only one report is known in the current literature which addresses the epoxidation of cyclopentenone [11]. By using chiral diketopiperazine-derived hydroperoxide **85** as stoichiometric oxidant, 2,3-epoxycyclopentanone (**48r**) was generated in 31% yield and 56:44 *er* (Fig. 4.4).

Eager to address the challenge that cyclopentenones pose to asymmetric catalysis, we reevaluated the catalytic potential of a variety of chiral primary amine salts for the epoxidation of cyclopentenone (Table 4.17). (*R*,*R*)-DPEN (**12**) as its TFA salt showed to be slightly less active and at the same time considerably less enantioselective than when employed in combination with (*S*)-TRIP (entries 1–2). Enantioselectivities attained with 9-amino(9-deoxy)*epi*cinchona alkaloid

---

[8] This is in accordance with our observation that 2-cyclopentenone (**46r**) dimerizes in the presence of the catalytic salt [**13** · 2 TFA] as detected by GC–MS and ESI-MS.

Table 4.16 Catalytic asymmetric epoxidation of 2-cyclopentenone (**46r**)

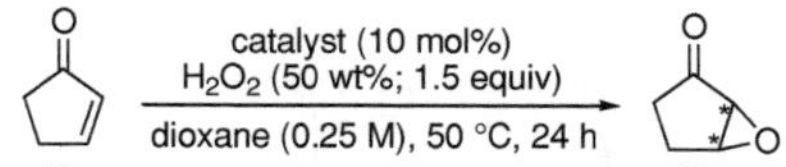

Entry	Catalyst	Product	Yield [%][a]	*er*[b]
1	[9-NH_2-*epi*Q (**13**) · 2 TFA]	(*S*,*S*)-**48r**	33	87:13
2	[(*R*,*R*)-DPEN · (*S*)-TRIP]	(*R*,*R*)-**48r**	33	90:10

[a] Determined by GC
[b] Determined by chiral GC

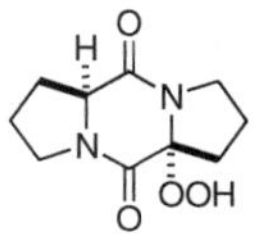

Fig. 4.4 Diketopiperazine-derived hydroperoxide **85** by Laschat et al.

derivatives **13**, **67–72** (as their diTFA salts) (cf. Table 4.4) all ranged between 84:16 and 86.5:13.5 *er* with cyclopentenone epoxide **48r** being generated in 27–45% yield. Re-screening different solvents for the cyclopentenone epoxidation confirmed dioxane as the solvent of choice. The use of mono TFA salt [**13** · TFA] increased the catalytic activity slightly and afforded the desired product in 44% yield with essentially equal optical purity (87:13 *er*) (entry 3), whereas a 1:3 amine/TFA ratio resulted in a drop of enantioselectivity to 79.5:20.5 *er* (entry 6). As for 2-cyclohexenone, the epoxidation proceeded with low enantioselectivity in the presence of amine **73** (9-NH_2-Q; cf. Table 4.4) featuring the natural (*R*)-configuration of quinine at C-9 (entry 7). We then had a look at different acid co-catalysts. Pleasingly, the use of diphenyl phosphoric acid (**66**) instead of TFA enhanced the catalytic activity, and moreover had a positive impact on the enantioselectivity. In the presence of [**13** · DPPOH], 2,3-epoxycyclopentanone was furnished in 52% yield with an enantiomeric ratio of 88.5:11.5 (entry 8). Unfortunately, replacing achiral diphenyl phosphate by chiral phosphate counteranions, e.g. TRIP (**7a**), could not improve the result (entries 10–13).

Beside improving the enantioselectivity of the process, we also strove to identify an exceedingly active primary amine salt catalyst with regard to the use of 3-substituted cyclopentenone derivatives (e.g. 3-methyl-2-cyclopentenone) which exhibit reduced reactivity compared to the parent compound (entry 14).[9]

With this aim, we re-evaluated truncated Cinchona alkaloid derivatives α-(aminomethyl)-quinuclidines **63** and **64** (Table 4.18). In particular, **64** effectively mediated the epoxidation reaction. In combination with (*S*)-TRIP in a 1:1 ratio, epoxide **48r** was formed in 91% yield after 24 h and with a promising enantioselectivity of 76.5:23.5 *er* (entry 11). However, screening of a wide range of

[9] This trend had previously been observed with the six-membered ring analogues.

Table 4.17 Evaluation of various primary amine salts for the catalytic asymmetric epoxidation of 2-cyclopentenone (**46r**)

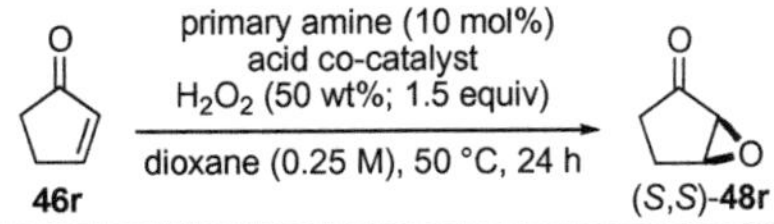

Entry	Amine	Acid co-catalyst	[mol%]	Yield [%][a]	*er*[b]
1	(*R*,*R*)-DPEN	TFA	10	19	17.5:82.5
2	(*R*,*R*)-DPEN	TFA	20	28	29:71
3	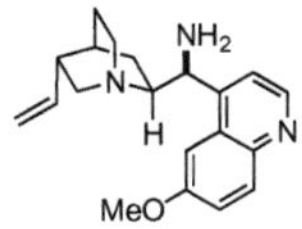 **13**	TFA	10	44	87:13
4		TFA	20	33	87:13
5[c]		TFA	20	45	87:13
6		TFA	30	34	79.5:20.5
7	**73**	TFA	20	42[d]	65:35
8	**13**	$(PhO)_2PO(OH)$ (**66**)	10	52	88.5:11.5
9	**13**	$(PhO)_2PO(OH)$ (**66**)	20	50[e]	72.5:27.5
10	**13**	(*S*)-TRIP	10	23	77:23
11	**13**	(*S*)-TRIP	20	22	56:44
12	**13**	(*R*)-TRIP	10	21	88:12
13	**13**	(*R*)-TRIP	20	29	69:31
14[c,f]	**13**	TFA	20	26	88:12

[a] Determined by GC
[b] Determined by chiral GC
[c] [**13** · 2 TFA] (20 mol%), 48 h
[d] 72 h
[e] Increased side product formation observed
[f] With 3-methyl-2-cyclopentenone as substrate

chiral BINOL-derived phosphoric acids bearing different substituents at the 3,3′-positions could not improve the result.

Encouraged by the results obtained with α-(aminomethyl)quinuclidine (**64**), we sought a way to modify the catalyst structure to enhance its stereoselectivity while at the same time retaining the high catalytic activity. We envisaged that α-(aminobenzyl)quinuclidine (**86a**), bearing an additional phenyl group at C-9 could meet these criteria (Fig. 4.5a). We speculated that such a de novo designed catalyst motif would ideally merge the high enantioselectivity of Cinchona alkaloid-derived catalysts such as 9-amino(9-deoxy)*epi*quinine (**13**) with the significantly more active truncated representative **64**, presumably due to reduced steric congestion surrounding the primary amine function.

α-(Amino-benzyl)quinuclidine (**86a**) was obtained in two steps from quinuclidine according to the retrosynthetic strategy presented in Fig. 4.5b. α-Lithiation of quinuclidine and subsequent reaction of the lithiated species with benzaldehyde gave α-(hydroxyl-benzyl)quinuclidine. The aminoalcohol was further converted into α-(amino-benzyl)quinuclidine (**86a**) via a Mitsunobu reaction with azide followed by a Staudinger reduction to afford the desired amine. A detailed presentation and discussion of this synthetic approach is provided in Sect. 4.7.1.2.

Table 4.18 Evaluation of truncated Cinchona alkaloid catalysts **63** and **64** in the catalytic asymmetric epoxidation of 2-cyclopentenone (**46r**)

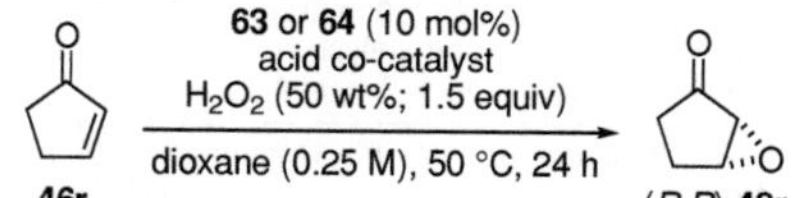

Entry	Diamine	Acid co-catalyst	[mol%]	Yield [%][a]	*er*[b]
1	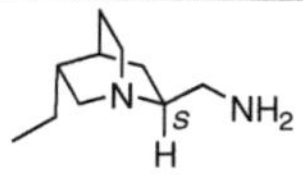	–	–	21[c]	*rac*
2		TFA	10	69	55:45
3		TFA	20	67	54.5:45.5
4	**63**	DPPOH (**66**)	10	57	*rac*
5		DPPOH (**66**)	20	31	*rac*
6		(*S*)-TRIP	10	59	63:37
7		(*R*)-TRIP	10	62	48:52
8	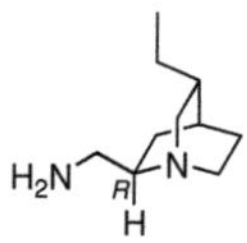	–	–	28[c]	46.5:53.5
9		TFA	10	80	44:56
10		TFA	20	82[d]	34.5:65.5
11		(*S*)-TRIP	10	91	23.5:76.5
12	**64**	(*R*)-TRIP	10	49	43:57
13		(*S*)-TRIP	20	91	29:71

[a] Determined by GC
[b] Determined by chiral GC
[c] 48 h
[d] 18 h

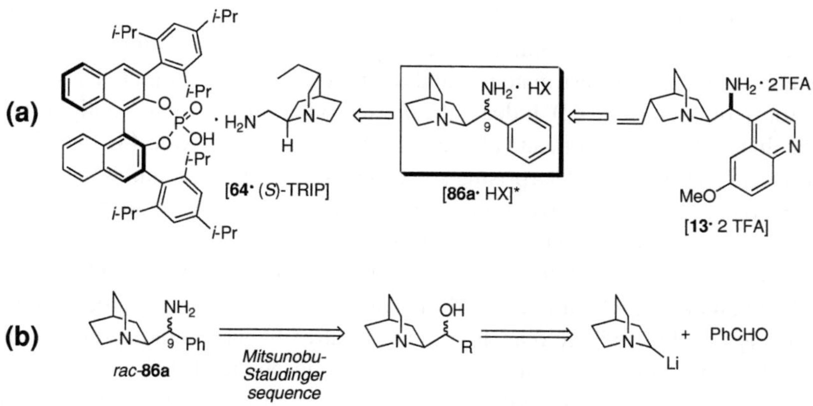

Fig. 4.5 **a** Design of α-(aminobenzyl)quinuclidine (**86a**) as new catalyst target structure, and **b** retro-synthetic analysis of **86a**

Separation of the enantiomers of racemic **86a** was attempted by chiral HPLC. Unfortunately, only the *erythro*- but not the *threo*-isomer of **86a** could be readily separated and tested as chiral catalyst in combination with different acid co-catalysts in the asymmetric epoxidation of 2-cyclopentenone (**46r**) (Table 4.19; entries 3–8).

Table 4.19 Catalytic asymmetric epoxidation of 2-cyclopentenone (**46r**) in the presence of α-(amino-benzyl)quinuclidine (**86a**) as catalyst

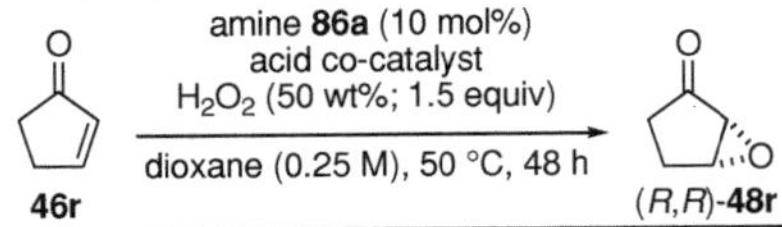

Entry	Amine	Acid co-catalyst	[mol%]	Yield [%][a]	*er*[b]
1	*rac-threo*-**86a**	TFA	10	65	–
2	*rac-threo*-**86a**	TFA	20	37	–
3[c]	(-)-*erythro*-**86a**	TFA	10	5	74:26
4[c]	(-)-*erythro*-**86a**	TFA	20	54	74:26
5[c]	(-)-*erythro*-**86a**	(*S*)-TRIP	10	9	64:36
6[c]	(-)-*erythro*-**86a**	(*S*)-TRIP	20	29	65:35
7[c]	(+)-*erythro*-**86a**	(*S*)-TRIP	10	5	32.5:67.5
8[c]	(+)-*erythro*-**86a**	(*S*)-TRIP	20	32	32:68

[a] Determined by GC
[b] Determined by chiral GC
[c] 30 h

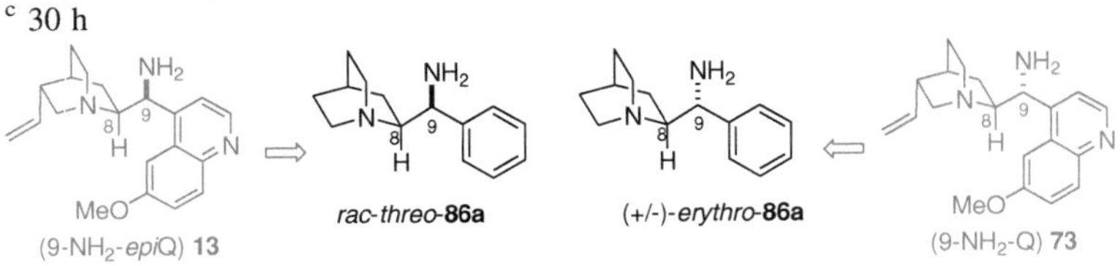

The best result was obtained by using *erythro*-**86a** as its di TFA salt (entry 4). Under these conditions, epoxycyclopentanone **48r** was generated in a maximum yield of 54% along with moderate enantioselectivity of 74:26 *er*. *threo*-**86a** was tested in racemic form to at least estimate its catalytic potential in the attempted asymmetric transformation (entries 1–2). Indeed, the *threo*-isomer turned out to be slightly more active than the *erythro*-form and afforded cyclopentenone epoxide **48r** in 65% yield at best (entry 1). This is not surprising since the *threo*-isomer resembles the powerful ***epi***cinchona alkaloid derivatives such as 9-NH_2-*epi*Q (**13**) with regard to the relative configuration at the carbon atoms C-8 and C-9. On the opposite, the *erythro*-diastereomer of **86a** just like 9-amino(9-deoxy)quinine (**73**) resembles natural Cinchona alkaloids, which in general display inferior catalytic potential compared with the ***epi***cinchona alkaloid analogues as has already been demonstrated in many cases including our epoxidation reaction of cyclic enones (cf. Table 4.4). Thus, future efforts in this area are aimed at providing the *threo*-**86a** in enantiomerically pure form to finally evaluate its catalytic activity and selectivity in the asymmetric epoxidation of 2-cyclopentenone (**46r**) among other asymmetric transformations.

During our screening studies, amino acid ester salts had turned out as highly active albeit poorly enantioselective catalysts. However, the encouraging results obtained with the DPEN-based catalyst [(*R*,*R*)-DPEN · (*S*)-TRIP] and the recent reports in the literature on the successful implementation of various chiral diamine catalysts, especially those bearing a primary amine moiety (some of which are readily availabe from naturally abundant amino acids), prompted us to further

Table 4.20 Screening of diamines **87-89** for the catalytic asymmetric epoxidation of 2-cyclopentenone (**46r**)

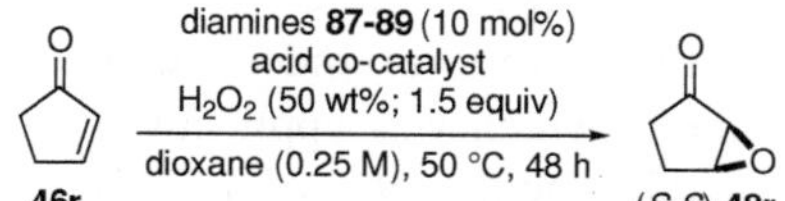

Acid\Diamine	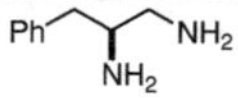 **87** **87**		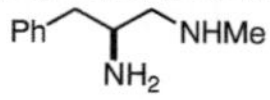 **88** **88**		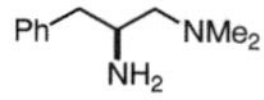 **89a** **89a**	
	Yield [%][a]	*er* [b]	Yield [%][a]	*er* [b]	Yield [%][a]	*er* [b]
TFA (10 mol%)	27	74:26	36	60.5:39.5	82	71:29
TFA (20 mol%)	28	69.5:30.5	51	59.5:40.5	60	73.5:26.5
(*S*)-TRIP (10 mol%)	28	82.5:17.5	36	62.5:37.5	47	66.5:33.5
(*R*)-TRIP (10 mol%)	31	85:15	38	41.5:58.5	48	74.5:25.5
–	6	69.5:30.5	10	58.5:41.5	35	62.5:37.5

[a] Determined by GC
[b] Determined by chiral GC

investigate on this [22, 23]. For this purpose, diamines exhibiting primary, secondary, and tertiary amine functionalities were quickly assembled through conventional synthetic methods (cf. Sect. 4.7.2) and tested in the epoxidation reaction (Table 4.20). Unfortunately, the original result obtained with [(*R*,*R*)-DPEN · (*S*)-TRIP] as catalyst could not be further improved.

In conclusion, during our screening studies for the catalytic asymmetric epoxidation of 2-cyclopentenone (**46r**), we have identified an alternative, highly active, yet moderately enantioselective catalyst system for the enantioselective epoxidation of five-membered cyclic enones: [**64**· (*S*)-TRIP] (cf. Table 4.18). Due to the increased catalytic activity of [**64**· (*S*)-TRIP] compared to the previously employed catalysts [9-NH_2-*epi*Q (**13**)· 2 TFA] and [(*R*,*R*)-DPEN · (*S*)-TRIP], the less reactive 3-substituted cyclopentenone derivative **90** could be converted to the corresponding epoxide 2,3-epoxy-3-phenethyl-cyclopentanone (**91**) in reasonable yield of 65% with an enantiomeric ratio of 85.5:14.5 *er*. As a comparison, after 48 h, the reaction catalyzed by [**13** · 2 TFA] had reached only 32% conversion as determined by GC. However, under those conditions the product was formed with superior enantioselectivity of 91:9 *er*.

4.1.3 Summary and Conclusions

In summary, we have successfully developed a method for the highly enantioselective epoxidation of simple cyclic enones catalyzed by chiral primary amine salts and employing aqueous hydrogen peroxide as a cheap and environmentally benign oxidant.

Prior to this work, and in spite of the wealth of enantioselective enone epoxidation methods known to date, cyclic α,β-epoxy ketones such as **48** have been difficult to access in optically active form. Indeed, there was no single asymmetric epoxidation method available which was applicable to a broad range of simple cyclic enones of the general structure **46** to furnish the corresponding epoxides with satisfying yields and enantioselectivities (cf. Sect. 2.2.6).

In the course of our screening studies we have identified two powerful and complementary chiral primary amine salt catalysts which efficiently mediated the desired transformation: the [(*R*,*R*)-DPEN (**12**) · (*S*)-TRIP (**7a**)] salt and 9-amino (9-deoxy)*epi*quinine (9-NH_2-*epi*Q; **13**) as its trifluoroacetic acid salt [9-NH_2-*epi*Q (**13**) · 2 TFA]. Both catalyst salts have in common the bifunctional nature of the amine component. Whereas the primary amine group was found to be crucial for catalytic activity, the second basic site seems to direct the attack of the hydrogen peroxide as depicted in Fig. 4.1, thus presumably accounting for the improved selectivities typically attained in reactions with diamine catalyst salts (cf. Table 4.2).

Using either catalyst [(*R*,*R*)-DPEN · (*S*)-TRIP] or [9-NH_2-*epi*Q · 2 TFA] at 10 mol% loadings, we achieved the highly enantioselective epoxidation of variously substituted cyclohexenone derivatives **46**. The (*R*,*R*)-DPEN-based catalyst system turned out to be particularly suitable for cyclohexenone derivatives **46** equipped with substituents at the 4- and 5-positions (Table 4.10). The 9-amino(9-deoxy)*epi*quinine salt [9-NH_2-*epi*Q · 2 TFA] provided in addition an array of 3-substituted 2,3-epoxycyclohexanones **48** with high yields along with excellent enantioselectivities of up to 99.5:0.5 *er*—including *inter alia* epoxides **48i-j** bearing sterically demanding isopropyl and even *tert*-butyl substituents at the 3-position (Table 4.12).

Moreover, both catalyst systems were applicable to cyclic enones of different ring sizes (Table 4.13). Macrocyclic enones such as 2-cyclododecenone (**46v**) and 2-cyclopentadecenone (**46x**) could be converted into the corresponding epoxides in high yields and outstanding enantioselectivity of 99.5:0.5 *er* in the presence of Cinchona alkaloid-derived catalyst [9-NH_2-*epi*Q (**13**) · 2 TFA]. Interestingly, seven and eight-membered cyclic enones **46s** and **46v**, respectively, provided mixtures of the corresponding α,β-epoxyketones along with the unanticipated bicyclic peroxyhemiketals **116** and **117** (Scheme 4.3), which will be the subject of further investigations described within this thesis (cf. Sect. 4.2.2.6).

Among the cyclic enone series, cyclopentenones proved to be particularly challenging substrates (cf. Sect. 4.1.2.5). In the reaction of 2-cyclopentenone (**46r**), both of our catalytic systems showed significantly reduced activity and selectivity compared with the reactions of the higher homologues. Therefore, we initiated a separate screening to identify a chiral primary amine salt with superior catalytic efficiency for the epoxidation of cyclopentenone and derivatives. Truncated Cinchona alkaloid **64** (cf. Table 4.18) equipped with a less sterically hindered primary amine group than the parent 9-amino(9-deoxy)*epi*quinine (**13**) displayed promising catalytic activity. Gratifyingly, the enantioselectivity of 65.5:34.5 *er* obtained with [**64** · 2 TFA] could be increased to 76.5:23.5 *er* by pairing amine **64** with the chiral BINOL phosphate TRIP (7a). At the same time, the high catalytic activity is retained, which also renders catalyst system [**64** · (*S*)-TRIP]

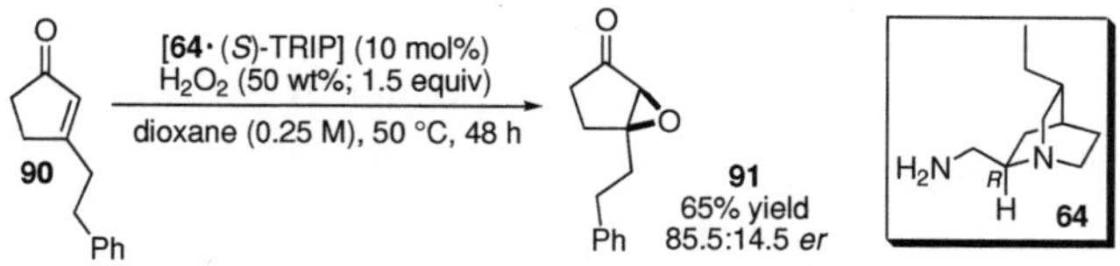

Scheme 4.10 Asymmetric epoxidation of 3-substituted cyclopentenone **90** catalyzed by [**64** · (*S*)-TRIP]

suitable for the epoxidation of less reactive 3-substituted cyclopentenone derivatives such as 3-phenethyl-2-cyclopentenone (Scheme 4.10). Although not yet optimal, we anticipate that these encouraging results may constitute a valid starting point for further developments, inspiring future research in this area.

4.2 Catalytic Asymmetric Epoxidation and Hydroperoxidation of Acyclic Enones

After having established a highly efficient and general method for the asymmetric epoxidation of cyclic enones catalyzed by chiral primary amine salts, we became interested in expanding the scope to acyclic substrates with a focus on aliphatic enolizable enones which still constitute challenging substrates for the currently available epoxidation methods.

4.2.1 Screening Studies

4.2.1.1 Initial Results

In a first attempt, we applied the optimized reaction conditions for the epoxidation of cyclic enones with 9-amino(9-deoxy)*epi*quinine (9-NH_2-*epi*Q, **13**) TFA salt as the catalyst system to the epoxidation of 3-decen-2-one (**92a**) as an example of a simple aliphatic acyclic enolizable enone. After 20 h, we obtained a product mixture consisting of the desired *trans*-epoxide **93a** in excellent enantioselectivity of >99.5:0.5 *er* along with the cyclic peroxyhemiketal (PHK) **94a** in a ratio of 43–57 as determined by ^{1}H NMR of the crude mixture (Scheme 4.11).

This result was surprising in view of the fact that peroxyhemiketals are typically only isolated in minor amounts (as by-products) in Weitz-Scheffer reactions, due to the overwhelming preference for epoxide formation under those conditions (cf. Sect. 2.3.2) [24–28]. Thus, we were highly intrigued by our observation of a peroxyhemiketalselective "epoxidation" reaction, even more so since a catalytic asymmetric hydroperoxidation of α,β-unsaturated ketones to obtain optically active peroxides such as **94a** was unprecedented in the literature.

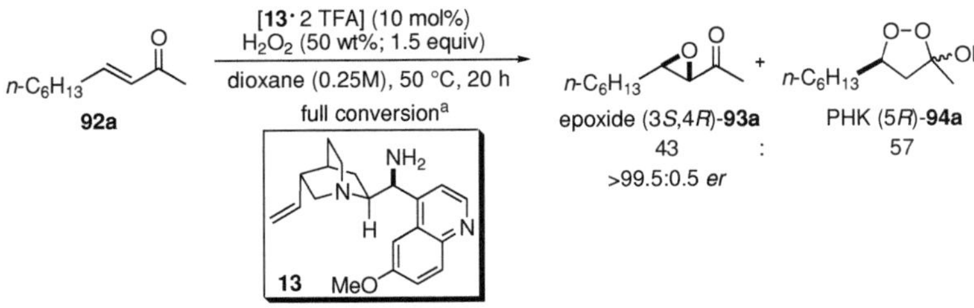

Scheme 4.11 Catalytic asymmetric epoxidation of 3-decen-2-one (**92a**): initial attempt. ([a]Determined by ^{1}H NMR)

Table 4.21 Evaluation of different catalysts for the catalytic asymmetric epoxidation of 6-phenyl-3-hexen-2-one (**92b**)

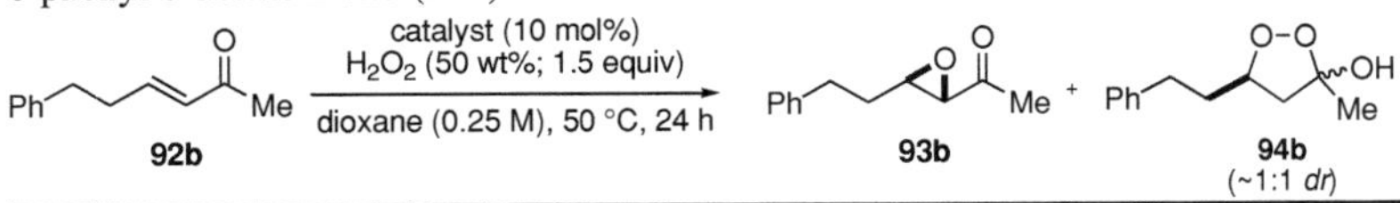

Entry	Catalyst	Conv. [%][a]	*er*(**93b**)[b]
1	[9-NH_2-*epi*Q · TFA]	57	98:2
2	[9-NH_2-*epi*Q · 2 TFA]	full	99.5:0.5
3	[(*R*,*R*)-DPEN · TFA]	61	13:87
4	[(*R*,*R*)-DPEN · 2 TFA]	92	6:94
5	[(*R*,*R*)-DPEN · (*S*)-TRIP]	70	9:91
6	[(*R*,*R*)-DPEN · (*R*)-TRIP]	52	25:75

Identical reactivity and selectivity trends were observed with 4-hexen-3-one (**92c**) as substrate
[a] Determined by GC
[b] Determined by chiral GC

At this stage, we speculated that the chiral amine salt catalyst, e.g. [**13** · 2 TFA], might not only render enantioselective the conjugate addition of hydrogen peroxide but also affect the product distribution between epoxide and peroxyhemiketal. Thus, we decided to optimize the catalyst system and reaction conditions towards increased peroxyhemiketal formation.

In this context, we tested alternative catalyst systems in the hydroperoxidation-epoxidation reaction of 6-phenyl-3-hexen-2-one (**92b**). (*R*,*R*)-DPEN salts such as [(*R*,*R*)-DPEN · TFA], [(*R*,*R*)-DPEN · 2TFA], [(*R*,*R*)-DPEN · (*S*)-TRIP], and [(*R*,*R*)-DPEN · (*R*)-TRIP] displayed moderate to high catalytic activity, yet gave inferior results in terms of enantioselectivity with acyclic compared to the previously studied cyclic substrates (Table 4.21, entries 3–6). Among all catalysts tested, 9-amino(9-deoxy)*epi*-quinine (9-NH_2-*epi*Q, **13**) coped best with the main challenge arising from the transition from cyclic to acyclic enones, which presumably is the control of the *s-cis* and *s-trans* geometry, and afforded excellent enantioselectivity also for acyclic substrates when employed together with an acid co-catalyst (TFA) in a ratio of 1:2 (entry 2). A 1:1 amine/TFA ratio led to only slightly lower enantioselectivity, but significantly reduced catalyst activity (entry 1).

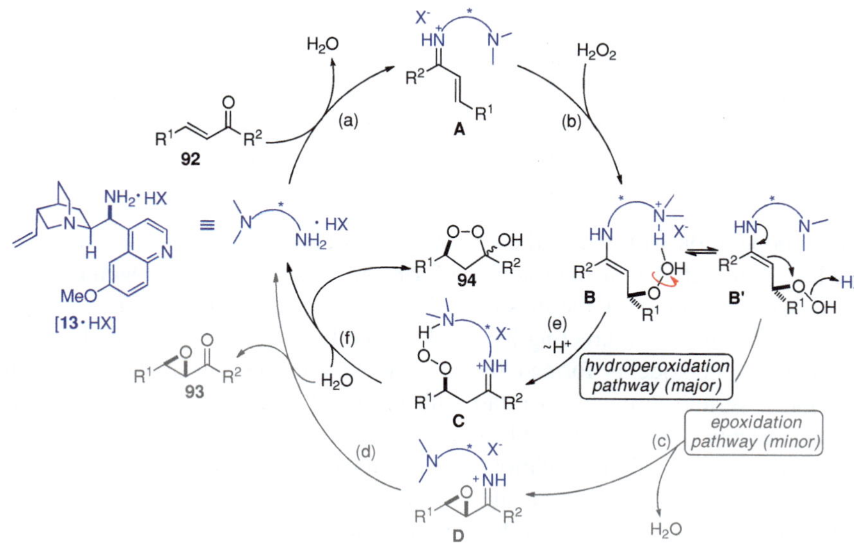

Scheme 4.12 Hypothetical catalytic cycle: hydroperoxidation versus epoxidation pathway in the reaction of acyclic enones (**92**)

Our proposed catalytic cycle for the reaction of acyclic α,β-unsaturated ketones **92** with hydrogen peroxide in the presence of the 9-amino(9-deoxy)*epi*quinine (**13**) TFA salt accounts for the formation of both peroxyhemiketals **94** and epoxides **93** (Scheme 4.12).

The initial steps are identical to those invoked for the epoxidation of 2-cyclohexenone (**46a**) (cf. Fig. 3.1). The catalytic cycle is triggered by the reversible formation of iminium ion **A** from the enone substrate **92** and the catalyst [9-NH_2-*epi*Q (**13**) · HX] (step a). Iminium ion formation may effectively lower the LUMO energy of the substrate and facilitate the enantioselective conjugate addition of hydrogen peroxide (step b) to provide β-peroxyenamine intermediate **B**. Subsequent intramolecular nucleophilic substitution may break the weak peroxide bond (step c), and afford α,β-epoxy ketone **93** after hydrolysis of α,β-epoxy iminium ion **D** (step d). However, the epoxidation pathway may be converted into a hydroperoxidation pathway if β-peroxyenamine **B** may be trapped by protonation (step e). β-Hydroperoxy iminium ion **C** may release a β-peroxyketone upon hydrolysis (step f), which in the case of acyclic enones such 3-decen-2-one (**92a**) exists in equilibrium exclusively as the cyclic peroxyhemiketal **94**.

Indeed, there are only isolated cases evident in the literature where the β-hydroperoxide generated by conjugate addition of hydrogen peroxide to α,β-unsaturated carbonyl compounds could be isolated [24–29]. Yet, none of them allows access to β-hydroperoxides in optically active form.

As for the epoxidation of cyclic enones, we reckoned that while the primary amine function of 9-amino(9-deoxy)*epi*quinine (**13**) activates the enone as iminium

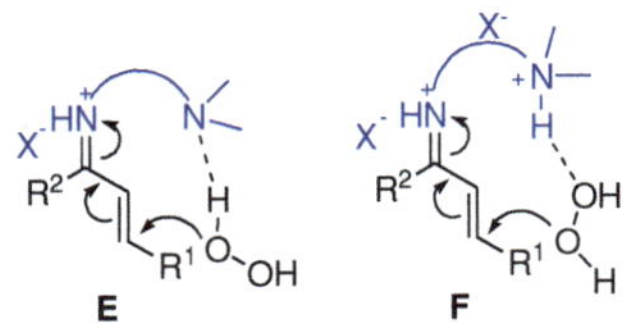

Fig. 4.6 Putative pre-transition state assemblies **E** and **F** for the conjugate addition of hydrogen peroxide to iminium ion **A**

ion in cooperation with the acid co-catalyst, the quinuclidine moiety either as the free base (Fig. 4.6, **E**), or in protonated form (**F**) may bind the hydrogen peroxide via hydrogen bonding interactions [30], thereby bringing it into close proximity to the reactive center and directing its attack to one face of the double bond.

When the epoxide is formed the O–O bond must be anti-periplanar to the enamine π-system to enable overlap with the O–O anti-bonding orbital. Taking this into account, we rationalized the preference for the hydroperoxidation pathway under our reaction conditions as an apparent failure of peroxyenamine intermediate **B** to adopt the requisite O–O bond orientation by which the enamine is optimally aligned for epoxide formation to take place such as in **B′** (Scheme 4.12). Presumably, secondary catalyst-substrate interactions via hydrogen bonding between the peroxyenamine moiety and the covalently linked Cinchona alkaloid scaffold might rigidify β-peroxyenamine intermediate **B**. This might impede its bond-rotational freedom (cf. equilibrium between **B** and **B′**). As a result, the peroxyenamine intermediate **B** may be stabilized and the barrier associated with epoxide formation may have become so high that peroxyenamine **B** is relegated to protonation and the reaction terminated at the hydroperoxidation stage.

On the basis of our mechanistic hypothesis, we reckoned that we could possibly further influence the peroxy hemiketal to epoxide ratio by the choice of the acid co-catalyst and by further fine-tuning of the reaction conditions. In this, we considered that an increased water content in the system might benefit the protonation and subsequent hydrolysis of β-peroxyenamine intermediate **B** prior to the irreversible epoxide ring closure. The acidity of the acid co-catalyst might influence the partitioning of the peroxyenamine intermediate **B** between the hydroperoxidation and epoxidation pathways insofar that a relatively strong acid co-catalyst might interfere with the multipoint binding interactions to the covalently linked quinine scaffold, with the result that the conformational rigidity might be reduced. Moreover, a strong acid co-catalyst should be able to bring forward epoxide formation by enhancement of the nucleofugacity of the hydroxyl moiety by protonation of the distal oxygen atom of the peroxy moiety in intermediate **B**. On the contrary, the use of a weaker acid might promote the hydroperoxidation pathway.

4.2.1.2 Optimization of the Catalytic System

Indeed, when we studied the influence of different acid co-catalysts on the reaction of enone substrate **92a** with hydrogen peroxide in the presence of 9-amino (9-deoxy)*epi*quinine (9-NH_2-*epi*Q; **13**) as its TFA salt [**13** · 2 TFA], we found a correlation between the pK_a value of the acid and the peroxyhemiketal to epoxide ratio (Table 4.22) [31]. Whereas a relatively strong acid such as *para*-toluene sulfonic acid gave predominately the epoxide (entries 1–3), the peroxyhemiketal was obtained as major product in presence of weaker acids (entries 4–29). Moreover, a decrease in temperature favoured PHK formation. Whereas the PHK-epoxide ratio was 57:43 at 50 °C, (entry 4) it could be increased to 73:27 by lowering the temperature to 10 °C albeit at the expense of reactivity (entry 8); 32 °C appeared to be optimal with regard to both selectivity and reactivitiy (entry 5). An increase in the amount of acid co-catalyst slightly enhanced the catalyst activity; however, the increased activity came along with reduced PHK selectivity (entries 7–8). The use of acids with a pK_a greater than 2 benefits the peroxyhemiketal-epoxide ratio but entails diminished catalyst efficiency. 2,4-Dinitrobenzoic acid (p$K_a \sim 2.17$) does promote the reaction giving a PHK-epoxide ratio of 75:25—albeit at a lower rate (entry 24). The same holds true for 3,3,3-trifluoropropanoic acid (pK_a 3.07) (entry 10). However, acetic acid (pK_a 4.76) as well as benzoic acid (pK_a 4.20) were both ineffective (entries 19–20, 23). The best compromise between selectivity and reactivity was exhibited by trichloroacetic acid (pK_a 0.65) with a PHK-epoxide ratio of 72:28 and virtually full conversion after 24 h at 28 °C (entry 15). Thus the 9-amino(9-deoxy)*epi*quinine (**13**) TCA salt [**13** · 2 TCA] was used in further optimization experiments.

4.2.1.3 Optimization of the Reaction Conditions

Further experiments were aimed at evaluating the dependency of the product distribution on the reaction time. To this end, samples were taken from an exemplary reaction at different time intervals (Table 4.23). The PHK-epoxide ratio remained unchanged between 4.5 and 28 h of reaction time (entries 1–3). Only upon extended stirring of the reaction for 10 days, did the peroxide to epoxide ratio drop from 72:28 to 66:34 (entry 4). Although this is only a slight decrease in peroxide, it is crucial to stop the hydroperoxidation reactions without delay once the starting material is consumed to ensure maximum yields of the PHK products.

Next, we were interested in reducing the catalyst loading as much as possible without affecting its efficiency (Scheme 4.13). However, with only 5 mol% [**13** · 3 TCA] the reaction rate dropped significantly and the reaction was complete only after 48 h.[10] Based on this result, 10 mol% of catalyst were used in further experiments.

A screening of different ethereal solvents validated 1,4-dioxane as the solvent of choice, as previously observed in the catalytic asymmetric epoxidation of cyclic enones (Table 4.24, entry 5; cf. Table 4.9). In Et_2O and MTBE the catalyst system [**13** · 2 TCA] showed comparable activity; yet the PHK to epoxide ratio was

[10] [**13** · 3 TCA] was chosen instead of [**13** · 2 TCA] since a 3:1 acid/9-amino(9-deoxy)*epi*quinine ratio seemed to have a beneficial effect on the catalyst activity (Table 4.22, entry 6 versus 7).

Table 4.22 Evaluation of the influence of acid co-catalysts (HX) on the PHK **94a** to epoxide **93a** ratio in the [**13** · HX]-catalyzed reaction of 3-decen-2-one (**92a**) with aqueous hydrogen peroxide

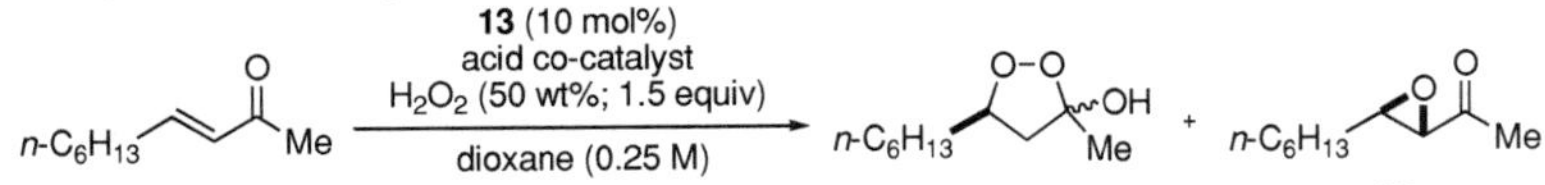

92a **94a** (~1:1 *dr*) **93a**

Entry	Acid co-catalyst (HX)	pK_a (H_2O)[a]	[mol%]	T [° C]	Conv. [%][b]	**94a**:**93a**[c]	*er*(**93a**)[d]
1	*p*TsOH · H_2O	−2.80	20	28	82 (48 h)	35:65	98:2
2			30	28	98 (48 h)[e]	31:69	92:8
3			30	10	65 (48 h)	35:65	94.5:5.5
4	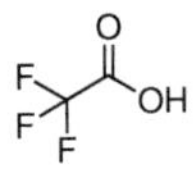 (TFA)	0.52	20	50	full (20 h)	57:43	>99.5:0.5
5			20	32	full (24 h)	68:32	>99.5:0.5
6			20	28	95 (24 h)	70:30	>99.5:0.5
7			30	28	97 (48 h)	72:28	>99.5:0.5
8	(TFA)		30	10	76 (48 h)	73:27	>99.5:0.5
9	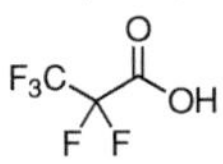	n.a.	20	32	97 (24 h)	69:31	>99.5:0.5
10	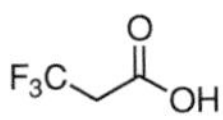	3.07	20	32	44 (28 h)	n.d.	99.5:0.5
11	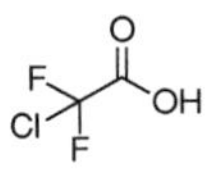	n.a.	20	32	full (24 h)	68:32	>99.5:0.5
12	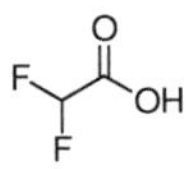	1.24	20	32	88 (48 h)	78:22	>99.5:0.5
13	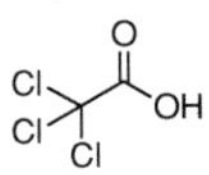 (TCA)	0.65	20	50	full (20 h)	60:40	>99.5:0.5
14			20	32	full (24 h)	70:30	>99.5:0.5
15			20	28	97 (24 h)	72:28	>99.5:0.5
16			20	20	90 (48 h)	75:25	>99.5:0.5
17	(TCA)		20	10	79 (48 h)	75:25	>99.5:0.5
18	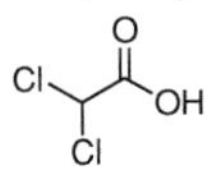	1.35	20	32	84 (48 h)	77:23	>99.5:0.5
19	HOAc	4.76	20	28	<10	n.d.	n.d.
20			30	28	<10	n.d.	n.d.
21	$EtCO_2H$	4.87	20	28	<10	n.d.	n.d.
22	*i*-$PrCO_2H$	n.a.	20	28	<10	n.d.	n.d.
23	$PhCO_2H$	4.20	20	28	15	n.d.	95.5:4.5
24	2,4-$(NO_2)_2$-$C_6H_4CO_2H$	2.17[f]	20	32	98% (48 h)	75:25	99.5:0.5
25	4-NO_2-$C_6H_4CO_2H$	3.44	20	32	45 (28 h)	n.d.	n.d.
26	2,6-F_2-$C_6H_3CO_2H$	n.a.	20	32	52 (28 h)	n.d.	n.d.
27	3,5-$(CF_3)_2$-$C_6H_3CO_2H$	n.a.	20	32	43 (28 h)	n.d.	n.d.

(continued)

Table 4.22 (continued)

Entry	Acid co-catalyst (HX)	pK_a (H_2O)[a]	[mol%]	T [° C]	Conv. [%][b]	**94a**:**93a**[c]	*er*(**93a**)[d]
28	4-Cl-$C_6H_4CO_2H$	3.99	20	32	23 (28 h)	n.d.	n.d.
29	$(PhO)_2PO(OH)$	1.29[g]	20	32	96 (24 h)	64:36	99.5:0.5

[a] The pK_a values are adopted from the pK_a compilations of Williams and Bordwell [31]
[b] Determined by GC
[c] Determined by ^1H NMR of the crude mixture
[d] Determined by chiral GC
[e] Extensive side product formation observed [32]
[f] pK_a of 2-NO_2-$C_6H_4CO_2H$
[g] pK_a of $(MeO)_2PO(OH)$

Table 4.23 Tracing the peroxyhemiketal to epoxide ratio over time

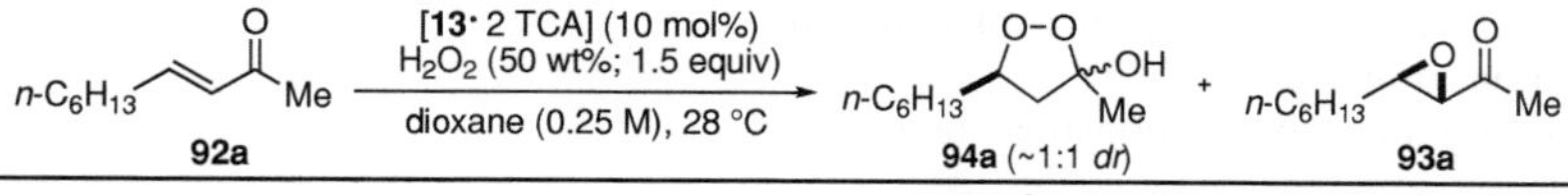

Entry	t [h]	Conv. [%][a]	**94a**:**93a** ratio[a]
1	4.5	42	71:29
2	24	97	72:28
3	28	full	72:28
4	10 d	full	66:34

[a] Determined by ^1H NMR of the crude mixture

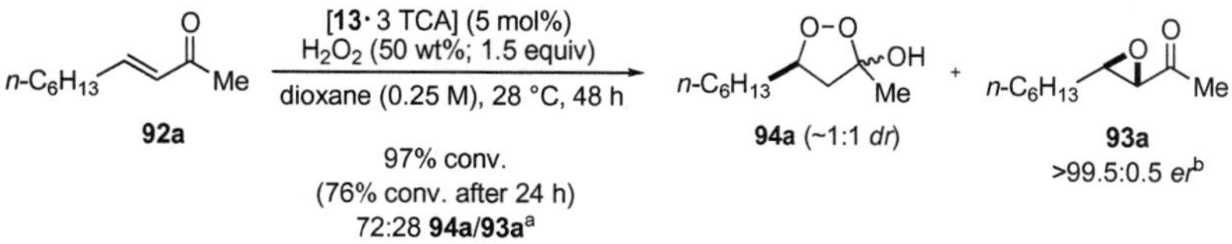

Scheme 4.13 Effect of the catalyst loading on the reaction efficiency. ([a]Determined by ^1H NMR of the crude mixture; [b]Determined by chiral GC)

unfavourably affected (entries 2 and 4). Nonetheless, both may be alternative solvents for applications which directly target epoxide synthesis rather than formation of peroxyhemiketals (cf. Sect. 4.2.2.2). Moreover, MTBE might also be a suitable substitute for 1,4-dioxane in large-scale applications (with the advantage of MTBE being for instance its lower hazardous potential).

Increasing the substrate concentration from 0.25 to 0.5 M had a positive effect on the PHK-epoxide ratio but was accompanied by increasing formation of unidentified side products (entry 6). As the concentration was lowered (0.125 M), proportionally more epoxide was obtained (entry 7). On the basis of these results, the screening studies were continued in dioxane at a substrate concentration of 0.25 M (entry 5).

Table 4.24 Screening of various ethereal solvents at different concentrations

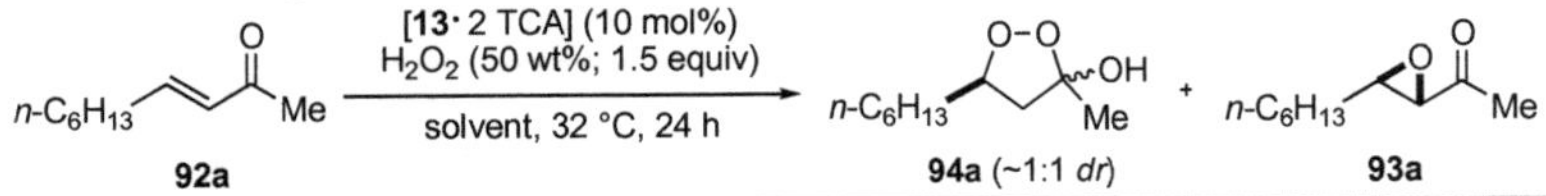

Entry	Solvent	Conc. [M]	Conv. [%][a]	**94a**:**93a** ratio[b]	*er*(**93a**)[c]
1	THF	0.25	43	n.d.	n.d.
2	Et_2O	0.25	99	65:35	99:1
3	DME	0.25	38	n.d.	n.d.
4	MTBE	0.25	full	68:32	99.5:0.5
5	dioxane	0.25	full	70:30	>99.5:0.5
6	dioxane	0.5	full[d]	71:29	>99.5:0.5
7	dioxane	0.125	89	67:33	>99.5:0.5

[a] Determined by GC
[b] Determined by ^1H NMR of the crude mixture
[c] Determined by chiral GC
[d] Increasing side product formation observed

Table 4.25 Optimization of the hydrogen peroxide concentration

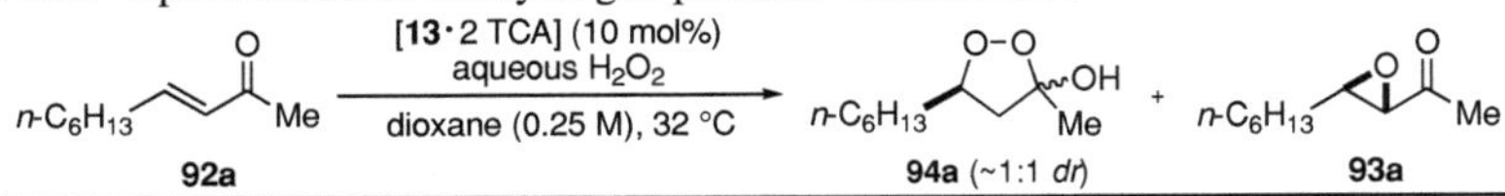

Entry	H_2O_2 equiv.	H_2O_2 conc. [wt%]	Conv. [%][a]	**94a**:**93a** ratio[b]	*er*(**93a**)[c]
1	1.2	50	full (24 h)	65:35	>99.5:0.5
2	1.5	50	full (24 h)	70:30	>99.5:0.5
3	2	50	full (24 h)	71:29	>99.5:0.5
4	5	50	full (24 h)	76:24	99.5:0.5
5	1.5	30	full (24 h)	76:24	>99.5:0.5
6[d]	1.5	30	full (36 h)	77:23	>99.5:0.5
7	3	30	full (24 h)	80:20	>99.5:0.5
8	1.5	25	full (30 h)	78:22	>99.5:0.5
9	3	25	full (30 h)	81:19	99.5:0.5
10	1.5	10	79 (6 d)	78:22	99:1
11	1.5	5	57 (6 d)	n.d.	98:2

[a] Determined by GC
[b] Determined by ^1H NMR of the crude mixture
[c] Determined by chiral GC
[d] At 28 °C

Next, we focused on the effect of the concentration of hydrogen peroxide (Table 4.25). Two trends emerged: The peroxyhemiketal to epoxide ratio increased as the amount of hydrogen peroxide increased (entries 1–4). Moreover, the use of less concentrated solutions of hydrogen peroxide (30 wt% instead of 50 wt%-solutions) had a beneficial effect (entry 2 versus. 5). A PHK-epoxide ratio of 80:20 was obtained by using 3 equivalents of a 30 wt% aqueous hydrogen peroxide solution (entry 7). Indeed, higher dilution (25 wt% aqueous hydrogen

Table 4.26 Screening of alternative oxidants: *tert*-butylhydroperoxide

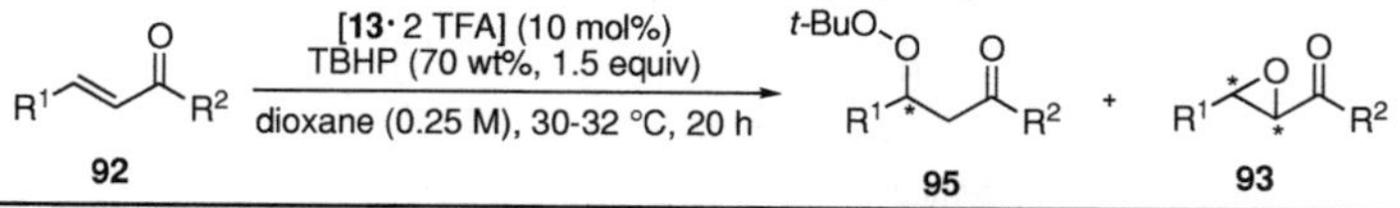

Entry	R^1	R^2	Enone	Peroxide	Epoxide	Yield (**95**:**93** ratio) [%][a]	*er*(**93**)[b]
1	Me	Et	**92c**	(5*S*)-**95c**	(4*R*,5*S*)-**93c**	88 (82:6)	14.5:85.5
2	*n*-C_6H_{13}	Me	**92a**	(4*R*)-**95a**	(3*S*,4*R*)-**93a**	91 (55)[c] (70:21)	99:1

[a] Determined by GC
[b] Determined by chiral GC
[c] Isolated yield of pure **95a** in parentheses

peroxide) further increased this value albeit at the expense of catalyst efficiency with regard to both activity and selectivity (entries 8–11).

4.2.1.4 *Tert*-butylhydroperoxide as the Oxidant

Although hydrogen peroxide is arguably the most attractive oxidant with respect to environ-mental and economic considerations it was interesting to find out how our system would respond to other oxidants. Initial experiments with aqueous *tert*-butylhydroperoxide (70 wt%) gave also promising, yet inferior results compared to the catalytic reaction with aqueous hydrogen peroxide as oxidant under otherwise identical conditions (Table 4.26).

Very much like in the catalytic reaction with hydrogen peroxide, α,β-epoxy ketone **93** was only formed in minor amounts. In contrast, the product of the conjugate addition of *tert*-butylhydroperoxide (TBHP)—*tert*-butylperoxide **95** was obtained as the major product with high selectivities. An intriguing feature was the inverted enantiofacial selectivity observed in the reaction of 4-hexen-3-one (**92c**) compared to the corresponding hydrogen peroxide reaction (entry 1, cf. Table 4.30, entry 18).

In general, we assume that acyclic α,β-unsaturated ketones **92** or the corresponding unsaturated iminium ions generated in the presence of catalyst [**13** · 2 TXA] (with X = Cl, F), respectively, react preferentially in the *s-trans* conformation which is in agreement with the observed absolute stereochemistry of both cyclic and acyclic enones (cf. Sect. 4.5.4). We speculate that the formation of the opposite epoxide enantiomer might indicate participation of the *s-cis* conformation. Moreover, epoxide **93c** was formed with relatively low enantioselectivity of 85.5:14.5 *er* which may support a competition of *s-cis* and *s-trans* reactive conformations.

The equilibrium distribution of *s-trans* and *s-cis* conformations of α,β-unsaturated ketones depends on the extent of van der Waals interaction between substituents (Table 4.27) [33]. In 3-penten-2-one ($R^1 = R^2 =$ Me) the mole fraction of the *s-cis* conformer in equilibrium amounts to 30% (entry 2). Yet, this value progressively increases as the size of the alkyl group R^2 increases beyond methyl. Although the situation in the corresponding iminium ions might differ

Table 4.27 Equilibrium distribution of *s-trans* and *s-cis* conformations of α,β-unsaturated ketones [33]

Entry	R^1	R^2	*s-trans* [%]	*s-cis* [%]
1	H	Me	73	27
2	Me	Me	70	30
3	Me	Et	55	45
4	Me	*i*-Pr	30	70
5	Me	*t*-Bu	0	100

Scheme 4.14 Catalytic asymmetric alkylperoxidation of α,β-unsaturated ketones according to Deng et al. [35]

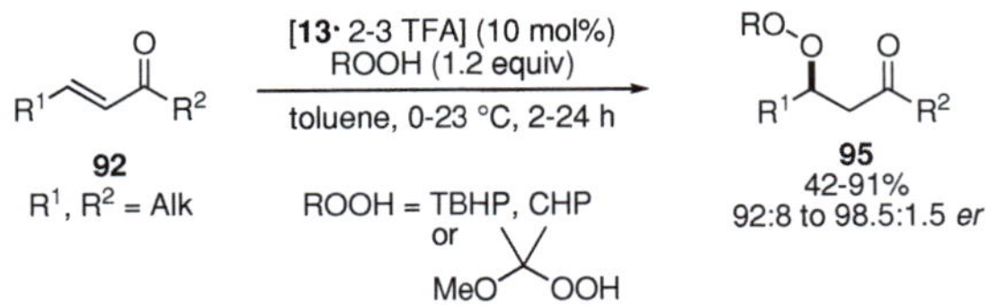

significantly, the *s-cis*/*s-trans* equilibrium should always be considered while analyzing the results obtained with acyclic enone substrates.

Nevertheless, it can not be ruled out that the reaction of 4-hexen-3-one (**92c**) proceeds through a distinct, competing mechanism which accounts for the inverted enantiofacial selectivity, potentially involving activation by hydrogen bonding interactions.

In fact, subsequent experiments validated the assumption of a substrate-specific effect in that our standard substrate 3-decen-2-one (**92a**) showed the same facial selectivity as previously observed in the reaction with hydrogen peroxide as terminal oxidant (Table 4.26, entry 2). With TBHP as oxidant under otherwise identical reaction conditions, *tert*-butylperoxide **95a** was obtained in good yield along with high enantioselectivity of 99:1 *er*.

Thus, *tert*-butyl hydroperoxide not only constitutes an alternative oxidant for the synthesis of epoxides but also provides concise access to interesting optically active β-alkylperoxy ketones **95** via conjugate addition. Yet, hydrogen peroxide meets all criteria for future applications potentially also on an industrial scale: it is inexpensive, readily available, and gives water as the only by-product. Thus, we continued our research with hydrogen peroxide as the oxidant of choice.

In 2008, after our group has introduced 9-amino(9-deoxy)*epi*quinine (**13**) (as its TFA salt) as powerful catalyst for the epoxidation of cyclic enones, the Deng group reported a catalytic asymmetric alkylperoxidation of α,β-unsaturated ketones using the same catalyst system [9-NH_2-*epi*Q (**13**) · 2-3 TFA] (Scheme 4.14) [34, 35].

Further oxidants were not tested since a broad screening of potential oxidizing reagents had already been carried out in the context of the asymmetric epoxidation of 2-cyclohexenone (cf. Table 4.7).

4.2.2 Reaction Scope and Discussion

Once we had identified the optimal catalyst system and established the optimal reaction conditions, we proceeded to investigate the scope of this novel asymmetric transformation.

The following protocol was applied: the catalytic asymmetric hydroperoxidation of α,β-unsaturated ketones **92** was conducted in dioxane (0.25 M) at 32 °C in the presence of 9-amino(9-deoxy)*epi*quinine TCA salt [**13** · 2 TCA] (10 mol%) and with 30 wt% aqueous hydrogen peroxide (3 equiv) as oxidant to afford optically active peroxyhemiketal diastereomers **94** virtually as a 1:1 mixture of C-1 hemiketal epimers in good yields along with high enantioselectivities. Details of the scope of the catalytic asymmetric hydroperoxidation are outlined in the following Sect. 4.2.2.1.

4.2.2.1 Hydroperoxidation of α,β-Unsaturated Ketones

The catalytic asymmetric hydroperoxidation of α,β-unsaturated ketones **92** proved to have a broad substrate scope. Various α,β-enones **92** were converted to the optically active peroxyhemiketals **94** in reasonable to good yields (30–72%) along with high enantioselectivities (96:4–98.5:1.5 *er*). In general, the only detected by-products were the corresponding epoxides **93**, which are easily separated from peroxides **94**. The absolute configuration of products **94** and **93** was established by reducing peroxide **94b** to the corresponding aldol-type product 4-hydroxy-6-phenyl-2-hexanone (**96b**) (see Sect. 4.3.2) and comparing its optical rotation with literature data [36]. 4-Hydroxy-6-phenyl-2-hexanone (**96b**) was obtained as the (*R*)-isomer; absolute configurations of other products **94** and **93** were assigned by analogy.

α,β-Unsaturated methyl ketones **92** (R^2 = Me) with linear as well as α'- and β'-branched alkyl residues R^1 at the β-position were suitable substrates (Table 4.28, entries 1–4). Compared with **92a-b**, and **d**, 4-cyclohexyl-3-buten-2-one (**92e**) proved to be slightly less reactive. Moreover, the percentage of peroxyhemiketal decreased as the degree of branching in the substituent R^1 increased. With 3-decen-2-one (R^1 = n-C_6H_{13}) it accounted for 80% of the PHK-epoxide product mixture, whereas a cyclohexyl substituent such as in enone **92e** reduced it to 68%. Moreover, our hydroperoxidation reaction features high functional group compatibility (entries 5–10). Enones equipped with an ester, keto, or protected hydroxyl group, a halide, or an olefin were well tolerated. Notably, the survival of acid labile acetal and silyl ether functionalities testifies the mildness of the adopted reaction conditions (Table 4.28, entries 7–8). The hydro-peroxidation reactions were generally conducted on a 0.5 or 1 mmol scale; yet a 3 mmol (~500 mg) scale affected neither the isolated yield nor the observed enantioselectivity (entry 1).

The hydroperoxidation product derived from keto-enone **92k** was isolated in 30% yield. ^{1}H NMR analysis of the purified peroxidic product **94k** revealed a mixture of three isomeric peroxyhemiketals consisting of the two 3-hydroxy-1,2-dioxolane

Table 4.28 Scope of the hydroperoxidation of α,β-unsaturated methyl ketones **92**

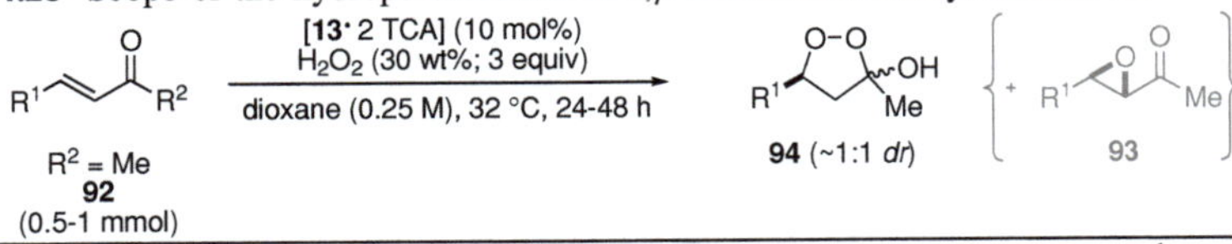

Entry	R^1	Enone	Product	**94**:**93** ratio[a]	Yield [%][b]	*er*(**94**)[c]
1[d]	n-C_6H_{13}	**92a**	**94a**	80:20	65	98.5:1.5
2	$PhCH_2CH_2$	**92b**	**94b**	78:22	68	97:3
3	*i*-Bu	**92d**	**94d**	71:29	61	97.5:2.5
4	Cy	**92e**	**94e**	68:32	54	98:2
5	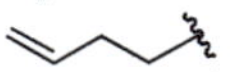	**92f**	**94f**	80:20	69	97.5:2.5
6		**92g**	**94g**	79:21	72	97:3
7	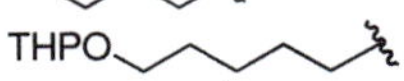	**92h**	**94h**	76:24	64	96.5:3.5
8	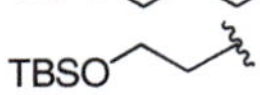	**92i**	**94i**	80:20	68	98:2
9	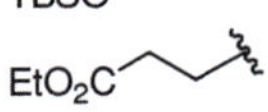	**92j**	**94j**	85:15	70	96.5:3.5
10[e]	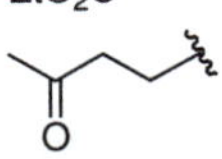	**92k**	**94k**	63:37	30	96:4
11[e]	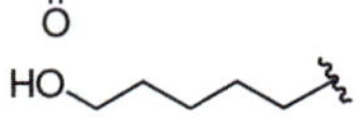	**92l**	**94l**	–	25[f]	(98:2)[g]

[a] Determined by ^{1}H NMR of the crude mixture
[b] Yields of pure, isolated products
[c] Determined by chiral GC after conversion to the corresponding epoxide with 1 N NaOH (1 equiv) in THF
[d] Reaction conducted on a 3 mmol scale
[e] With [**13** · 2 TFA] (10 mol%), H_2O_2 (50 wt%; 1.5 equiv)
[f] 69% conv. determined by ^{1}H NMR
[g] *er* of the initially formed epoxide **93l**

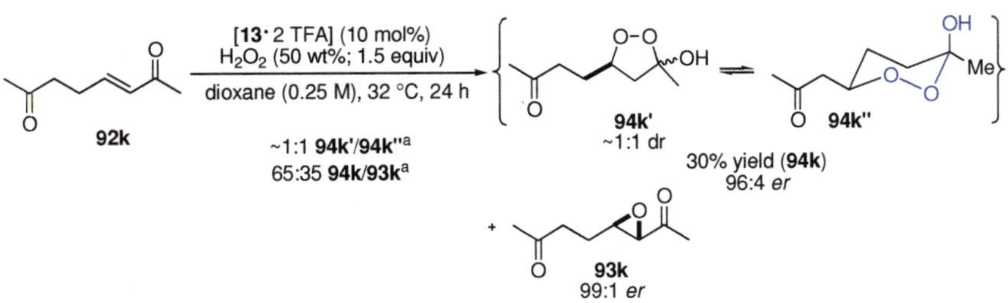

Scheme 4.15 Hydroperoxidation of keto-enone **92k**: formation of regioisomeric peroxyhemiketals **94k′** and **94k″**. ([a]Determined by ^{1}H NMR)

diastereomers **94k′** and 3-hydroxy-1,2-dioxane **94k″** which was formed via hemiketalization of the remote keto group and obtained as a single diastereomer. The relative configuration of compound **94k″** depicted in Scheme 4.15 was tentatively assigned on the basis of anomeric stabilization.

Table 4.29 Influence of the substituent R^2 on the peroxide to epoxide ratio

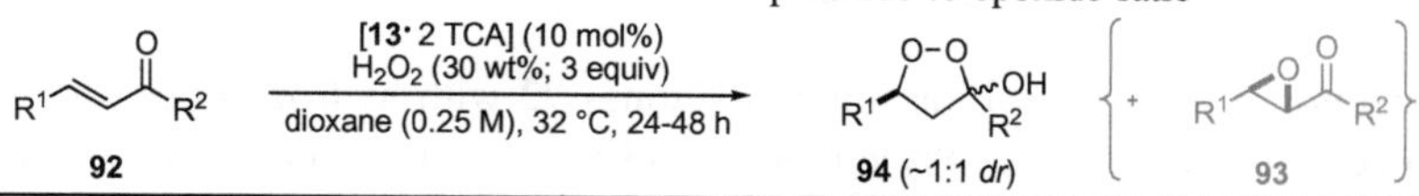

Entry	R^1	R^2	Enone	Product	**94**:**93** ratio[a]	Yield [%][b]	*er*(**94**)[c]
1[d]	*n*-C_6H_{13}	Me	**92a**	**94a**	80:20 (57:43)	65	98.5:1.5
2	*n*-C_5H_{11}	*n*-C_5H_{11}	**92m**	**94m**	45:55 (25:75)	40	97:3
3	*n*-C_5H_{11}	*i*-Bu	**92n**	**94n**	5:95 (<1:> 99)	–[e]	–
4[f]	Me	*i*-Pr	**92o**	**94o**	50:50	39[g]	96:4
5	Me	Et	**92c**	**94c**	74:26 (67:33)	56	97:3
6	*n*-C_9H_{19}	Et	**92p**	**94p**	63:37 (47:53)	48	98:2

[a] Determined by ^{1}H NMR of the crude mixture. Values in parentheses correspond to the reaction with [**13** · 2 TFA] (10 mol%), H_2O_2 (50 wt%; 1.5 equiv) at 50 °C
[b] Yields of pure, isolated products
[c] Determined by chiral GC after conversion to the corresponding epoxide with 1 N NaOH (1 equiv) in THF
[d] Conducted on a 3 mmol scale
[e] 90% conv. determined by ^{1}H NMR
[f] With [**13** · 2 TFA] (20 mol%), H_2O_2 (50 wt%; 1.5 equiv) at 50 °C
[g] 96% conv. determined by ^{1}H NMR

Table 4.29 displays the results obtained with α,β-unsaturated ketones **92** bearing substituents R^2 other than methyl ($R^2 \neq$ Me). All enones **92c**, **m-o** smoothly participated in the reaction and alterations of the steric properties of the aliphatic ketone substituent did not impact negatively on the enantioselectivity. However, the desired peroxide was obtained in varying yields depending on the steric bias of the substrate (entries 2–6). Increasing the steric demand of residue R^2 impeded the formation of peroxyhemiketals in favour of the corresponding epoxides. For instance, replacing the methyl group in enone **92a** (R^2 = Me) by an *n*-pentyl group decreased the peroxide to epoxide ratio from 80:20 to 45:55 (entry 1 versus 2). Somewhat surprisingly, with a β'-branched isobutyl group as ketone substituent R^2, essentially only the epoxide was isolated (entry 3). On the contrary, sterically crowded isopropyl ketone **92o** exhibited greatly reduced reactivity but still allowed for a reasonable yield of 39% of peroxide **94o** in slightly lower enantioselectivity of 96:4 *er* (entry 4). Increasing the steric demand of substituent R^1 also affected the reaction, yet to a minor extent: Whereas 3-hexen-4-one (**92c**) gave 74% of peroxyhemiketal in the crude product mixture, this value was reduced to 63% in 3-tetradecen-4-one (**92p**) (entry 5 versus 6).

The influence of the reaction conditions on the product distribution between PHK and epoxide is dependent on the nature of the substrate as highlighted by the **94**:**93** values given in parentheses for the reaction catalyzed by [**13** · 2 TFA] at 50 °C with 50 wt% hydrogen peroxide (1.5 equiv). With 7-tridecen-6-one (**92m**) an increase from 25:75 to 45:55 could be achieved by optimizing the reaction conditions (cf. Sect. 4.2.1) which then allows for synthetically useful yields of peroxyhemiketal **94m** (entry 2).

4.2.2.2 Epoxidation of Acyclic α,β-Unsaturated Ketones

The reaction of acyclic α,β-unsaturated ketones **92** with hydrogen peroxide in the presence of a 9-amino(9-deoxy)*epi*quinine (**13**) salt can further serve as entry to optically active α,β-epoxy ketones **93**.

To this end, [**13** · 2 TFA] (10 mol%) was used as catalyst system at 50 °C in dioxane (0.25 M) together with 50 wt% hydrogen peroxide (1.5 equiv) as oxidant. These conditions bring forward epoxide formation with the result that the epoxide content of the crude product mixture is proportionally higher than under the conditions previously established for the hydroperoxidation reaction (cf. Sect. 4.2.1.3).

Yet under the "epoxidation conditions", substantial amounts of peroxides **94** are still detected as by-products in the crude product mixture. Neither prolonged reaction times nor elevated temperatures were sufficient to significantly increase the epoxide yield (Table 4.30, entry 2). Cyclic peroxyhemiketals of the general type **94** are known to be transformed to the corresponding epoxides under basic conditions (Scheme 4.16a) [37, 38]. Thus, basic work-up of the crude product mixture will always enable quantitative epoxide formation independent of the initial product distribution between peroxyhemiketal **94** and epoxide **93** (Scheme 4.16b).

Peroxide **94c** was cleanly converted to α,β-epoxy ketone **93c** in the presence of catalytic amounts of aqueous NaOH (Scheme 4.17). However, the reaction was greatly accelerated without affecting the optical purity of the product when stoichiometric amounts of base were used.

Initial attempts to conduct the basic rearrangement following the catalytic reaction in the sense of a sequential one-pot process gave good to high yields of pure, isolated α,β-epoxy ketones **93** albeit with slightly reduced enantioselectivities. Thus, we decided to carry out an aqueous work-up once the hydroperoxidation-epoxidation reaction was complete, and then treat the crude product containing peroxide **94** and epoxide **93** with 1 N aqueous NaOH (1 equiv) in THF or diethyl ether at room temperature until TLC analysis indicated full conversion to the epoxide (generally within less than 20 min). According to this protocol, a wide range of α,β-epoxy ketones **93** were obtained in good to high yields (50–90%) as a single diastereomer along with excellent enantioselectivities (96.5:3.5–99.5:0.5 *er*) (Table 4.30).

Various α,β-enones **92** bearing linear and branched alkyl groups both at the β-position and adjacent to the carbonyl were suitable substrates giving optically active epoxides **93** regardless of the steric congestion in >98.5:1.5 *er* in most of the cases. Substrates which stood out were 6,6-dimethyl-3-hepten-2-one (**92r**) and 2-methyl-5-undecen-4-one (**92n**) (entries 8 and 21). Both furnished the corresponding epoxides **93r** and **n**, respectively, with an exceptional enantiomeric ratio of >99.5:0.5 and in high yields. Moreover, with enone **92n** the epoxide was the only product detected whereas formation of the corresponding peroxyhemiketal was not observed.

Gratifyingly, pseudoenantiomeric 9-amino(9-deoxy)*epi*quinidine (9-NH_2-*epi*QD; **67**) as its TFA salt afforded opposite enantiomeric epoxides with only slightly lower enantioselectivity, as exemplified by the epoxidation of enones **92a**-**b** in presence of [**67** · 2 TFA] (entries 3 and 5). Pseudoenantiomeric catalyst

Table 4.30 Substrate scope of the catalytic asymmetric epoxidation of α,β-unsaturated ketones **92**

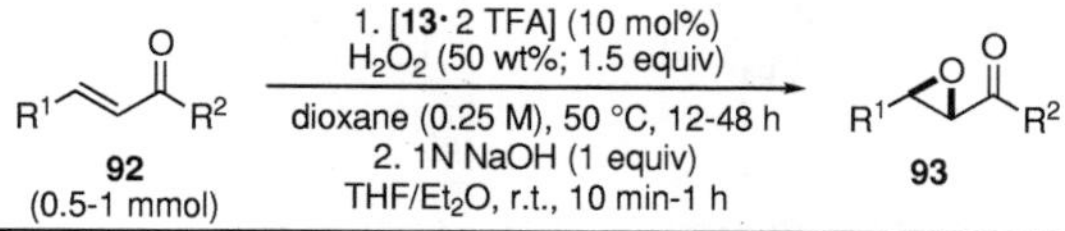

Entry	R^1	R^2	Enone	Product	**94**:**93** ratio[a]	Yield [%][b]	*er*[c]
1	n-C_6H_{13}	Me	**92a**	**93a**	57:43	75	98.5:1.5
2[d]	n-C_6H_{13}	Me	**92a**	**93a**	51:49	n.d.	n.d.
3[e]	n-C_6H_{13}	Me	**92a**	*ent*-**93a**	70:30	81	96:4
4	$PhCH_2CH_2$	Me	**92b**	**93b**	56:44	85	98.5:1.5
5[e]	$PhCH_2CH_2$	Me	**92b**	*ent*-**93b**	72:28	90	95:5
6[i]	$PhCH_2$	Me	**92q**	**93q**	47:53	70[a,f]	99:1
7	*i*-Bu	Me	**92d**	**93d**	71:29	77	98.5:1.5
8	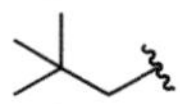	Me	**92r**	**93r**	45:55	81	>99.5:0.5
9	*i*-Pr	Me	**92s**	**93s**	52:48	50[g]	98:2
10	Cy	Me	**92e**	**93e**	42:58	84	98.5:1.5
11	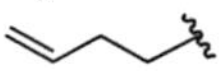	Me	**92f**	**93f**	63:37	76	98.5:1.5
12		Me	**92g**	**93g**	67:33	75	98.5:1.5
13	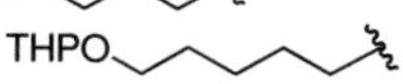	Me	**92h**	**93h**	58:42	88	98:2
14	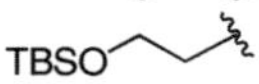	Me	**92i**	**93i**	66:34	72	98.5:1.5
15[h]	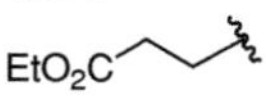	Me	**92j**	**93j**	68:32	81	98:2
16[i]	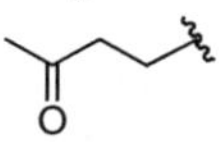	Me	**92k**	**93k**	58:42	40	98.5:1.5
17[i]		Me	**92l**	**93l**	50:50	51[j]	98:2
18	Me	Et	**92c**	**93c**	67:33	55[g]	98.5:1.5
19	n-C_9H_{19}	Et	**92p**	**93p**	47:53	82	99:1
20[k]	n-C_5H_{11}	n-C_5H_{11}	**92m**	**93m**	25:75	76	99:1
21[k,i]	n-C_5H_{11}	*i*-Bu	**92n**	**93n**	<1:>99	81	>99.5:0.5
22[k]	Me	*i*-Pr	**92o**	**93o**	50:50	60[g]	96.5:3.5

[a] Determined by ^1H NMR of the crude mixture
[b] Yields of pure, isolated products
[c] Determined by chiral GC
[d] 7 d
[e] With [**67** · 2 TFA] (10 mol%)
[f] Sum of **93q** and the corresponding PHK **94q** (at 94% conv. determined by ^1H NMR)
[g] Reduced yield due to the high volatility of the product
[h] NaOEt (1 equiv) was used instead of NaOH
[i] Second step was omitted
[j] Sum of **93l** and the corresponding PHK **94q** (at 69% conv. determined by ^1H NMR)
[k] With [**13** · 2 TFA] (20 mol%)

[**67** · 2 TFA] emerged as slightly more active than its quinine-derived analogue [**13** · 2 TFA] and the corresponding epoxides **93a-b** were obtained in high yields along with 96:4 and 95:5 *er*, respectively.

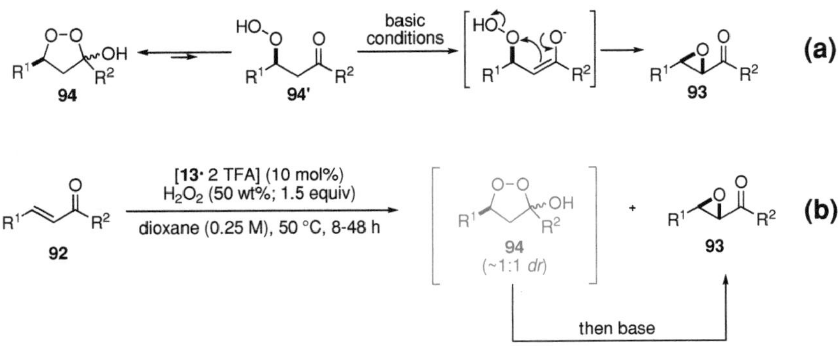

Scheme 4.16 **a** Mechanism of the peroxyhemiketal-epoxide rearrangement. **b** General strategy for the epoxidation of α,β-unsaturated ketones **92**

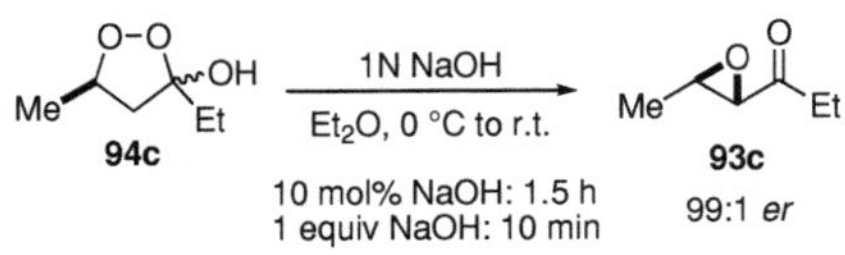

Scheme 4.17 Peroxyhemiketal-to-epoxide rearrangement mediated by aqueous NaOH

In contrast to the hydroperoxidation, the procedure for the epoxidation requires basic conditions for the peroxide-epoxide rearrangement which creates an additional challenge in terms of functional group compatibility (entries 11–17). Enones **92f-i** corresponded well to the two-step protocol and gave the desired functionalized α,β-epoxy ketones **93f-i** in good yields and high enantioselectivities. Clean conversion of dienone **92f** to α,β-epoxy ketone **93f** illustrated the selectivity for electron-deficient olefins. Despite the short exposure time, the ethyl ester group of enone **92j** was partly saponified in the presence of 1 N NaOH (1 equiv) resulting in only a moderate 47% yield of epoxide **93j**, yet in high enantioselectivity of 98:2 *er*. The use of NaOEt (1 equiv) as base in ethanol cleared that issue and α,β-epoxy ketone **93j** was isolated in 81% yield and identical optical purity (entry 15). In case of substrate **92k**, the second step was omitted since the base treatment did not improve the yield of epoxide **93k** (26%, 96:4 *er*). Product loss might arise from competing base-catalyzed aldol reaction. Thus, epoxide **93k** was isolated from the peroxide-epoxide product mixture in 40% yield along with 98.5:1.5 *er* (entry 16).

The reaction of 9-hydroxy-3-nonen-2-one (**92l**) equipped with an unprotected hydroxyl group stopped at 69% conversion as determined by ^{1}H NMR of the crude product and did not progress even upon extended runtime (entry 17). Epoxide **93l** could not be easily separated by column chromatography from the remaining starting material and was obtained in a moderate yield of 20% along with the corresponding peroxyhemiketal **94l** (25%), albeit with high enantioselectivity (98:2 *er*).

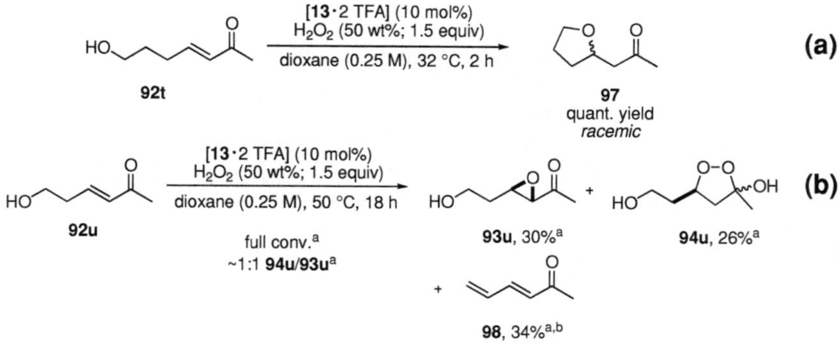

Scheme 4.18 Specific side reaction patterns with substrates **92t** and **92u**. ([a]Determined by ^1H NMR; [b]Determined by GC)

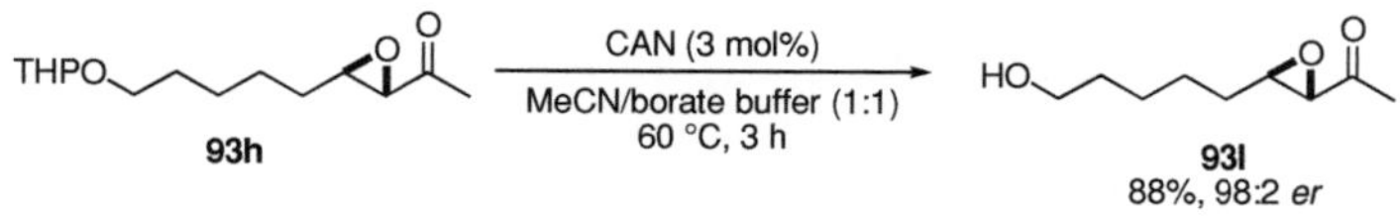

Scheme 4.19 CAN-catalyzed THP ether deprotection of epoxide **93h**

Unprotected hydroxyl enones **92t-u** were less suitable substrates for our catalytic reaction (Scheme 4.18). The nature of the competing pathway was dependent on the chain length between the hydroxyl group and the enone moiety. Whereas 7-hydroxy-3-hepten-2-one (**92t**) furnished exclusively racemic substituted tetrahydrofuran **97** via intramolecular conjugate addition of the pendant hydroxyl group (Scheme 4.18a), 6-hydroxy-3-hexen-2-one (**92u**) suffered from elimination as major side reaction which generated hexa-3,5-dien-2-one (**98**)[11] as side product (Scheme 4.18b).

Enones equipped with free hydroxyl groups constitute problematic substrates giving either low conversion or diverse side reactions. However, CAN-catalyzed deprotection of the THP ether functionality in epoxide **93h** cleanly provided 3, 4-epoxy-9-hydroxy-2-nonanone (**93l**) in 88% yield (Scheme 4.19) [39, 40]. Deprotection of the THP ether under standard conditions using catalytic *para*-toluenesulfonic acid in methanol was impeded by competing epoxide ring opening entailing a moderate yield of product **93l**.

4.2.2.3 Effect of the Enone Geometry

We also investigated the influence of the olefin geometry on the [**13** · 2 TFA]-catalyzed epoxidation reaction. Remarkably, enones (*E*)-**92b** and (*Z*)-**92b** furnished the same enantiomer of *trans*-epoxide **93b** in very high enantioselectivity

---

[11] Determined by GC-MS.

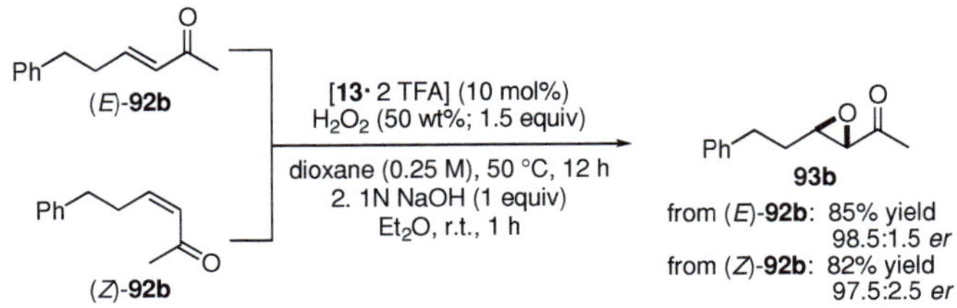

Scheme 4.20 Effect of the enone geometry on the catalytic asymmetric epoxidation reaction: illustration of the stereoconvergency

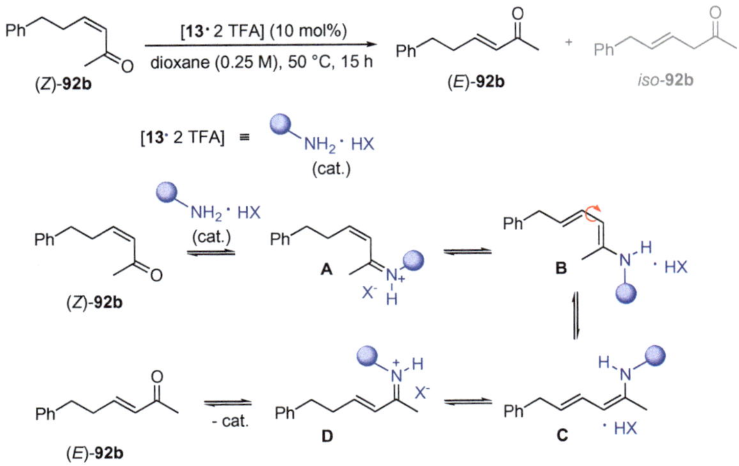

Scheme 4.21 Isomerization of (*Z*)-enone (*Z*)-**92b** in the presence of [**13** · 2 TFA] via dienamine intermediates **B** and **C**

evidencing its complete stereoconvergency (Scheme 4.20). Thus, this transformation does not require the use of pure enone isomers **92**; *E*/*Z*-mixtures can also be employed (if necessary) without any problems.

GC-MS analysis of samples taken from the epoxidation of pure (*Z*)-enone **92b** revealed the formation of the (*E*)-enone **92b** indicating a rapid isomerisation of the (*Z*)-enone to the corresponding (*E*)-isomer under reaction conditions prior to enantioselective epoxidation. (*Z*)-enone **92b** also isomerizes in the presence of only catalytic amounts of the primary amine salt [**13** · 2 TFA]—and without assistance of hydrogen peroxide. This suggests that isomerization takes place via a dienamine intermediate which allows free rotation about the carbon–carbon single bound as illustrated in Scheme 4.21.

The observation that *trans*-epoxide **92b** was formed from (*Z*)-enone **92b** with slightly lower optical purity than in the reaction of the corresponding (*E*)-enone may be rationalized by a minor contribution of conjugate addition of hydrogen

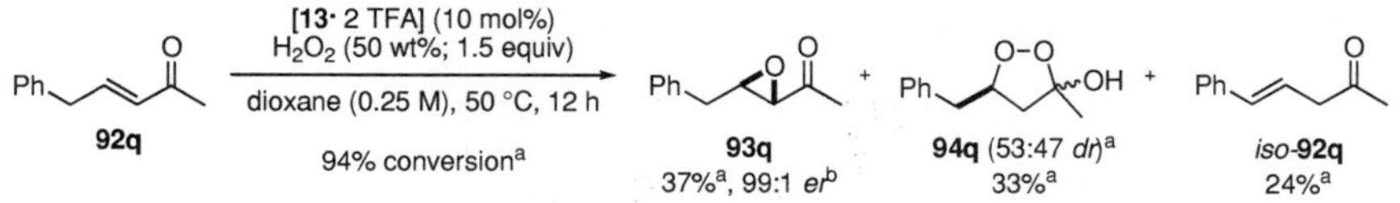

Scheme 4.22 Catalytic asymmetric epoxidation of 4-benzyl-3-buten-2-one (**92q**): double bond isomerization as side reaction. ([a]Determined by ^{1}H NMR of the crude mixture; [b]Determined by chiral GC)

peroxide prior to isomerization to the iminium ion **A** with (*Z*)-configuration which should deliver the opposite enantiomeric product (cf. Fig. 4.16 *vide infra*).

The progress of the isomerization was monitored by GC-MS. Samples were taken from the reaction mixture at different time intervals and GC-MS analysis revealed that the starting (*Z*)-enone **92b** was isomerized after 20, 40, and 120 min to 35, 50 and 75%, respectively. After 15 h, the (*E*)-isomer accounted for 99% of enone **92b**. Only trace amounts of (*Z*)-enone **92b** were present in the reaction mixture at this stage. Moreover in absence of hydrogen peroxide, regioisomeric enone *iso*-**92** was detected by GC-MS after extended reaction times. Whereas γ-protonation of dienamine **C** gives rise to the formation of (*E*)-**92b** (via iminium ion **D**), regioisomeric enone *iso*-**92** presumably is generated via competing α-protonation.,[12] [41] An alternative mechanism for the double bond isomerisation may be taken into considerations which is the reversible conjugate addition of the amine base 9-amino(9-deoxy)*epi*quinine (**13**) [42].

However, the observation of deconjugated starting material *iso*-**92** in the reaction mixture supports the involvement of dienamine intermediates in the isomerisation process. Dienamine intermediates may according to the present mechanistic rationale explain best the observation of a typically thermodynamically unfavourable isomerization of an conjugated double bond out of conjugation (cf. Scheme 4.21).

Undesired deconjugation was also observed in the epoxidation of 4-benzyl-3-buten-2-one (**92q**) (Scheme 4.22). With enone **92q**, "deconjugation" becomes a favourable pathway since in contrast to other acyclic substrates **92** the double bond in the isomerized product *iso*-**92q** benefits from conjugation with the phenyl ring. The isomerisation hampered the isolation of epoxide **93q** since it could not be readily separated by column chromatography from the isomerized starting material *iso*-**92q**.

4.2.2.4 Current Limitations

Currently, the catalytic asymmetric epoxidation is limited to aliphatic enones (Table 4.31). Among those, β-monosubstituted enones are more reactive than β, β-disubstituted analogues (entry 1; cf. Sect. 4.2.2.5). 4-Oxoenoates did not undergo our epoxidation reaction although their use in related transformations via

[12] The sterically demanding 9-amino(9-deoxy)*epi*quinine (**13**) catalyst may somewhat shield the α-position and thus hinder α-protonation in favour of competing γ-protonation.

Table 4.31 Limitations of the catalytic asymmetric epoxidation-hydroperoxidation of α,β-unsaturated ketones

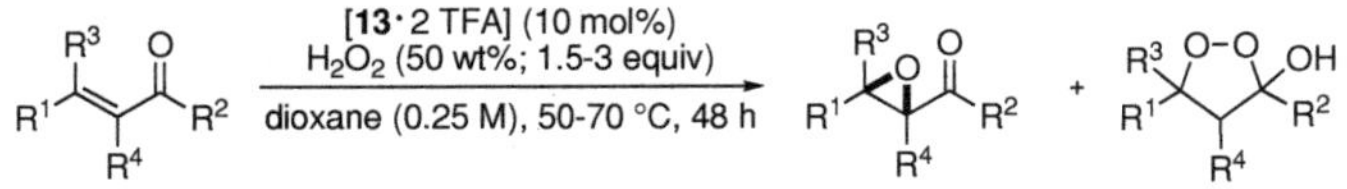

Entry	R^1	R^4	R^3	R^2	Enone	Conv. [%][a]	*er*[b]
1	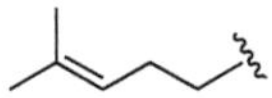	H	Me	Me	**100**	43 (23)	–
2	CO_2Et	H	H	Me	**101**	–	–
3	CO_2Et	H	Me	Me	**102**	–	–
4	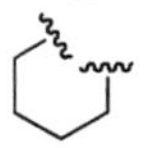		H	Me	**103**	–	–
5	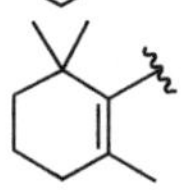	H	H	Me	**104**	<10	n.d.
6	$BnOCH_2$	H	H	Me	**105**	28 (20)	n.d.
7	Ph	H	H	Me	**106a**	15 (5)	98.5:1.5
8[c]	Ph	H	H	Me	**106a**	<10 (3)	99.5:0.5
9	4-NO_2-C_6H_4	H	H	Me	**106b**	–	–
10	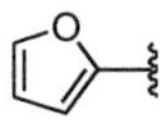	H	H	Me	**107**	<5	n.d.
11	Me	H	H	Ph	**108**	–	–
12	Ph	H	H	Ph	**109**	–	–
13	H	H	H	*n*-C_5H_{11}	**110a**	–	–

[a] Determined by GC. Values in parentheses correspond to the total yield of PHK and epoxide determined by GC

[b] Determined by chiral GC

[c] [**64** · (*S*)-TRIP] (10 mol%) was used

asymmetric iminium ion catalysis is well-precedented in the literature (entries 2–3). To date, neither substitution at the α-position, nor extended conjugation of the double bond is tolerated, as illustrated by the reaction of 1-acetylcyclohexene (**103**) (entry 4) and β-ionone (**104**) (entry 5), respectively. Moreover, a benzyloxy group at the γ-position impeded the reaction (entry 6), whereas enones with remote protected hydroxyl functionalities readily reacted with hydrogen peroxide (*vide supra*). Benzylideneacetone (**106a**) gave the corresponding epoxide in less than 10% yield as detected by GC-MS albeit with high enantioselectivity of 98.5:1.5 *er* (entry 7). Even catalyst [**64** · (*S*)-TRIP] (cf. Table 4.2), which exhibited superior catalytic activity compared with [**13** · 2 TFA] in the asymmetric epoxidation of 2-cyclopentenone (**46r**), was unable to efficiently mediate the reaction (entry 8). In both cases, considerable amounts of the starting material **106a** decomposed to phenylacetaldehyde[13] (up to 12%) presumably via a Dakin pathway. The use of a benzylideneacetone derivative bearing electron withdrawing substituents at the phenyl ring (4′-nitrophenyl-3-buten-2-one, **106b**), and thus constituting a more

[13] Determined by GC–MS.

Scheme 4.23 Reversible acid-catalyzed conjugate addition of 9-amino(9-deoxy)*epi*quinine (**13**) to 3-octenone (**110a**)

potent acceptor substrate, suppressed this side reaction, but overall did not improve the results (entry 9). Other aromatic enones such as furfurylidenacetone (**107**), 1-phenyl-2-buten-1-one (**108**), or chalcone (**109**) proved to be unreactive under the present reaction conditions (entries 10–12). Thus, both substituents R^1 and R^2 are limited to alkyl groups [43].'[14] Indeed, the same substrate scope limitation has previously been detected in several iminium-catalytic transformations of α, β-unsaturated ketones employing 9-amino(9-deoxy)*epi*cinchona alkaloid catalysts. Indeed, the substituent R^2 adjacent to the ketone only very rarely tolerates aromatic rings [44, 45]. In this, our catalytic epoxidation of α,β-unsaturated ketones very much complements conventional methods such as metal-catalyzed or Juliá-Colonna epoxidations.

Somewhat surprisingly, terminal enones were unsuitable substrates (entry 13), in the presence of catalyst [**13** · 2 TFA] as well as [**64** · (*S*)-TRIP]. We speculate that in the case of more active Michael acceptors, conjugate addition of 9-amino (9-deoxy)*epi*quinine (**13**) to form β-amino ketones is favoured over the 1,2-addition to the ketone carbonyl group, even more so since hydrogen peroxide is not a particular strong nucleophile to compete with (Scheme 4.23).

To support our hypothesis, we conducted the epoxidation of 3-decen-2-one (**92a**) in the presence of the terminal enone 7-octen-6-one (**110a**). Epoxide **93a** derived from 3-decen-2-one (**92a**) was still generated, yet at lower rates. This is consistent with the assumption of the catalyst being captured as β-amino ketone and only partly released from this "catalyst resting state" keeping its absolute concentration low [46].

4.2.2.5 Application to β,β-Disubstituted α,β-Unsaturated Ketones

The scope of the catalytic asymmetric hydroperoxidation-epoxidation reaction was further explored by using β,β-disubstituted α,β-unsaturated ketones as potential substrates.

This substrate class had turned out to be considerably less reactive than β-monosubstituted analogues and only moderate conversions of up to 43% were achieved with citral-derived enone **100** (Table 4.32). Under the reaction conditions, *E*/*Z*-isomerization of the starting material **100** took place—presumably via a

---

[14] Deng and co-workers reported identical substrate scope limitations in their [**13** · 3 TFA]-catalyzed asymmetric alkylperoxidation of α,β-unsaturated ketones with TBHP and related oxidants.

Table 4.32 Screening studies: hydroperoxidation-epoxidation of 4,8-dimethyl-3,7-nonadien-2-one (**100**)

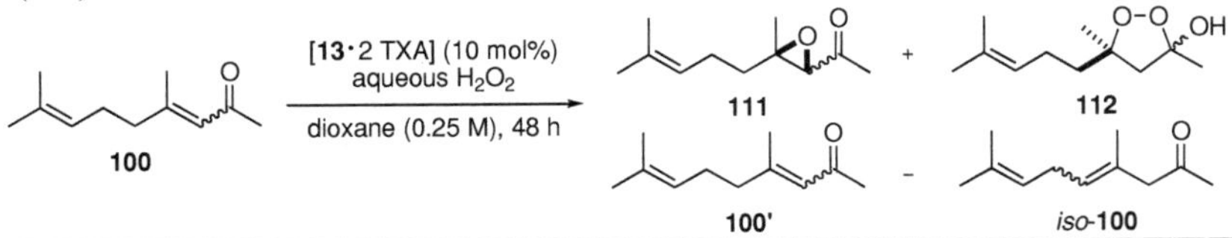

Entry	*E*/*Z*(**100**)[a]	TXA	H_2O_2 [equiv]	T [° C]	conv. [%][a]	*E*/*Z*(**100′**)[a]	Yield (*trans*/*cis*) **111** [%][a]	*er*(**111**)[c]	Yield **112** [%][a]
1	4:96	TCA	3	32	14	17:83	4	n.d.	5
2	4:96	TFA	3	50	30	26:74	7 (77:23)	n.d.	14
3	95:5	TCA	3	32	29	83:17	5	n.d.	8
4	95:5	TFA	3	50	42	76:24	5 (85:15)	n.d.	17
5	95:5	TFA	1.5[d]	50	43	74:26	3	n.d.	20
6	95:5	TFA	1.5[d]	70	65[b]	64:36	12[b] (81:19)	90.5:9.5 *er*(*trans*) 57:43 *er*(*cis*)	40[b]
7[e]	73:27	TFA	3[d]	70	85[b]	66:34	10[b] (88:12)	95.5:4.5 *er*(*trans*) 59.5:40.5 *er*(*cis*)	69[b]

[a] Determined by GC
[b] Determined by ^1H NMR of the crude mixture
[c] Determined by chiral GC
[d] 50 wt% aqueous H_2O_2 was used
[e] [**13** · 2 TFA] (20 mol%) was used

dienamine intermediate (cf. Scheme 4.21), and from both the *E*-and *Z*-isomer of citral-derived enone **100**, the *trans*-isomer of epoxide **111** was generated predominantly. In the case of *β*,*β*-disubstituted enones, the *E*/*Z*-interconversion of enone **100** also seems to proceed at higher rates than the conjugate addition of hydrogen peroxide which allowed us to employ a 73:27 mixture of *E* and *Z* isomers of **100** without noticeably affecting the reaction outcome. Detection of dienone *iso*-**100** by ^1H NMR and GC analysis of the crude product mixture denoted the isomerisation of *α*,*β*-enone **100** to the *β*,*γ*-unsaturated derivative as competing side reaction. This behaviour has been observed earlier with 4-benzyl-3-buten-2-one (**92q**) (cf. Scheme 4.20).

Forcing reaction conditions comprising increased catalyst and hydrogen peroxide loadings at elevated temperature proved essential to achieve reasonable conversions (entries 6–7). Under not yet fully optimized reaction conditions, peroxyhemiketal **112** was obtained in 56% yield with an enantiomeric ratio of 78:22 *er*. Epoxide **111** was formed concomitantly in 10% yield and 88:12 *dr*(*trans*/*cis*) with 95.5:4.5 *er* for the *trans*- and 59.5:40.5 *er* for the *cis*-isomer, respectively (Scheme 4.24).

As illustrated by the group of *Dussault*, 3-hydroxy-1,2-dioxolanes such as **112** may function as key intermediates in the synthesis of plakinic acid natural products (cf. Fig. 2.6) which display promising antitumor and antifungal activity [47].

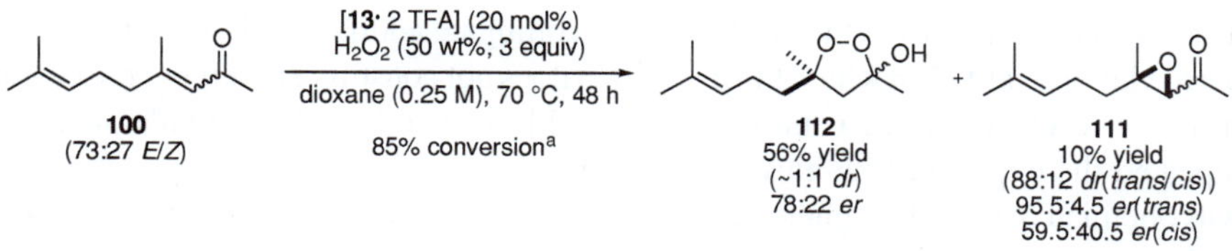

Scheme 4.24 Catalytic asymmetric hydroperoxidation of 4,8-dimethyl-3,7-nonadien-2-one (**100**) (73:27 *E*/*Z*). ([a]Determined by ^{1}H NMR)

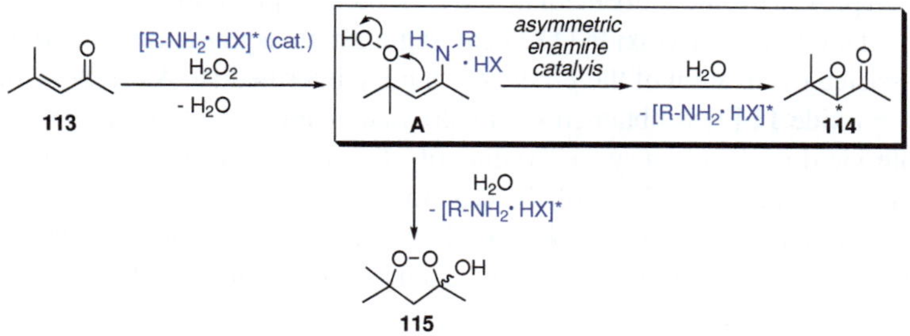

Scheme 4.25 Catalytic asymmetric Weitz-*Scheffer* epoxidation of mesityloxide (**113**)

Table 4.33 Screening studies: epoxidation of mesityloxide (**113**)

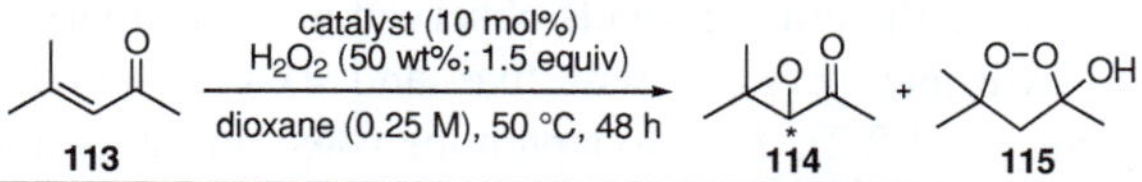

Entry	Catalyst	Conv. [%][a]	Yield (**114**)[a]	Yield (**115**)[a]	*er* (**114**)[b]
1	[**13** · 2 TFA]	66	28	38	19:81
2	[(*R*,*R*)-DPEN · TFA]	10	7	3	74.5:25.5
3	[(*R*,*R*)-DPEN · (*S*)-TRIP]	66	34	32	93.5:6.5

[a] Determined by ^{1}H NMR of the crude mixture
[b] Determined by (chiral) GC

Another β,β-disubstituted α,β-unsaturated ketone which was examined in the catalytic asymmetric epoxidation reaction was mesityloxide. Mesityloxide (**113**) is in contrast to citral-derived enone **100**, symmetrically substituted at the β-position. The enantio-determining step is not the conjugate addition but the epoxide ring closure (like in the case of terminal enones) (Scheme 4.25). Thus, the intended transformation may proceed via iminium ion activation but is probably classified correctly as (intramolecular) asymmetric **enamine** catalysis rather than asymmetric iminium ion catalysis [1].

Table 4.33 summarizes the attempts toward a catalytic asymmetric epoxidation of mesityloxide (**113**). In the presence of catalyst [**13** · 2 TFA] at 50 °C, epoxide **114** was obtained in only moderate yield and enantioselectivity (28%, 81:19 *er*) (entry 1).

The greater part (38%) of the product of the conjugate addition of H_2O_2 remained as peroxyhemiketal **115** in which the *β*-position does not constitute a stereogenic center. In order to carry a higher amount of peroxide forward to the epoxide stage we repeated the experiment at elevated temperature. At 70 °C, *α*,*β*-epoxy ketone **114** was obtained in slightly higher yield of 32%, albeit in lower optical purity (57:43 *er*).

Recently our group solved the problem of highly enantioselective epoxidation of symmetrically *β*,*β*-disubstituted *α*,*β*-unsaturated aldehydes, e.g. 3-methyl-2-butenal, by using a chiral catalyst [**10** · (*S*)-TRIP] featuring TRIP as chiral counteranion (cf. Scheme 2.30). Inspired by this work, we tested catalyst [(*R*,*R*)-DPEN · (*S*)-TRIP] which had proven to efficiently mediate the asymmetric epoxidation of cyclic enones in the epoxidation of mesityloxide (**113**) envisaging high asymmetric induction from a TRIP-assisted cyclization of the *β*-peroxyenamine intermediate **A** (cf. Scheme 4.25). Indeed, epoxide **114** was obtained with increased enantioselectivity of 93.5:6.5 *er* in moderate yield of 32% (entry 3). To quantify the impact of the chiral counteranion TRIP on the enantioselectivity of the reaction, we conducted the epoxidation with catalytic amounts of [(*R*,*R*)-DPEN · TFA]. Yet, this catalyst furnished epoxide **114** with significantly reduced optical purity, and moreover, exhibited dramatically lower catalytic activity (entry 2).

4.2.2.6 Medium and Large Ring Enones Revisited

Whereas epoxides were the only products observed in the primary amine salt-catalyzed reaction of hydrogen peroxide with five- and six-membered as well as large ring enones (cf. Sect. 4.1.2.2), 2-cycloheptenone (**46s**) and 2-cyclooctenone (**46v**) provided under identical reaction conditions a product mixture consisting of the expected epoxides **48s** and **v** along with peroxidic by-products **116** and **117**—both in optically active form—as previously described for acyclic enones (Scheme 4.26). Yet, in contrast to monocyclic peroxyhemiketals **94** derived from acyclic enones, varying amounts of cyclic *β*-hydroperoxyketones **116** and **117b**, which in solution exist in equilibrium with the corresponding bicyclic peroxyhemiketals (**116** and **117a**), could be detected by NMR.

Scheme 4.26 displays the results obtained for the reaction of 2-cycloheptenone (**46s**) and 2-cyclooctenone (**46v**) with the 9-amino(9-deoxy)*epi*quinine salt [**13** · 2 TFA] as catalyst (cf. Table 4.10). 2-Cycloheptenone (**46 s**) afforded predominantly the expected *α*,*β*-epoxy ketone **48s**, which was isolated in 65% yield with excellent enantioselectivity of >99.5:0.5 *er* (Scheme 4.26a). In contrast, *β*-hydroperoxycyclooctanone **117b** was obtained as the major product (30:70 **48v**/**117**) in the reaction of eight-membered ring enone **46v**. Moreover, significant amounts of deconjugated 3-cyclooctenone *iso*-**46v** could be detected by GC and ^{1}H NMR of the crude product mixture.[15] Yet, as previously seen for acyclic enones

[15] *Iso*-**46v** was detected in the crude product mixture prior to treatment with base. Deconjugation was also noticed in the reaction of 2-cycloheptenone (**46s**), yet only to a negligible extent (~1%).

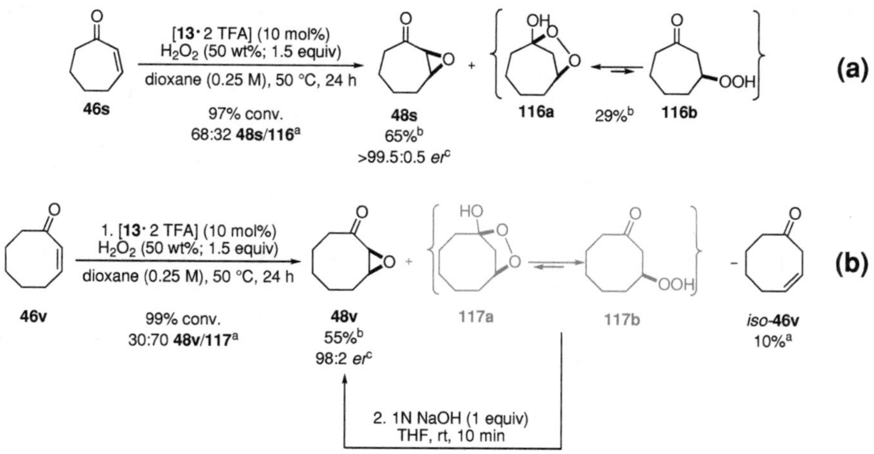

Scheme 4.26 Catalytic reaction of 2-cycloheptenone (**46s**) and -octenone (**46v**): formation of peroxidic by-products. ([a]Determined by ^{1}H NMR of the crude mixture; [b]Yields of pure, isolated products; [c]Determined by chiral GC)

92 (cf. Scheme 4.16a), peroxide **117** can be transformed to the corresponding epoxide under basic conditions and thus, 2,3-epoxycyclooctanone (**48v**) was finally obtained in 55% yield with 98:2 *er* over two steps (Scheme 4.26b).[16]

The peroxidic compounds **116** and **117** piqued our interest since both substances had not been previously described in the literature. Notably, the bicyclic peroxidic structure (of **116** and **117b**) resembles potent antimalarial agents. With the goal of optimizing the yield of the peroxidic products at the expense of the corresponding epoxides, we evaluated various catalyst systems in the reaction of 2-cycloheptenone (**46s**) and 2-cyclooctenone (**46v**).

Since the peroxidic compound **116** was only isolated in 29% yield under the original reaction conditions (cf. Scheme 4.26a), we were pleased to see that the previously observed product distribution was inverted by conducting the reaction with [**13** · 2 TCA] as catalyst and 30 wt% hydrogen peroxide (3 equiv) at 32 °C (Table 4.34, entries 1–2), as previously optimized for the hydroperoxidation of acyclic enones. Under those conditions, peroxyhemiketal **116** accounted for 77% of the product mixture as detected by ^{1}H NMR and was generated with only a slight loss of enantioselectivity (99:1 *er*). Catalyst [(*R*,*R*)-DPEN· (*S*)-TRIP] displayed the same selectivity as [**13** · 2 TFA] and provided antipodal epoxide (2*R*,3*R*)-**48s** as the major product with high enantioselectivity (99:1 *er*) (entry 3, cf. Table 4.13, entry 6). DPEN-based catalyst systems lacking the chiral counteranion TRIP gave inferior results with regard to both yields and enantioselectivities (entries 4–5).

[16] The same two-step protocol was applied to 2-cycloheptenone (**46s**). Yet, the yield was not significantly improved to compensate for the loss of enantioselectivity (96.5:3.5 *er*).

Table 4.34 Evaluation of different catalyst systems for the reaction of 2-cycloheptenone (**46s**)

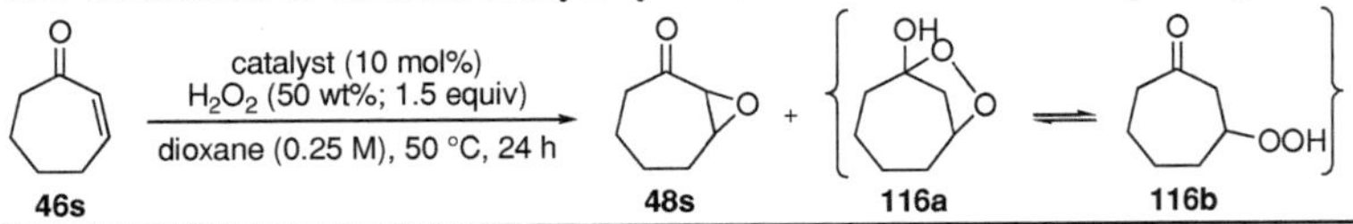

46s **48s** **116a** **116b**

Entry	Catalyst	Conv. [%][a]		**48s**:**116**[a]	*er*(**48s**)[b]
1	[**13** · 2 TFA]	97	(2*S*,3*S*)-**48s**	68:32	>99.5:0.5
2[c]	[**13** · 2 TCA]	89	(2*S*,3*S*)-**48s**	23:77	99:1
3	[(*R*,*R*)-DPEN · (*S*)-TRIP]	98	(2*R*,3*R*)-**48s**	70:30	1:99
4	[(*R*,*R*)-DPEN · TFA]	66	(2*R*,3*R*)-**48s**	41:59	4:96
5	[(*R*,*R*)-DPEN · 2 TFA]	87	(2*R*,3*R*)-**48s**	55:45	2:98

[a] Determined by ^1H NMR of the crude product
[b] Determined by chiral GC
[c] At 32 °C with H_2O_2 (30 wt%; 3 equiv)

Table 4.35 Evaluation of different catalyst systems for the reaction of 2-cyclooctenone (**46v**)

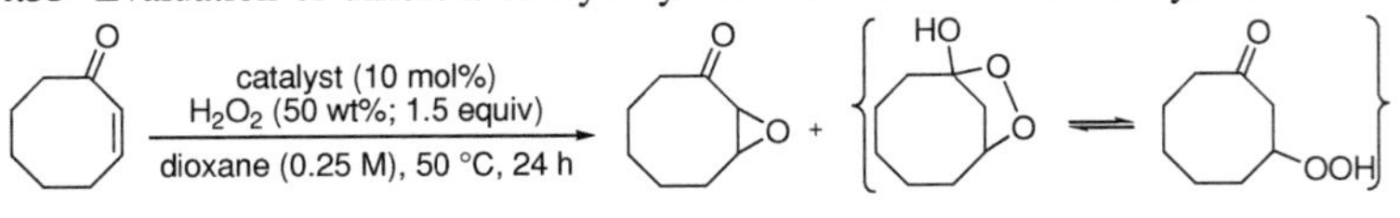

46v **48v** **117a** **117b**

Entry	Catalyst	Conv. [%][a]	Product	**48v**:**117**[a]	*er*(**48v**)[b]
1	[**13** · 2 TFA]	99	(2*S*,3*S*)-**48v**	30:70	>99.5:0.5
2[c]	[**13** · 2 TCA]	91	(2*S*,3*S*)-**48v**	12:88	>99.5:0.5
3	[(*R*,*R*)-DPEN · (*S*)-TRIP]	80	(2*R*,3*R*)-**48v**	37:63	1:99
4	[(*R*,*R*)-DPEN · TFA]	52	(2*R*,3*R*)-**48v**	26:74	10.5:89.5
5	[(*R*,*R*)-DPEN · 2 TFA]	71	(2*R*,3*R*)-**48v**	19:81	9:91

[a] Determined by ^1H NMR of the crude product
[b] Determined by chiral GC
[c] At 32 °C with H_2O_2 (30 wt%; 3 equiv)

Similar trends were observed with 2-cyclooctenone (Table 4.35). The amount of peroxidic product **117** could be further increased by applying the optimized hydroperoxidation conditions (entries 1–2). Catalyst [(*R*,*R*)-DPEN · (*S*)-TRIP] gave comparable results, yet at slightly reduced catalyst activity (entry 3). Moreover, simple TFA salts [(*R*,*R*)-DPEN · 1-2 TFA] exhibited lower catalytic activity, and again their use brought about an erosion of enantioselectivity (entries 4–5). Interestingly, in contrast to five- and six-membered cyclic enones for which DPEN-mono-TFA salt afforded superior results than DPEN-di-TFA salt, their catalytic efficiency is inverted with cycloheptenone and cyclooctenone.

Under the optimum conditions, peroxidic compounds **116** and **117** were isolated in synthetically useful yields of 55 and 59%, respectively (Scheme 4.27).

Interestingly, hydroperoxidation product **116** exists in equilibrium preferentially as bicyclic peroxyhemiketal **116a** rather than as hydroperoxyketone **116b** (Scheme 4.28). The precise ratio is depending on the solvent with the concentration

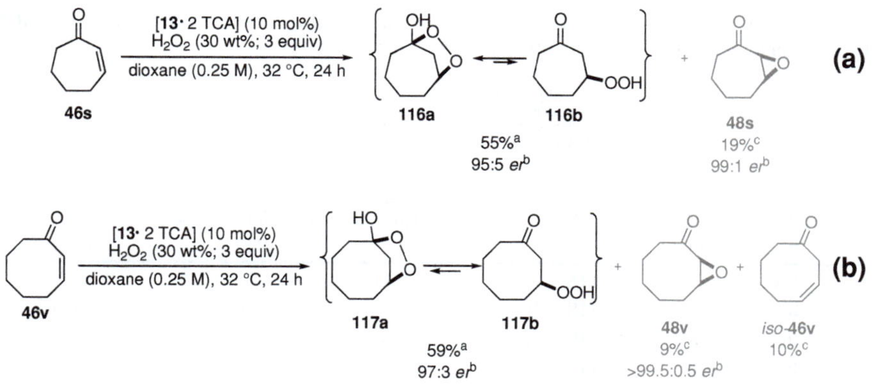

Scheme 4.27 Catalytic asymmetric hydroperoxidation of **a** 2-cycloheptenone (**46s**), and **b** 2-cyclooctenone (**48v**). ([a]Yields of pure, isolated products; [b]Determined by chiral GC; [c]Determined by ^{1}H NMR of the crude product)

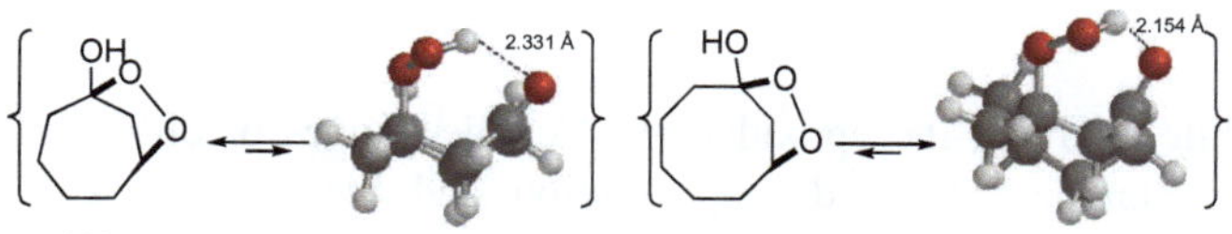

	116a	116b	117a	117b
THF:	80	20	30	70
$CHCl_3/CH_2Cl_2$:	82	18	34	66

Scheme 4.28 Mole fractions of bicyclic peroxyhemiketals and β-hydroperoxyketones in different solvents at ambient temperature

of peroxyhemiketal being slightly increased in non-coordinating solvents such as $CHCl_3$. In contrast, peroxidic product **117** derived from 2-cyclooctenone (**46v**) exists predominantly as hydroperoxyketone **117b**. DFT calculations[17] revealed the existence of an internal hydrogen bond which may account for the stabilization of hydroperoxyketones **116** and **117b** as illustrated in Scheme 4.28; though, due to geometric considerations less so in the case of the seven-membered ring.

Alike in the case of acyclic enones, the use of *tert*-butyl hydroperoxide as oxidant instead of hydrogen peroxide generated predominantly the conjugate addition product in the reaction catalyzed by [**13** · 2 TFA], thereby offering an alternative entry to peroxidic compounds related to **116** and **117** (cf. Sect. 4.2.1.4).

In contrast to 2-cycloheptenone (**46s**) and 2-cyclooctenone (**46v**), 3-substituted cyclo-heptenone derivatives afforded exclusively the corresponding epoxide. Thus, our approach to such bicyclic peroxyhemiketals is complementary to that of Blanco et al. [48, 37]—the oxidation of bicyclic cyclopropanols—which is only feasible for the synthesis of 3-substituted bicyclic peroxides.

[17] Density functional theory (DFT) computation at the B3LYP/6-31G* level of theory (Vacuum) was performed using Spartan'08 Windows from Wavefunction, Inc.

Table 4.36 Catalytic asymmetric epoxidation of macrocyclic enones: cyclododecenone (**46w**) (n = 3) and cyclopenta-decenone (**46x**) (n = 6)

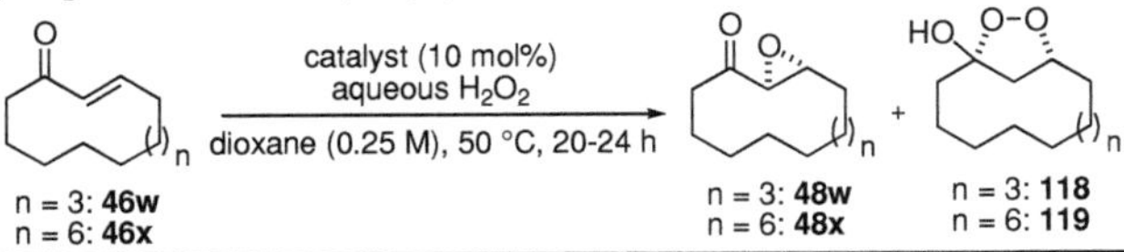

Entry	(*E*)-enone	(*E*)-epoxide	PHK	Conditions[a]	Epoxide:PHK[b]	Yield (epoxide) [%]	*er* (epoxide)[c]
1	**46w**	**48w**	**118**	A	100:0	92	99.5:0.5
2	**46w**	**48w**	**118**	B	97:3	n.d.	99.5:0.5
3	**46x**	**48x**	**119**	A	96:4	87	99.5:0.5
4	**46x**	**48x**	**119**	B	91:9	n.d.	99.5:0.5

[a] Conditions A: [**13** · 2 TFA] (10 mol%), H_2O_2 (50 wt%; 1.5 equiv), dioxane (0.25 M), 50 °C, 20 h; Conditions B: [**13** · 2 TCA] (10 mol%), H_2O_2 (30 wt%; 3 equiv), dioxane (0.25 M), 32 °C, 20 h

[b] Determined by ^{1}H NMR of the crude mixture

[c] Determined by chiral GC

While studying the epoxidation of cyclic enones of different ring sizes, macrocyclic enones attracted our attention and we came across very intriguing observations: in contrast to both acyclic enones and medium ring enones, macrocyclic enones give almost exclusively epoxides in the reaction with hydrogen peroxide and [**13** · 2 TFA] (or [(*R*,*R*)-DPEN · (*S*)-TRIP]) as catalyst (Table 4.36, entries 1 and 3; cf. Sect. 4.1.2.2). Moreover, these epoxides are formed with an outstanding optical purity of 99.5:0.5 *er*. Even under reaction conditions which usually promote the hydroperoxidation pathway (conditions B), the maximum yield of bicyclic peroxyhemiketal remained below 10% (entries 2 and 4).

We presume that conformational constraints in the seven- and eight-membered ring systems, in contrast to 2-cyclohexenone, may hamper the peroxyenamine intermediate in adopting the proper conformation for epoxide ring closure which requires the antiperiplanar alignment of the π-system of the enamine with the O–O single bond. Those difficulties might be reflected in the accumulation of hydroperoxidation product in the course of the reaction. Large rings are very flexible, and show complex conformational behaviour. Presumably, as a result, cyclododecenone and cyclopentadecenone alike cyclohexenone provide almost exclusively the corresponding α,β-epoxyketones both under epoxidation (A) and hydroperoxidation (B) conditions.

It might be interesting to study the cyclic enone series continuously to identify a cross-over point from which on the amount of peroxyhemiketal would start decreasing whereas the epoxide formation begins rising again (Fig. 4.7). A correlation with the inherent ring strain seems likely.

In contrast, acyclic enones possess enhanced conformational freedom. Yet, at the transition to acylic enones, other conformational considerations as for instance the allylic strain might become decisive.

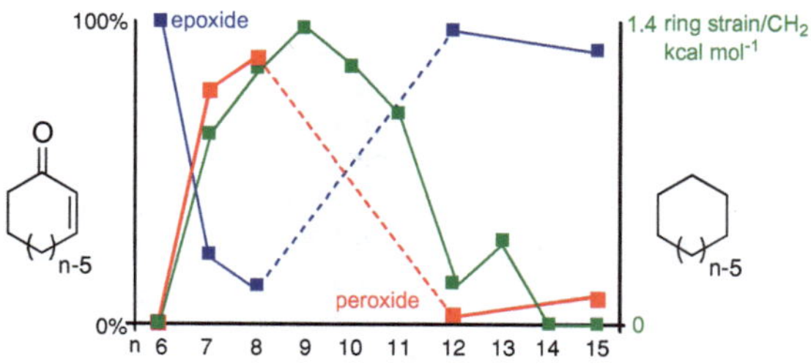

Fig. 4.7 Cyclic enone series: concentration of hydroperoxide and epoxide against ring size. (*blue*: epoxide; *red*: peroxide; *green*: ring strain/CH_2 of the corresponding cycloalkanes) [33]

4.2.3 Summary and Conclusions

After having developed a method for the asymmetric epoxidation of cyclic α,β-unsaturated ketones (cf. Sect. 4.1), we became further interested in expanding the scope of this powerful transformation. We therefore turned our attention toward acyclic substrates. In particular, salts of the readily available quinine-derived primary amine 9-amino(9-deoxy)*epi*quinine (9-NH_2-*epi*Q; **13**) retained their high catalytic activity and selectivity for the reaction with hydrogen peroxide when facing acyclic rather than cyclic substrates (Table 4.21).

Our work with acyclic enones is marked by two major achievements, namely: the development of a catalytic asymmetric hydroperoxidation to furnish optically active 3-hydroxy-1,2-dioxolanes; and the implementation of a highly enantioselective epoxidation protocol applicable to a broad array of acyclic aliphatic, enolizable enones, a substrate class which has long eluded existing catalytic systems.

Interestingly, both transformations share a common origin. Initial experiments with acyclic enones **92** had revealed that those substrates (in contrast to cyclic enones) furnish product mixtures comprising the expected α,β-epoxyketones **93** along with unanticipated five-membered cyclic peroxyhemiketals **94** upon exposure to aqueous hydrogen peroxide in the presence of a primary amine catalyst (Scheme 4.11). Notably, such an asymmetric hydroperoxidation reaction has not previously been reported in the literature. Optimization studies revealed factors affecting the product distribution between peroxyhemiketals and the corresponding epoxides. By adjusting the amount and concentration of the hydrogen peroxide used (30 wt%; 3 equiv; Table 4.25) and the choice of the acid co-catalyst (TCA instead of TFA; Table 4.22), we were able to identify reaction conditions which allowed the obtention of good isolated yields (up to 72%) of highly enantiomerically enriched 3-hydroxy-1,2-dioxolanes (up to 98.5:1.5 *er*). The asymmetric hydroperoxidation is applicable to a range of enones **92** and high enantiomeric ratios are obtained irrespective of the substitution pattern as far as β-substituted enones are concerned (Table 4.28). The yields vary only slightly, and typically decrease to some extent as the steric demand of the enone substituents increases (Table 4.29). The mild reaction conditions are testified by the high functional group compatibility. Enone substrates **92h** and **i** equipped with acid labile functional groups are well tolerated and cleavage of the TBS or THP ethers was not observed.

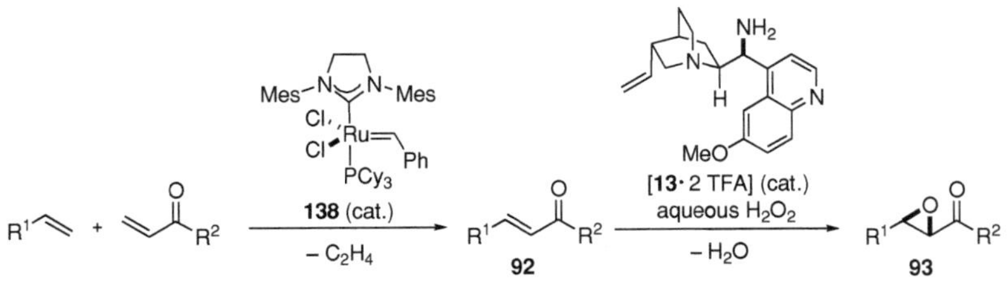

Scheme 4.29 Concise access to optically active α,β-epoxyketones via a CM-epoxidation sequence

3-Hydroxy-1,2-dioxolanes **94** can furthermore serve as valuable precursors to optically active aldol products **96** and 1,2-dioxolanes such as **127** as will be illustrated within the next Chapter (cf. Sect. 4.3).

Importantly, they can be cleanly converted to the corresponding epoxides **93** under basic conditions. Based on this, we have converged initially obtained product mixtures (of peroxides and epoxides) selectively to the epoxide by treatment with 1 N NaOH once the starting material was consumed. According to this protocol, we prepared a broad range of *trans*-epoxides **93** in good to high yields (up to 90%) and excellent enantioselecivities of up to >99.5:0.5 as demonstrated with substrates **92n** and **r** (Table 4.30). Once again functional groups are well-tolerated, and notably, the process proceeds with full stereoconvergency which allows the use of E/Z isomeric enone mixtures as starting materials without affecting the reaction outcome (Scheme 4.20).

Encouraging results have also been obtained with β,β-disubstituted enones which are generally less reactive than their β-monosubstituted counterparts when using primary amine salt catalysts such as [9-NH_2-*epi*Q (**13**) · 2 TFA] (cf. Sect. 4.2.2.5). Remarkably, citral-derived enone **100** provided the corresponding peroxyhemiketal **112** in good yield and encouraging enantioselectivity of 78:22 *er* when reacted with aqueous hydrogen peroxide in the presence of [**13** · 2 TFA] (Scheme 4.24). Peroxyhemiketals similar to compound **112** might play a pivotal role toward the enantioselective synthesis of members of the plakinic acid natural product family (cf. Fig. 2.6). Moreover, structurally interesting optically active bicyclic peroxidic compounds of potentially biological importance were obtained in the reaction of seven- and eight-memberd cyclic enones as substrates (cf. Sect. 4.2.2.6).

Gratifyingly, the use of the pseudoenantiomeric catalyst as [9-NH_2-*epi*QD (**67**) · 2 TFA] derived from naturally abundant quinidine (instead of quinine), provides an entry to the antipodal product series with only slightly reduced enantioselectivity as has been highlighted with the syntheses of α,β-epoxyketones *ent*-**93a** and *ent*-**93b** (Table 4.30).

Overall, our asymmetric epoxidation process significantly advances the area of chemical space formed by highly enantioenriched α,β-epoxyketones. Indeed, and thanks to the groundbreaking advances in olefin cross-metathesis, such compounds can now be fashioned in only two steps (Scheme 4.29).

Cross metathesis of any functionalized terminal olefin with the appropriate vinyl ketone (cf. Table 4.41, *vide infra*), followed by organocatalytic epoxidation delivers the highly valuable epoxyketone building blocks. This synthetic manifold can be contrasted with the more "classical" approach to those compounds, relying on preparation of a suitable allylic alcohol (typically employing Wittig and related olefinations), followed by Sharpless epoxidation and finally adjustment of the oxidation state [49]. It is evident that both the low atom and redox economy of this "classical" blueprint render it fairly unattractive from the standpoint of synthetic efficiency. Conversely, the merits of the approach described in Scheme 4.29 to these valuable intermediates rest precisely on both its high atom economy and the conceptual and practical advantages of employing two catalytic transformations to rapidly increase molecular complexity.

4.3 Synthetic Transformations of Optically Active α,β-Epoxy Ketones and 3-Hydroxy-1,2-dioxolanes

4.3.1 One-Pot Synthesis of Aldol Products From α,β-Unsaturated Ketones

The aldol reaction is arguably one of the most important methods for the formation of carbon–carbon bonds. It creates β-hydroxy carbonyl compounds of high interest as key intermediates in construction of a variety of important natural products and drugs. Extensive research is therefore devoted toward the development of highly efficient direct catalytic enantioselective aldol reactions employing either synthetic metal complexes, enzymes, or purely organic molecules as chiral catalysts [50–54].

Despite the progress made in recent years mainly within the realm of organocatalysis, the use of α-unsubstituted, and thus readily enolizable aldehydes **120** is generally problematic. That is the case also in aldol reactions catalyzed by proline or proline derivatives with this substrate class affording inferior enantioselectivities compared to aromatic aldehydes and low yields due to competing self-aldolization and the formation of condensation products (Scheme 4.30) [55–58].

With the catalytic asymmetric hydroperoxidation/epoxidation of α,β-unsaturated ketones **92**, we hold a powerful reaction that affords peroxides and epoxides with excellent enantioselectivities of up to >99.5:0.5 *er* for a range of enones bearing aliphatic enolizable substituents at the β-position and adjacent to the ketone group (cf. Sect. 4.2.2).

We became intrigued by the idea of developing a one-pot synthesis of optically active aldol-type products **96** from α,β-unsaturated ketones **92** which capitalizes on this highly enantioselective transformation. Notably, such a process would overall represent a formal olefin hydration (Scheme 4.31).

3-Hydroxy-1,2-dioxolanes **94** are readily converted to the corresponding β-hydroxy ketones by a variety of reducing agents such as trivalent phosphorus

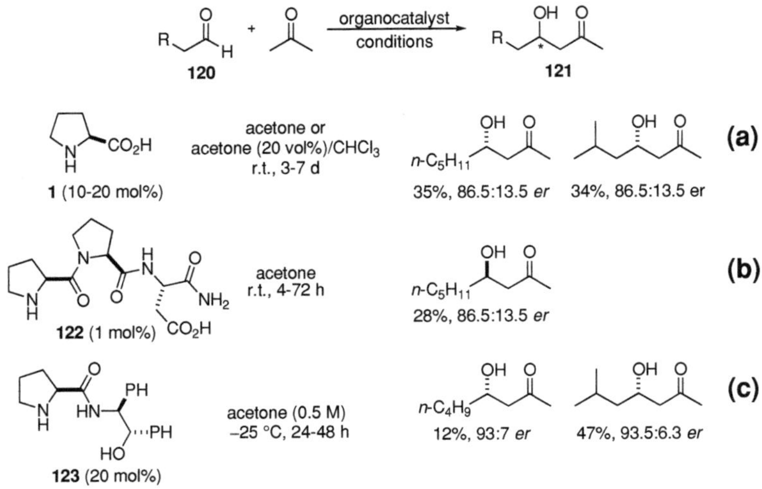

Scheme 4.30 Aldol addition of acetone to α-unbranched aliphatic aldehydes **120** catalyzed by (*S*)-proline (**1**)[58] and proline derivatives (**122–123**) [55–57]

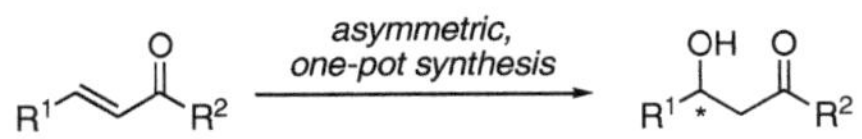

Scheme 4.31 Asymmetric, one-pot synthesis of aldol products from α,*β*-enones

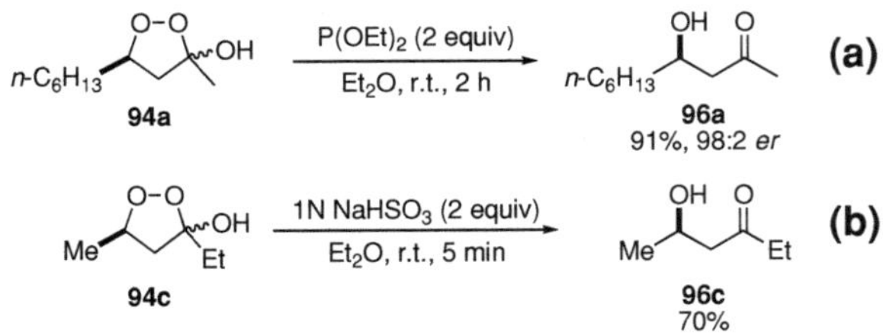

Scheme 4.32 Reduction of peroxyhemiketals **94** to aldol-type products **96** by **a** $P(OEt)_3$, and **b** $NaHSO_3$

compounds [59–61], sulfur-based reducing agents [62], or by catalytic hydrogenation (Scheme 4.32) [25].

For the one-pot synthesis of *β*-hydroxy ketones **96**, triethyl phosphite was added directly to the reaction mixture once the hydroperoxidation of α,*β*-unsaturated ketones **92** was complete as indicated by TLC analysis. The best results were obtained via the portionwise addition of 5 equivalents of $P(OEt)_3$. According to this protocol, aldol-type products **96** were isolated in good yields along with high enantioselectivities of up to 97:3 *er* (Table 4.37).

The reduction of 3-hydroxy-1,2-dioxolanes **94** with trivalent phosphorus reagents PX_3 is considered to proceed via a sequence involving biphilic insertion of PX_3 into

Table 4.37 Scope of the **one-pot** synthesis of aldol-type products **96** from α,β-enones **92**

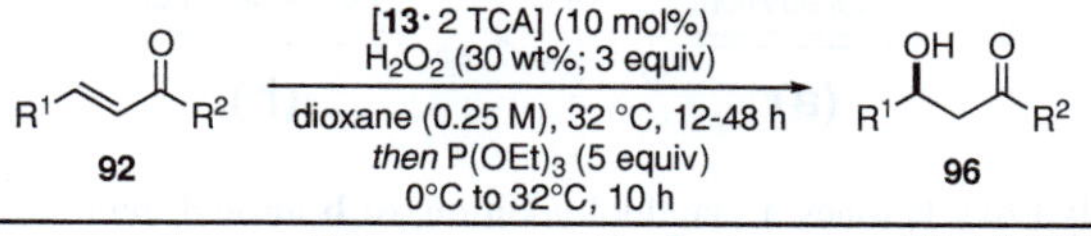

Entry	R	Enone	Product	Yield [%][a]	*er*[b]
1	n-C_6H_{13}	**92a**	**96a**	59	97:3
2	$PhCH_2CH_2$	**92b**	**96b**	53	96.5:3.5
3	*i*-Bu	**92d**	**96d**	56	96.5:3.5
4	Cy	**92e**	**96e**	46	96:4
5	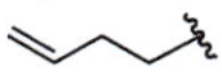	**92f**	**96f**	55	96:4

[a] Yields of pure, isolated products
[b] Determined by chiral HPLC/GC

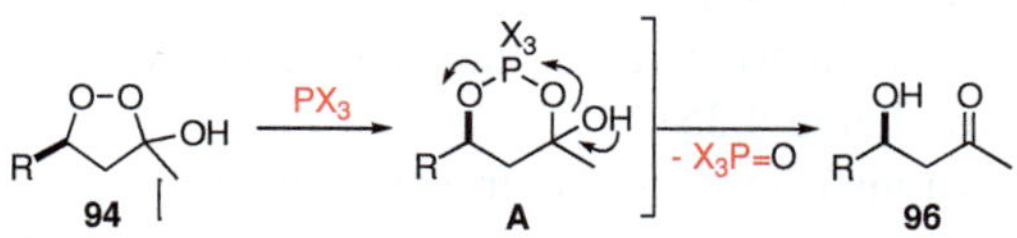

Scheme 4.33 Mechanism of the reduction of 3-hydroxy-1,2-dioxolanes **94** by trivalent phosphorus reagents PX_3

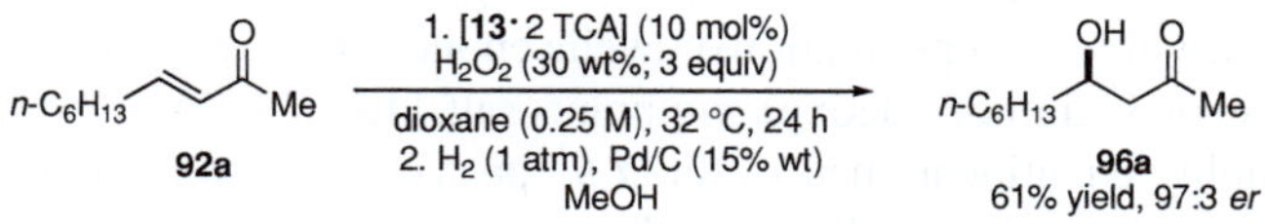

Scheme 4.34 Alternative protcol: aldol synthesis from α,β-enones through a hydroperoxidation-hydrogenation sequence (This reaction was carried out by Lifchits.)

the O–O bond followed by scission of the phosphorane intermediate **A** and elimination of $X_3P = O$ to yield directly β-hydroxy ketones **96** (Scheme 4.33) [60, 61].

Alternatively, catalytic hydrogenation effected the desired transformation to give aldol products **96** with the same level of enantioselectivity as illustrated with the reaction of 3-decen-2-one (**92a**) (Scheme 4.34).[18]

Employing comparatively simple starting materials—α,β-unsaturated ketones **92** and hydrogen peroxide, this approach nicely complements aldol reactions catalyzed by proline and its derivatives since it affords high enantioselectivities also for β-hydroxy ketones **96** formally derived from α-unbranched aldehydes **120** (cf. Scheme 4.30). This asymmetric hydroperoxidation-reduction sequence also

[18] Due to the incompatibility of the reaction conditions of the hydroperoxidation and the hydrogenation, this transformation was carried out in a stepwise fashion simply interrupted by an aqueous work-up alike in the epoxidation protocol.

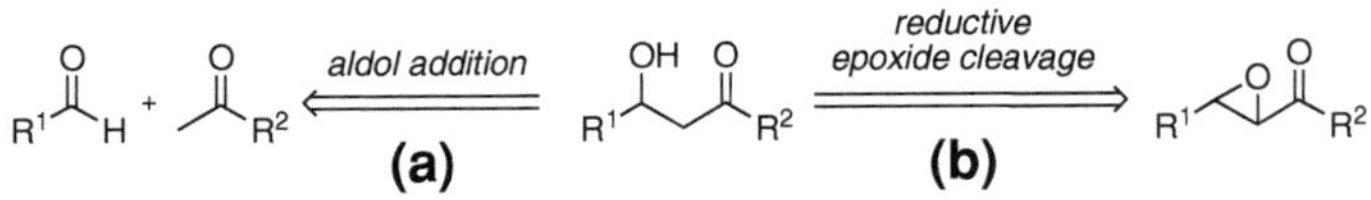

Scheme 4.35 *β*-Hydroxy ketones **a** via aldol addition, or **b** through reductive epoxide cleavage

represents an attractive and simple solution to the long-standing challenge of enantioselectively adding water to *α*,*β*-unsaturated ketones [36, 63, 64].

4.3.2 Synthesis of Aldol Products by Reductive Cleavage of α,β-Epoxy Ketones

Regioselective *α*-reduction of *α*,*β*-epoxyketones is a synthetically useful transformation providing an alternative to the aldol addition for the formation of *β*-hydroxy ketones (Scheme 4.35).

Nevertheless, this transformation is far from trivial due the presence of a reducible ketone functional group combined with the intrinsic instability of *β*-hydroxy ketone systems (which tend to give rise to the formation of enones and other by-products). Several reducing agents have been investigated for this particular process with the most prominent being single electron-transfer reagents such as aluminium amalgam [65] and samarium diiodide [66], or organoselenium reagents [67, 68].

Catalytic asymmetric epoxidation of cyclic enones **46** with hydrogen peroxide in the presence of 9-amino(9-deoxy)*epi*quinine salt [**13** · 2 TFA] provides concise access to highly enantioenriched cyclic *α*,*β*-epoxyketones **48** as outlined in Sect. 4.1.2. These **cyclic** *α*,*β*-epoxyketones **48** may function as precursors to optically active cyclic aldol products **124** which are difficult to access by any other means [69]. In addition, the reductive cleavage of the epoxide ring of **acyclic** *α*,*β*-epoxyketones **93** complements the hydroperoxidation-reduction sequence (see Sect. 4.3.1) for those acyclic enone substrates **92** which give only minor amounts of peroxyhemiketal (or none at all) (cf. Table 4.29).

Reductive epoxide cleavage of *α*,*β*-epoxycyclohexanone (**48a**) with samarium diiodide afforded 3-hydroxycyclohexanone (**124a**) in 52% yield. In the event, the formation of several by-products was concomitantly observed. In contrast, treatment of **48a** with phenyl selenide anion (generated from diphenyl diselenide by reduction with sodium borohydride) was more effective and furnished cyclic aldol product **124a** in excellent yield of 94% within a very short reaction time (<5 min) (Scheme 4.36) [67]. Most importantly, the optical purity of the starting material (96.5:3.5 *er*) was retained under the reaction conditions and 3-hydroxycyclohexanone (**124a**) was isolated for the first time in enantiomerically enriched form with 96:4 *er*.

Remarkably, tertiary cyclic aldol product **124b** was obtained from 2,3-epoxy-3-methylcyclohexanone (**48f**) (98:2 *er*) in 87% yield with only a slight loss of

Scheme 4.36 Reductive cleavage of optically active cyclic α,β-epoxyketones

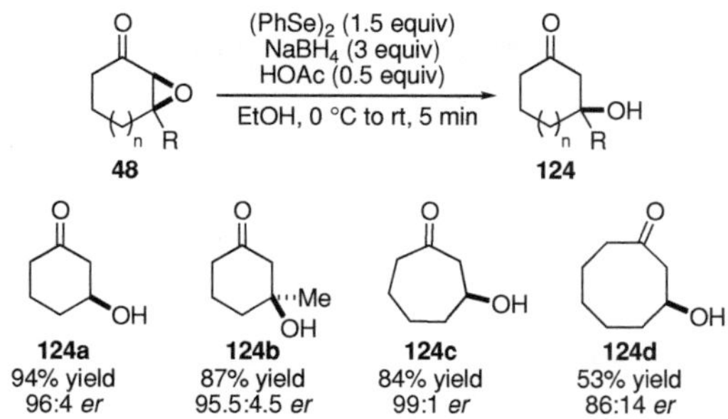

enantiopurity (95.5:4.5 *er*). Whereas seven-membered ring epoxide **48s** (n = 2, R = H) afforded 3-hydroxycycloheptanone (**124c**) in analogous fashion with a high yield and excellent enantioselectivity (84%, 99:1 *er*), the reaction of 2, 3-epoxycyclooctanone (**48v**) proceeded rather sluggishly, and after 60 h, GC-MS analysis indicated a conversion of 95%.[19] At this stage, 3-hydroxycyclooctanone (**124d**) was isolated in a yield of 53%. Moreover, 3-hydroxy-2-(phenylselanyl)cyclooctanone, an intermediate which is formed upon epoxide opening by phenyl selenide anion via S_N2-substitution at the α-carbon atom (cf. intermediate **A**, Scheme 4.37) [70], was recovered from the reaction mixture in 15% yield. Somewhat surprisingly, an erosion of optical purity took place in the course of the reaction: whereas the starting material **48v** exhibited an optical purity of 97:3 *er*, eight-membered cyclic aldol **124d** was obtained with only 86:14 *er*.

Notably, cyclic aldol products **124a-d** have not, to the best of our knowledge, been reported in the literature in optically active form to this date. The synthesis of seven- and eight-membered cyclic aldol products **124c** and **d** is of particular value in light of the fact that they can not (easily) be accessed via aldol disconnection. Intramolecular aldol reactions of keto-aldehydes of the general type **125** give rise to six-membered cyclic aldol products **126** through *enolexo*-aldolizations rather than to 3-hydroxycycloheptanone (**124c**) or -octanone (**124d**) through *enolendo*-aldolizations (Schemes 4.38, 4.39).

We speculate that the reactivity differences between six- and eight-membered α,β-epoxyketones **48a** and **v** arise from their different conformations. Minimized energy structures and LUMO orbitals of α,β-epoxyketones *ent*-**48a** and *ent*-**48v** are depicted in Scheme 4.40.[20] Whereas the α-position of α,β-epoxycyclohexanone (**48a**) appears accessible and the LUMO orbital at the carbonyl group ideally aligned for stabilizing interactions with the benzeneselenolate nucleophile (anticipating an S_N2-displacement of the C_α-O bond), the α-position in the corresponding

[19] We discovered a catalytic effect of the eight-membered ring peroxide **117** on the reductive cleavage of 2,3-epoxycyclooctanone **48v** which is to date unknown in nature. Full conversion to 3-hydroxycyclooctanone **124d** was achieved within 10 min at ambient temperature albeit once again with erosion of optical purity.

[20] Density functional theory (DFT) computation at the B3LYP/6-31G* level (Vacuum) was performed using Spartan '08 Windows from Wavefunction, Inc.

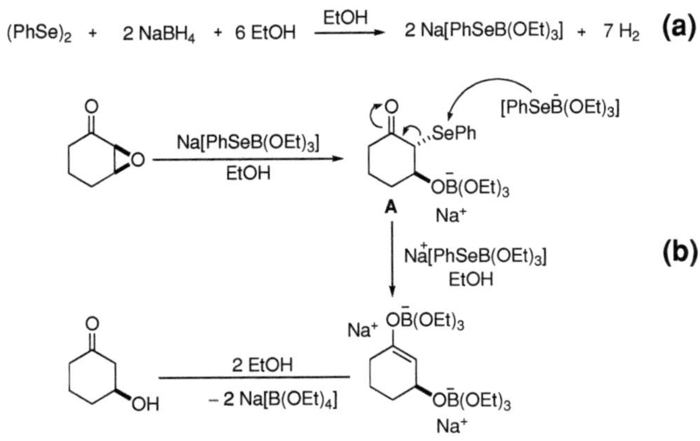

Scheme 4.37 **a** Generation of sodium phenylseleno(triethyl)borate complex Na[PhSeB(OEt)$_3$]. **b** Mechanism of the reductive cleavage of α,β-epoxyketones by benzeneselenolate

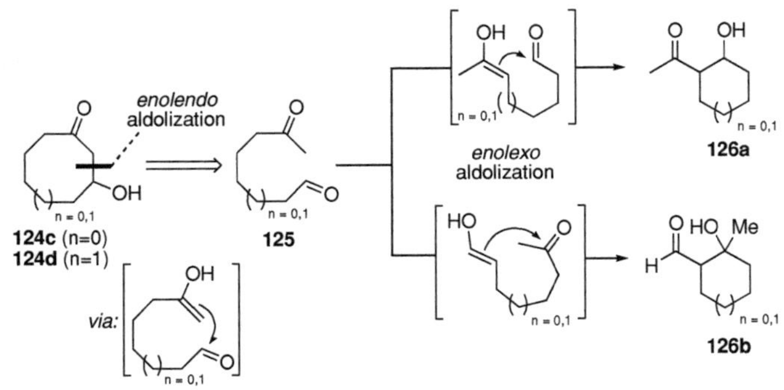

Scheme 4.38 *Enolendo*- versus *enolexo*-aldolization of keto-aldehydes of the general type **125**

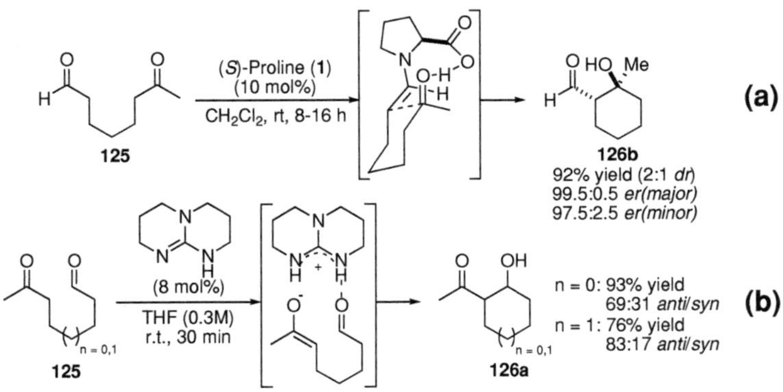

Scheme 4.39 *Enolexo*-aldolizations of keto-aldehydes of the general type **125** according to **a** List et al. [71], and **b** Baati et al. [72]

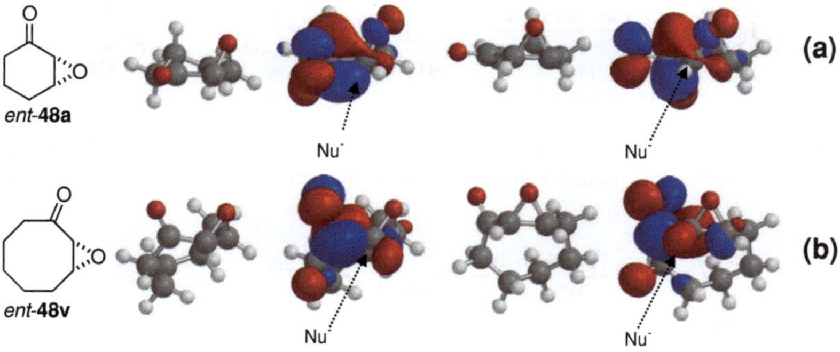

Scheme 4.40 Minimized energy structures and LUMO orbitals of **a** α,β-epoxycyclohexanone **48a**, and **b** α,β-epoxycyclooctanone **48v** (*grey*: C, *red*: O, *white*: H; O-C_α-C = O: 48a: −111.02°, 48v: −34.15°; C_β-C_α-C = O: 48a: −177.04°, 48v: 116.95°)

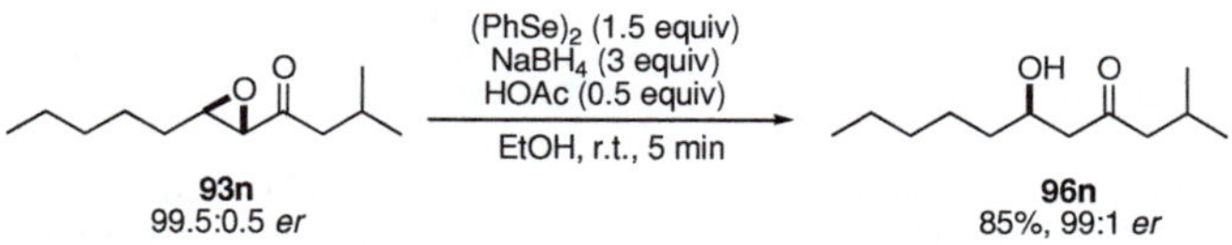

Scheme 4.41 Synthesis of aldol product **96n** by reductive cleavage of α,β-epoxyketone **93n** (This experiment was carried out by Lifchits.)

eight-membered ring epoxyketone **48v** seems to be sterically encumbered. Moreover, according to the model the trajectory of the incoming nucleophile lies in the node plane of the LUMO orbital. This should minimize the possibility for stabilizing orbital interactions during the S_N2 displacement and hamper the reductive cleavage.

The reaction of hydrogen peroxide and acyclic enone **92n** equipped with an *iso*-butyl group as ketone substituent afforded in the presence of 9-amino (9-deoxy)*epi*quinine salt [**13** · 2 TCA] essentially exclusively α,β-epoxyketone **93n** and only traces of the desired peroxyhemiketal **94n** (**93n**:**94n** = 95:5; cf. Table 4.29, entry 3). Thus, the asymmetric hydroperoxidation-reduction sequence (see Sect. 4.3.1) is not feasible to access the aldol product in this particular case. However, 6-hydroxy-2-methyl-4-undecanone (**96n**) was readily obtained in excellent yield and enantioselectivity (99:1 *er*) via reductive cleavage of the corresponding epoxyketone **93n** (Scheme 4.41).

4.3.3 Synthesis of 1,2-Dioxolanes From Peroxyhemiketals

The noticeable lack of asymmetric syntheses of 1,2-dioxolanes in the chemical literature stimulated us to seek for approaches to this substrate class based on our methodology [73]. We envisaged optically active 3-hydroxy-1,2-dioxolanes **94** as

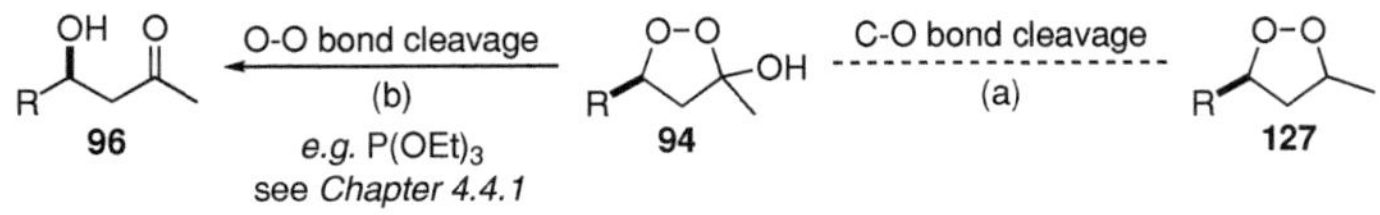

Scheme 4.42 Complementary reduction pathways of 3-hydroxy-1,2-dioxolanes **94**: **a** via O–O bond cleavage to give aldol products **96**, and **b** via C–O bond cleavage to afford 1,2-dioxolanes **127**

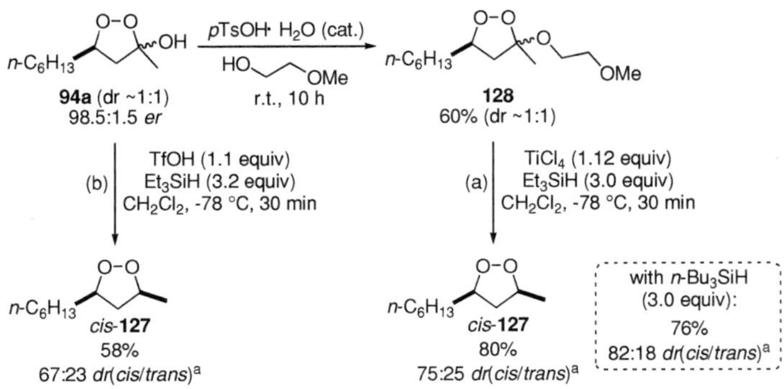

Scheme 4.43 Silane reduction of **a** 3-alkoxy-1,2-dioxolane **128** mediated by $TiCl_4$, and **b** 3-hydroxy-1,2-dioxolane **94a** mediated by TfOH. ([a]Determined by ^{1}H NMR of the crude product)

potential precursors which may be converted into the corresponding 1,2-dioxolanes **127** upon reduction (Scheme 4.42a), a transformation which has not been described to date. Whereas reduction to aldol products **96** proceeds with cleavage of the O–O single bond (cf. Sect. 4.3.1, Scheme 4.42b), the synthesis of 1,2-dioxolane **127** requires a chemoselective reduction which does not affect the O–O linkage.

Several Lewis acids are known to mediate the ionization of peroxyacetals and hemiacetals to provide peroxocarbenium ions which can be trapped by various nucleophiles [47, 74–78]. Applying this strategy, Dussault et al. [47] accomplished the synthesis of 1,2-dioxolanes (in racemic form) from 3-alkoxy-1,2-dioxolanes and allylsilanes, silyl enol ethers, and silyl ketene acetals as nucleophiles.

Guided by this precedent, we carried out the reduction of 3-alkoxy-1,2-dioxolane **128** in the presence of titanium tetrachloride as Lewis acid and triethylsilane as reducing agent to obtain 1,2-dioxolane **127** in 80% yield and a 75:25 diastereomeric ratio in favor of the *cis*-isomer (Scheme 4.43a) [79]. Changing to *n*-tributylsilane under otherwise identical conditions slightly improved the diastereomeric ratio to 82:18.

Moreover, 1,2-dioxolane **127** could be directly obtained in good yield and only slightly lower diastereoselectivity (67:23 *dr*) via reduction of 3-hydroxy-1,2-dioxolane **94a** with triethyl-silane in the presence of stoichiometric amounts of triflic acid (Scheme 4.43b), which presumably generates triethylsilyl triflate as catalytically active species under the reaction conditions. Whereas related Brønsted acid-mediated reductions of hemiacetals are well-precedented, a corresponding transformation with peroxyhemiketals has not been described to date. Notably, this is also the first report on the direct conversion of peroxyhemiketals into the corresponding dioxolanes, obviating the need for prior formation of a peroxyketal.

4.4 Summary and Conclusions

In summary, the synthetic usefulness of the peroxyhemiketal and epoxide products obtained could be demonstrated. In retrospect, the former intermediates allowed the greatest versatility due to the combination of their high-energy O–O bond and the hemiketal linkage. Not only could they lead to their epoxide counterparts (cf. Sect. 4.2.2.2), but also to aldol products **96** (Sect. 4.3.1) and 1,2-dioxolanes such as **127** (Sect. 4.3.3) simply through appropriate choice of reaction conditions. While the thus obtained aldols are actually generated with higher enantioselectivities than those currently afforded by state-of-the-art aldol technology (e.g. proline-catalyzed aldol reaction), the enantioenriched 1,2-dioxolanes represent intriguing and biologically relevant substructures in their own right.

In addition, the epoxides **48** derived from cyclic enones afforded, through reductive epoxide cleavage, a new class of chiral enantioenriched cyclic aldols **124** which (to the best of our knowledge) have not been reported prior to this work (Sect. 4.3.2).

That the long sought-after products of our methodology are themselves easily converted to additional, highly valuable chiral building blocks with essentially no loss in enantiopurity is but further demonstration of the plethora of possible applications of this synthetic method in a preparative context.

4.5 Mechanistic Considerations

4.5.1 Evaluation of Modified Catalysts: Structure-Activity Relationships

Mechanistically, we propose the epoxidation/hydroperoxidation reaction to proceed via activation of α,β-unsaturated ketones **46** and **92** as iminium ions in the presence of chiral amine salt catalysts [9-NH_2-*epi*Q · 2 TXA] (X = F, Cl) or [(*R*,*R*)-DPEN · (*S*)-TRIP]. The amine catalysts which proved optimal in terms of both activity and selectivity, 9-amino(9-deoxy)*epi*quinine (**13**) and (*R*,*R*)-diphenylethylenediamine (**12**), both possess, in addition to a primary amine functionality, (at least) one additional basic site being either another primary or a tertiary amine moiety. Thus, we suggest that the second basic amine site may bring in hydrogen peroxide via hydrogen bonding interactions, pre-organize the transition state, and direct its attack toward one enantiotopic face of the double bond of cyclic enones **46** or acyclic enones **92** (Fig. 4.8)

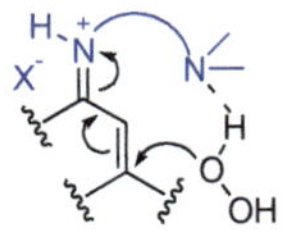

Fig. 4.8 Hypothetical pre-transition state assembly for the catalytic epoxidation/hydroperoxidation of α,β-unsaturated ketones **46** or **92**

To provide experimental support for our proposed mechanism based on iminium ion catalysis we performed the epoxidation reaction of 2-cyclohexenone **46a** in the presence of various modified catalyst systems (Table 4.38). These experiments revealed that both primary amine functionality and acid co-catalyst are crucial for high catalyst efficiency.

Whereas [**13** · 2 TFA] efficiently promotes the reaction giving epoxycyclohexanone **48a** in high yield and excellent enantiomeric ratio (82%, 96:4 *er*; entry 1), the use of 9-amino(9-deoxy)*epi*quinine (**13**) in absence of TFA entailed somewhat lower enantioselectivity and a significantly reduced reaction rate (45%, 92:8 *er*; entry 2). Moreover, derivatives of 9-amino(9-deoxy)*epi*quinine which lack the primary amine group such as monomethylated **129**, thiourea- and sulfonamide-functionalized **131** and **132**, and quinine (**130**) itself showed very low catalytic activity (with or without acid co-catalyst) or were not catalytically active at all (entries 3–10). Similar trends were detected upon modifying the catalyst system [(*R*,*R*)-DPEN · (*S*)-TRIP] (entry 11). When only (*R*,*R*)-DPEN was present in the reaction mixture, epoxide **48a** was generated in modest yield and essentially without stereocontrol (43%, 48:52 *er*; entry 13). Blocking one of the primary amine groups by tosylation, which concomitantly decreases its basicity, preserved the catalytic activity but led to inferior enantiocontrol (92%, 17:83 *er*; entry 14). This may suggest that amine **133**—due to the absence of the second, presumably directing basic site—does not achieve the same high degree of transition state organization, which then translates to a lower enantioselectivity of product **48a**. Tetramethylated DPEN derivative **134** did not promote the reaction, which nicely demonstrates once more that the primary amine group is a vital prerequisite for catalyst activity (entries 15–17). Similar structure selectivity relationships were also identified for other catalyst motifs such as 1,2-diaminocyclohexane (DACH; **60**) (cf. Sect. 4.1.1.1).

In the presence of molecular sieves, catalyst turnover was inhibited and 2,3-epoxycyclohexanone (**48a**) was generated in only 10% yield, albeit with essentially equally high enantioselectivity of 95:5 *er*.

Overall, these experiments provide strong support for the catalytic asymmetric epoxidation of 2-cyclohexenone (**46a**) (as a representative of other α,β-unsaturated ketones) proceeding via iminium ion catalysis, and not via general base catalysis or phase-transfer catalysis. In addition, the substrate scope commonly observed in phase transfer catalysis complements the one of our reaction. Whereas PTC consistently affords higher yields of the desired epoxide with 5- and 6-substituted cyclohexenones compared to 4-mono- or disubstituted derivatives, the trend in our reaction is vice versa. This indicates that steric hindrance in proximity to the

Table 4.38 Evaluation of modified catalyst systems for the catalytic asymmetric epoxidation of 2-cyclohexenone (**46a**)

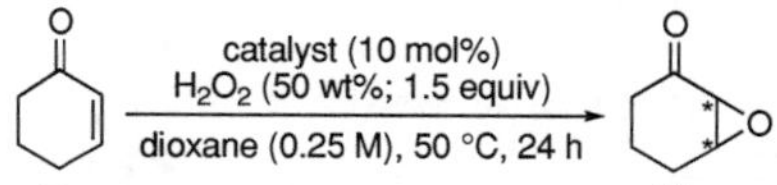

46a **48a**

Entry	Catalyst	Yield [%][a]	*er*[b]
1	[**13** · 2 TFA]	82	96:4
2	**13**	45	92:8
3	[**129** · 2 TFA]	13	67:33
4	quinine (**130**)	0	–
5[c]	quinine (**130**)	0	–
6	[**130** · TFA]	0	–
7	**131**	12	*rac.*
8	[**131** · 2 TFA]	18	76.5:23.5
9	**132**	<5	*rac.*
10	[**132** · 2 TFA]	<5	77:23
11	[(*R*,*R*)-DPEN · (*S*)-TRIP]	92	4:96
12[d]	[(*R*,*R*)-DPEN · (*S*)-TRIP]	10	5:95
13	(*R*,*R*)-DPEN	43	48:52
14	[**133** · TFA]	92	17:83
15	**134**	0	–
16	[**134** · TFA]	0	–
17	[**134** · (*S*)-TRIP]	0	–

[a] Determined by GC
[b] Determined by chiral GC
[c] 20 mol% quinine was used
[d] With 4 Å MS

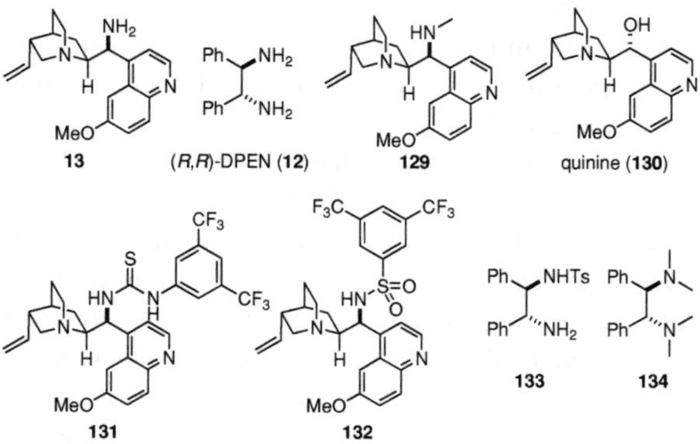

carbonyl group has a detrimental effect on our reaction, which supposedly proceeds via iminiun ion catalysis, but not so in phase-transfer catalysis [80].

The interruption of catalyst turnover in the presence of molecular sieves neatly testifies to the crucial role of water in the reaction (entry 12). This observation is consistent with the assumption of a catalytic cycle based on iminium ion activation of α,β-unsaturated ketones, insofar as water is essential for catalyst turnover since the catalyst is regenerated by hydrolysis of intermediate iminium ions.

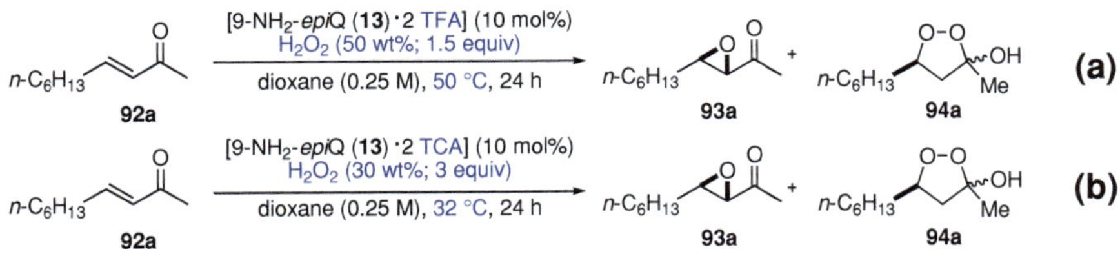

Scheme 4.44 Test reactions for ESI-MS analysis: reaction of 3-decen-2-one (**92a**) with hydrogen peroxide under **a** epoxidation, and **b** hydroperoxidation conditions

4.5.2 ESI-MS Studies

To gain further insight into the reaction mechanism of the epoxidation-hydroperoxidation process we monitored the progress of the reaction by using Electrospray-Ionization Mass Spectroscopy (ESI-MS). For this purpose we carried out the reaction of 3-decen-2-one (**92a**) with hydrogen peroxide in the presence of catalytic amounts of the 9-amino(9-deoxy)*epi*quinine catalyst salt [9-NH_2-*epi*Q (**13**) · 2 TXA] (X = F, Cl): once under the standard conditions of the epoxidation reaction (Scheme 4.44a), and the other time according to the optimized hydroperoxidation protocol (Scheme 4.44b).

Samples were taken from the reaction mixtures at different time intervals over 24 h and submitted to ESI-MS analysis. Spectra recorded prior to hydrogen peroxide addition and after 1 and 12 h of reaction time, respectively, are depicted in the following figures (Fig. 4.9, 4.12, 4.15).

It should be noted that 3-decen-2-one (**92a**), its corresponding epoxide **93a** and peroxide **94a** could not be detected by ESI-MS, as the spectrometer utilized could only detect molecules with molecular weights higher than 200 g mol^{-1}. Therefore, neither signals corresponding to the starting material nor to the products appear in Figs. 4.9, 4.10, 4.11, 4.12 and 4.13.

The ESI-MS spectra of the samples taken prior to hydrogen peroxide addition (Fig. 4.9) showed a signal at *m/z* 460 which corresponds to the iminium ion **A** formed by condensation of 9-amino(9-deoxy)*epi*quinine (**13**) with 3-decen-2-one (**92a**). Moreover, the signal of 9-amino(9-deoxy)*epi*quinine at *m/z* 324 [(**13** + $H)^+$] could be identified. The signal at *m/z* 596 could not be assigned at this stage.

The fact that the signal at *m/z* 460 could not be detected any longer after addition of hydrogen peroxide (Fig. 4.10), indicates that iminium ion **A** is consumed by hydrogen peroxide, and thus validates our assumption that iminium ions such as **A** are indeed intermediates in our epoxidation/hydroperoxidation reaction. The signal corresponding to 9-amino(9-deoxy)*epi*-quinine at *m/z* 324 [(**13** + $H)^+$] is still present. Yet, after 1 h of reaction time, two new signals have appeared at *m/z* 476 and 494, which match the masses of the iminium ion **B** derived from 2, 3-epoxy-2-decanone (**93a**) and iminium ion **C** derived from the corresponding *β*-hydroperoxyketone which upon hydrolysis affords peroxyhemiketal **94a**. The fact that the signal of the unsaturated iminium ion **A** at *m/z* 460 cannot be

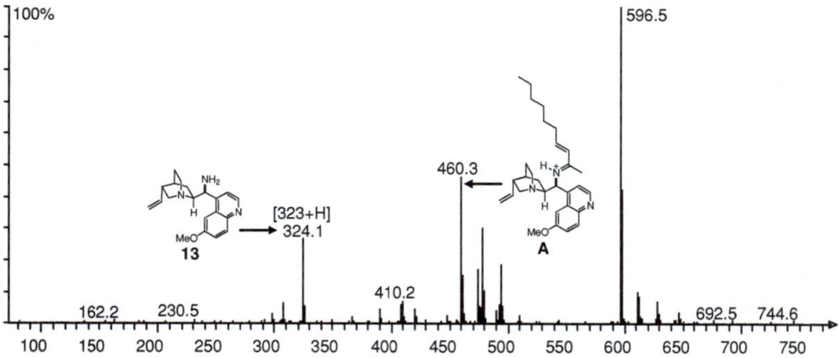

Fig. 4.9 ESI-MS spectrum of (**a**) refers to reaction (a) in Scheme 4.44 recorded prior to H_2O_2-addition. (*MW* molecular weight, *EM* exact mass)

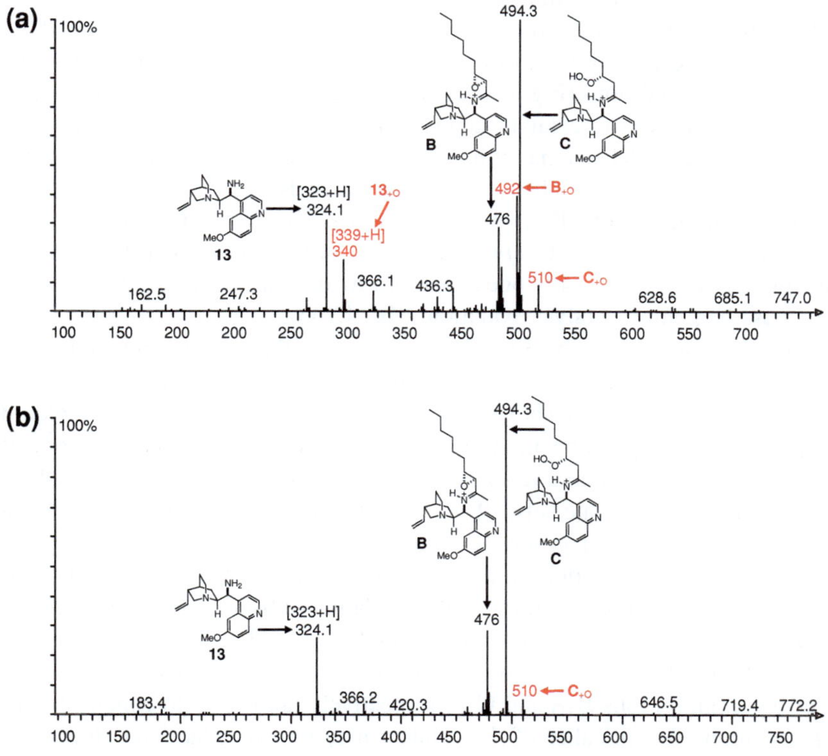

Fig. 4.10 ESI-MS spectra after 1 h from the reaction, **a** with [**13** · 2 TFA] at 50 °C, and **b** [**13** · 2 TCA] at 32 °C (cf. Scheme 4.44)

detected in the presence of hydrogen peroxide suggests that its concentration remains low under reaction conditions. Thus, based on these results, we may propose that neither the conjugate H_2O_2-addition nor the intramolecular epoxide

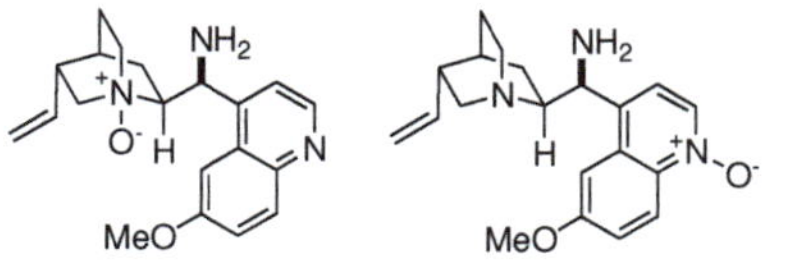

$13_{+O,1}$ $13_{+O,2}$

Fig. 4.11 Proposed structures of regioisomeric 9-amino(9-deoxy)*epi*quinine *N*-oxides $\mathbf{13}_{+O,1}$ and $\mathbf{13}_{+O,2}$

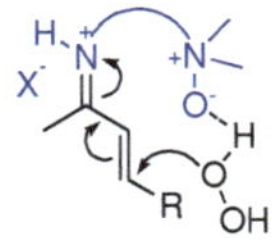

Fig. 4.12 *N*-oxide as directing element

closure, but indeed the formation of the iminium ion intermediate **A** represents the rate-limiting step of these transformations.

In particular in reaction (a) run at 50 °C, additional signals at *m/z* 340, 492, and 510 can be detected which presumably arise from the oxidation of the catalyst under reaction conditions (Fig. 4.10a). Since an excess of oxidant was used and the amine was employed in catalytic amount, it is reasonable to expect oxidation of the amine to occur. The signal at *m/z* 340 may correspond to monooxygenated 9-amino(9-deoxy)*epi*quinine $\mathbf{13}_{+O}$. Moreover, signals at *m/z* 492 and 510 were assigned to iminium ions $\mathbf{B}_{+O}$ and $\mathbf{C}_{+O}$, analogous to **B** and **C**, yet incorporating oxygenated 9-amino(9-deoxy)*epi*quinine $\mathbf{13}_{+O}$ instead of **13**. On the contrary, in reaction (b) run at 32 °C only one signal of low intensitiy at *m/z* 510 was detected to witness beginning catalyst oxidation (Fig. 4.10b).

After 12 h at 50 °C, the ESI-MS spectrum of the sample from reaction (a) indicates extensive catalyst oxidation (Fig. 4.13a). Strong signals at *m/z* 340 and 492 match (as mentioned above) monooxygenated 9-amino(9-deoxy)*epi*quinine $\mathbf{13}_{+O}$ and its corresponding iminium ion formed with α,β-epoxy decanone $\mathbf{B}_{+O}$, respectively. On the contrary, signals at *m/z* 324, 476, and 494 have suffered a considerable loss of intensity. The epoxidation of decenone at 50 °C with [**13** · 2 TFA] (10 mol%) generally requires 10 h to go to completion. Partially oxidized amine $\mathbf{13}_{+O}$ could already be detected after 1 h of reaction time (at ~20% conversion). Intriguingly, the oxidation of 9-amino(9-deoxy)*epi*quinine **13** to $\mathbf{13}_{+O}$ seems to not shut down the catalytic cycle. The fact that we can detect signals at *m/z* 492 and 510, which presumably correspond to α,β-epoxy and β-peroxy iminium ions $\mathbf{B}_{+O}$ and $\mathbf{C}_{+O}$, respectively, allows us to conclude that the primary amine functionality of the catalyst may remain intact and able to effect iminium ion activation of α,β-unsaturated ketones. We assume that amine **13** gets oxidized to the *N*-oxide at either the tertiary quinuclidine or chinoline nitrogen and suggest the structures depicted in Fig. 4.11 for compound $\mathbf{13}_{+O}$. Due to its higher basicity and nucleophilicity, the quinuclidine nitrogen atom should get oxidized preferentially to form the *N*-oxide $\mathbf{13}_{+O,1}$ [81]. However, in addition, the molecular confor-mation of amine **13** and the precise reaction conditions may play a critical role [82]. In the presence of two equivalents of the acid co-catalysts (TFA or

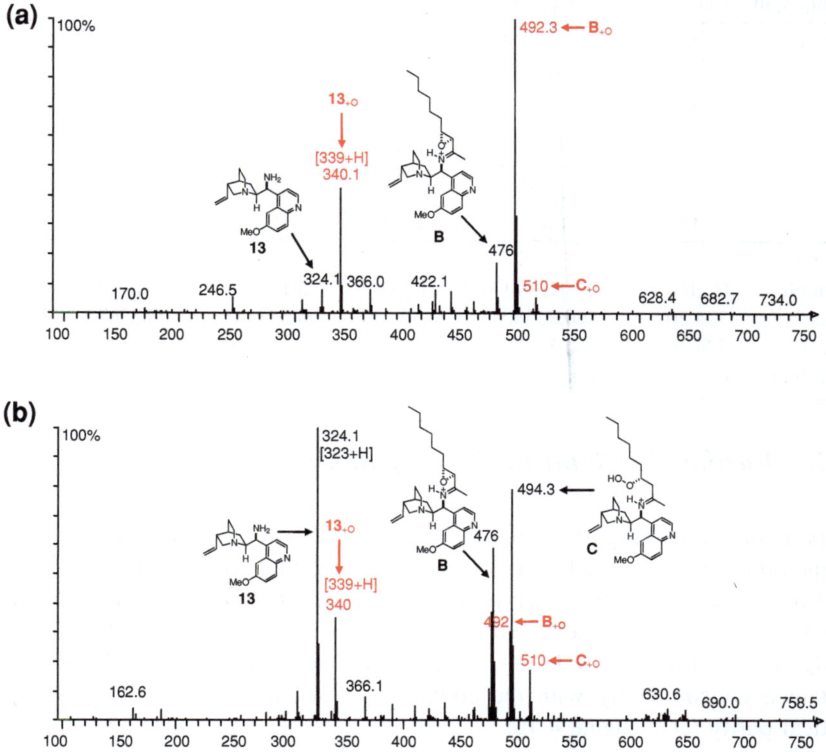

Fig. 4.13 ESI-MS spectra after 12 h from the reaction, **a** with [**13** · 2 TFA] at 50 °C, and **b** [**13** · 2 TCA] at 32 °C (cf. Scheme 4.44)

TCA), the quinuclidine nitrogen may mostly exist in protonated form what should impede its oxidation. Double oxidation of amine **13**, in other words a signal at *m/z* 356, could not be perceived throughout the ESI-MS experiments.

What is intriguing about these results is, that the epoxide **93a** continues to be formed in outstanding enantioselectivities in the presence of $\mathbf{13}_{+O}$ as competing catalytic species. Thus, we suspect that $\mathbf{13}_{+O}$ likewise mediates the epoxidation reaction via a highly organized transition state despite. In the α,β-unsaturated iminium ion derived from *N*-oxide $\mathbf{13}_{+O,1}$, which lacks the quinuclidine nitrogen as directing element, the *N*-oxide itself may bring in hydrogen peroxide for the conjugate addition (Fig. 4.12).

Signals attributed to catalyst oxidation were also perceived with increasing intensity in the reaction with [**13** · 2 TFA] at 32 °C (Fig. 4.13b). Yet in this case, the signals at *m/z* 324, 476 and 494 still dominate the picture which reveals the higher reaction temperature of 50 °C compared to 32 °C as a major cause of catalyst oxygenation [83].

The complete data of the putative intermediates of the hydroperoxidation-epoxidation process found during the ESI-MS investigation is summarized in Table 4.39.

Table 4.39 Potential intermediates found during ESI-MS studies

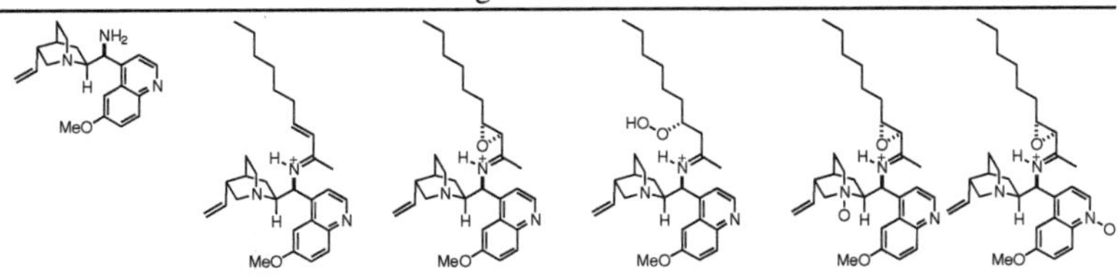

Species	13+H	A	B	C	B_{+O}
HRMS					
Formula	$C_{20}H_{26}N_3O$	$C_{30}H_{42}N_3O$	$C_{30}H_{42}N_3O_2$	$C_{30}H_{44}N_3O_3$	$C_{30}H_{44}N_3O_3$
Found	324.207039	460.332803	476.327879	494.338147	492.322675
Calcd	324.206741	460.332238	476.327154	494.337720	494.322065
Error[ppm]	0.92	1.23	1.52	0.86	1.24

4.5.3 Further Mechanistic Investigations

In particular, one intriguing feature of our hydroperoxidation-epoxidation reaction piqued our interest: namely, the truly outstanding enantioselectivity of >99.5:0.5 *er* (in most of the cases) observed for the **initially formed**[21] epoxides **93** in the reaction of acyclic α,β-enones **92** in the presence of catalytic amounts of [9-NH_2-*epi*Q (**13**) · 2 TXA] (X = F, Cl). However, the peroxyhemiketals **94** which were obtained concomitantly with the epoxides generally exhibited a slightly lower optical purity (cf. Scheme 4.45).

This observation suggests that a kinetic resolution is associated with the epoxide ring closure step, as verified through a control experiment. In the event, cyclization of racemic peroxyhemiketal *rac*-**94a** took place upon treatment with catalyst salt [**13** · 2 TFA] in dioxane at 50 °C, giving scalemic epoxide (3*S*,4*R*)-**93a** and PHK (5*S*)-**94a** at 44% conversion (Scheme 4.46) [1, 84].

Moreover, optically active peroxyhemiketal (*R*)-**94a** with an enantiomeric ratio of 98.5:1.5 *er* was resolved by converting it into the epoxide in the presence of catalyst [**13** · 2 TCA] (Scheme 4.47).

Both experiments clearly demonstrate that chiral catalyst [**13** · 2 TXA] (X = F, Cl) is indeed involved in the conversion of peroxyhemiketals **94** to epoxides **93** under our reaction conditions with the effect of a moderately efficient kinetic resolution accompanying this reaction step. This kinetic resolution may well contribute to the truly outstanding enantioselectivities observed for epoxides **93** generated under reaction conditions. Yet, for the same reason, peroxides **94** are obtained with slightly lower enantiomeric ratios. The "true" enantioselectivity of the conjugate addition of hydrogen peroxide for the reaction depicted in Scheme 4.45 may thus lie at about 99:1 *er*.

[21] Initially formed refers to the epoxide fraction which is formed directly under the reaction conditions, and not through conversion of the corresponding peroxyhemiketal to the epoxide while treating the crude mixture with base (NaOH).

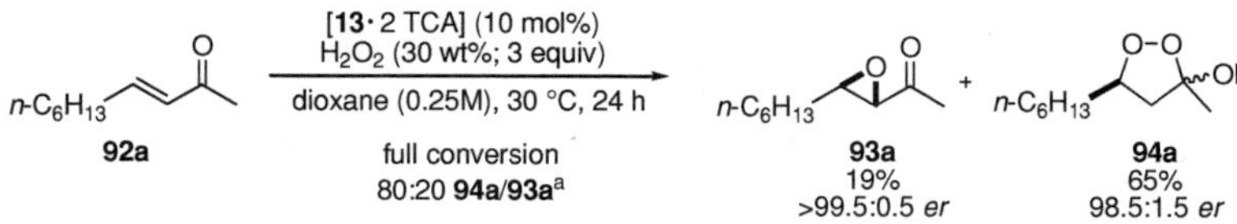

Scheme 4.45 Catalytic asymmetric hydroperoxidation-epoxidation of 3-decen-2-one (**92a**): discrepancy in optical purity between epoxide **93a** and peroxide **94a**. ([a]Determined by ^{1}H NMR of the crude mixture)

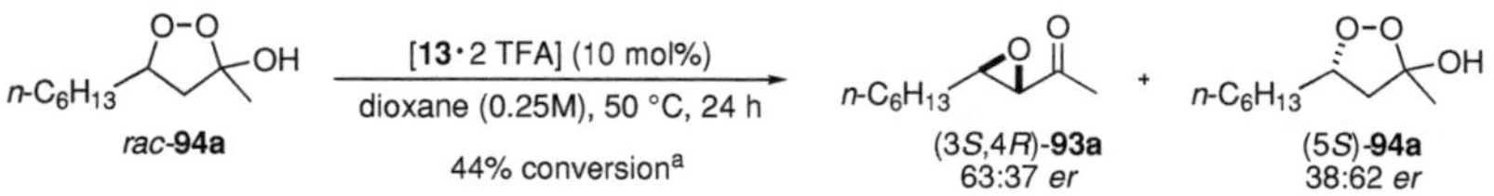

Scheme 4.46 Kinetic resolution I: racemic peroxyhemiketal in the presence of catalyst [**13** · 2 TFA]. ([a]Determined by ^{1}H NMR)

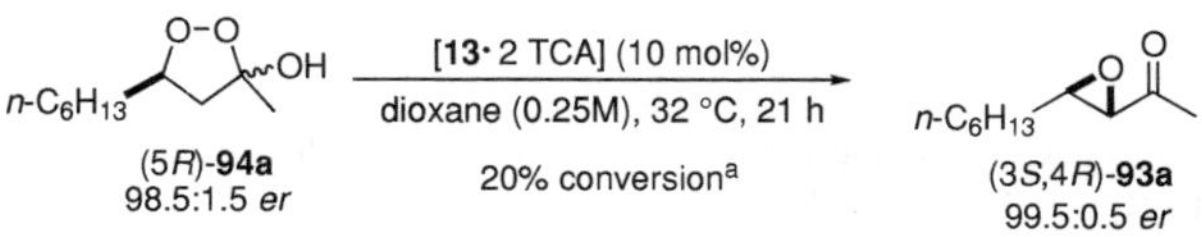

Scheme 4.47 Kinetic resolution II: optically active peroxyhemiketal in the presence of catalyst [**13** · 2 TFA]. ([a]Determined by GC/MS)

Another explanation could be put forward to account for the observation of slightly lower enantioselectivity values for peroxyhemiketals: under the basic conditions of the peroxyhemiketal-epoxide rearrangement (carried out prior to *er*-determination)[22] the conjugate addition of hydrogen peroxide could exhibit (to some extent) reversibility since those conditions resemble classic Weitz-Scheffer reaction conditions. Nevertheless, since enone formation was not perceived under those conditions while rearranging purified PHKs, we believe that the slightly reduced optical purity is indeed a result of the resolution of PHKs by epoxide formation.

4.5.4 Summary and Conclusions

4.5.4.1 Proposed Catalytic Cycle

Based on experimental evidence and the information gained from the ESI-MS studies, we propose the catalytic cycle, which is depicted for clarity again in Fig. 4.14, for the catalytic asymmetric hydroperoxidation and epoxidation of α,

[22] Enantiomeric ratios were determined after converting the peroxyhemiketals to the corresponding epoxides.

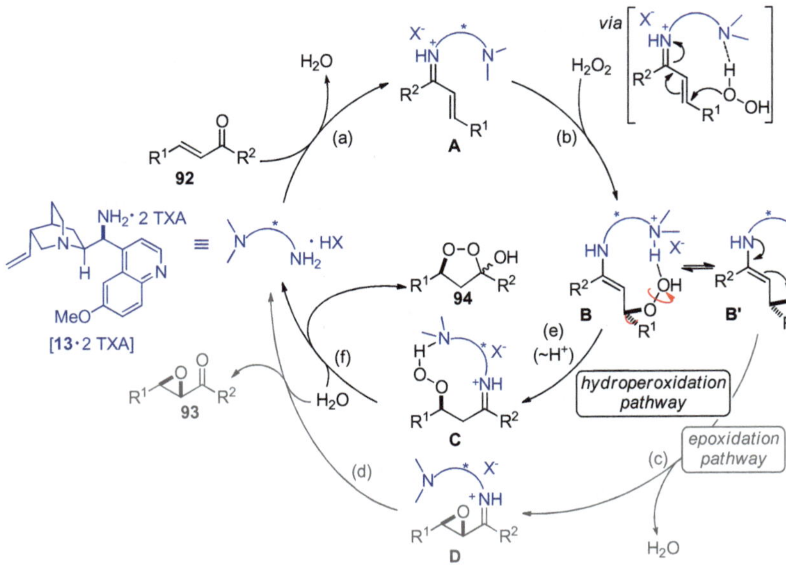

Fig. 4.14 Proposed catalytic cycle for the epoxidation and hydroperoxidation in the reaction of α,β-unsaturated ketones **92** and cyclic enones **46s** and **v**. (TXA with X = F, Cl)

β-unsaturated ketones **92** in the presence of catalytic amounts of [**13** · 2 TXA] (X = F, Cl).[23] In the case of cyclic enones **46** (except for 2-cycloheptenone **46s** and -octenone **46v**), only the epoxidation pathway is relevant.

The catalytic cycle is initiated by the reversible formation of α,β-unsaturated iminium ion **A** from the enone and the catalyst (step a) to activate the substrate, by lowering its LUMO energy. Conjugate nucleophilic addition of hydrogen peroxide then takes place in the following step (step b).

In *Weitz-Scheffer* epoxidations under basic conditions, hydroperoxide anion is the active nucleophile and the addition step is considered to be fast and reversible as confirmed by kinetic data [85, 86]. Under our reaction conditions in the presence of catalyst salts [**13** · 2 TXA] (X = F, Cl) including two equivalents of either TFA or TCA, neutral hydrogen peroxide rather than the hydroperoxide anion may be the active nucleophile,[24] and the conjugate addition is most likely irreversible and constitutes the enantiodetermining step of the catalytic cycle. The same holds true for the epoxidation in the presence of catalyst system [(*R*,*R*)-DPEN · (*S*)-TRIP].[25] An alternative scenario would be a reversible addition of hydrogen peroxide followed by an enantioselective irreversible epoxide ring closure.

---

[23] The reaction catalyzed by [(*R*,*R*)-DPEN · (*S*)-TRIP] supposedly proceeds in an analogous fashion.

[24] pK_a(H_2O_2) 11.8, for comparison: pK_a(quinuclidine) 11.1, pK_a(*i*-PrNH_2) 10.7, pK_a(benzyl amine) 9.34, pK_a(quinoline) 4.85.

[25] pK_a(2,3-diaminobutane, *rac.*) 10.00, 6.91.

However, this possibility could be excluded by the experiment presented in Scheme 4.46. That is to say, the epoxide should have been furnished with the same enantioselectivity as under reaction conditions if the latter assumption was valid.[26] Besides, β-peroxycyclooctanone (**117**) was isolated in 97:3 *er* (cf. Scheme 4.27). If the addition step was reversible and not irreversible and under kinetic control, this β-peroxy compound would have been inevitably obtained in racemic form.

The conjugate addition of hydrogen peroxide generates β-peroxyenamine intermediate **B** (step b). This intermediate can either pursue the epoxidation pathway and afford epoxide **93** via intramolecular nucleophilic substitution at the proximal oxygen atom (step c) followed by hydrolysis (step d) or remain at the hydroperoxidation stage to deliver 3-hydroxy-1,2-dioxolane **94** after protonation (step e) and hydrolysis (step f). It is the β-peroxyenamine intermediate **B** which cyclizes to provide epoxide **93** and not the corresponding enol form. Beside the enamine being more nucleophilic, cyclization of the corresponding β-peroxyenol would not result in an enhancement of the enantioselectivity as revealed in independent experiments (cf. Scheme 4.47). Cyclic peroxyhemiketals **94** derived from acyclic α,β-unsaturated ketones **92** exist in equilibrium with the acylic form. However, the equilibrium distribution favours the peroxyhemiketal whereas the acyclic form cannot be perceived by NMR. On the contrary, with seven- and eight membered cyclic enones **46s** and **v**, the β-peroxyketone form significantly contributes to the equilibrium mixture of hydroperoxidation products **116** and **117** as detected by NMR. Both the epoxidation and the hydroperoxidation pathway are terminated by hydrolysis of an iminium ion **D** or **C** to release the product and regenerate the catalyst (step d and f, respectively).

The epoxidation of cyclic enones in the presence of catalytic amounts of either [**13** · 2 TFA] or [(*R*,*R*)-DPEN · (*S*)-TRIP] are assumed to proceed accordingly. Presumably, due to conformational constraints, in five- and six-membered as well as in macrocyclic enones, only epoxide formation is observed. Thus, the hydroperoxidation manifold is not operative in those cases (cf. Fig. 3.1).

Based on the ESI-MS investigations, we may take into account an alternative, simultaneously operating catalytic cycle with *N*-oxide $\mathbf{13}_{+O}$ (as its TFA salt) as the catalytically active species.[27]

4.5.4.2 Rationalizing the Absolute Stereo-Chemistry

Of major interest is the identification of a plausible model for the transition state of the H_2O_2-conjugate addition to the α,β-enone, which constitutes the enantiodetermining step of the hydro-peroxidation-epoxidation process. Such an endeavour may allow us to rationalize the observed absolute stereochemistry with both

[26] Alternatively, a cross-over experiment could have been carried out by mixing a peroxyhemiketal with a different enone substrate under reaction condition in the absence of H_2O_2.

[27] In particular for reactions conducted at elevated temperatures of 50 °C or higher.

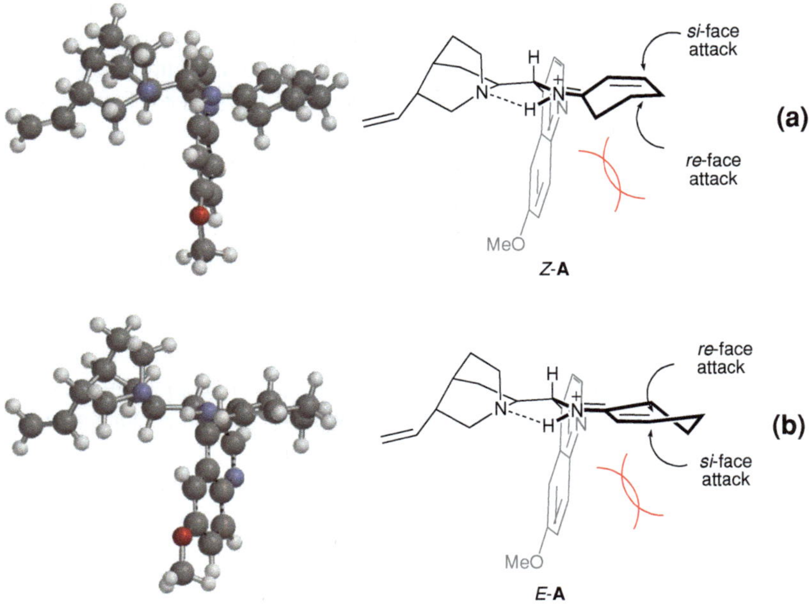

Fig. 4.15 Calculated structures of **a** *Z*-configured α,β-unsaturated iminium ions *Z*-**A**, and **b** *E*-configured α,β-unsaturated iminium ions *E*-**A**

primary amine salt catalyst systems [9-NH_2-*epi*Q (**13**) · 2 TFA] and [(*R*,*R*)-DPEN · (*S*)-TRIP].

DFT calculations[28] at the B3LYP/6-31G* level of theory (Vacuum) on the structure of the relevant α,β-unsaturated iminium ions *Z*-**A** and *E*-**A** (cf. Fig. 4.15; with 2-cyclohexenone (**46a**) as the substrate) revealed a steric shielding of the *re*-face by the quinoline moiety of 9-amino(9-deoxy)*epi*quinine (**13**) in the *Z*-configured iminium ion *Z*-**A** (Fig. 4.15a). A preferred attack at the *si*-face of the cyclohexenone is in accordance with the absolute stereochemistry observed with 9-amino (9-deoxy)*epi*quinine-based catalyst systems (cf. Table 4.5). On the contrary, the *E*-configured iminium ion, which is higher in energy, would suggest a *re*-face attack of hydrogen peroxide in contradiction with the experimental observations (Fig. 4.15b). Thus, *Z*-configured iminium ion *Z*-**A** might be invoked as a working model for the transition state structure.

At this stage, a directing effect of the second basic amine site of catalyst 9-NH_2-*epi*Q (**13**) could not be testified by calculations. DFT Calculations conducted on structures of the corresponding iminium ions *Z*-**A** and *E*-**A** generated from (*R*,*R*)-DPEN (**12**) with 2-cyclohexenone gave similar results. However, it might be desirable to calculate structures of the pivotal α,β-unsaturated iminium ions **A** in

[28] DFT calculations were performed using Spartan'08 Windows from Wavefunction, Inc.

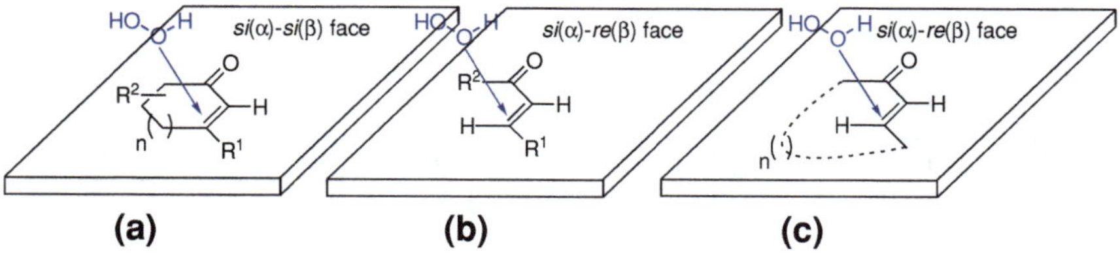

Fig. 4.16 **a** *si(β)*-attack in the [**13** · 2 TFA]-catalyzed epoxidation of cyclic enones (n = 0–3), and *re(β)*-attack in the [**13** · 2 TXA]-catalyzed hydroperoxidation/epoxidation of, **b** acyclic enones, and **c** macrocyclic enones (n = 7–10). (X = F, Cl)

solution and at a higher level of theory. Furthermore, it might be critical to take the influence of the counteranion into account.

DFT calculations need not, however, be the single tool used in order to gain access to structural data of those intermediates. For instance, it would be particularly interesting to synthesize imines resulting from the condensation of an array of enone substrates and the primary amines 9-NH_2-*epi*Q (**13**) or (*R*,*R*)-DPEN (**12**). These imines should, at the outset, be carefully analyzed for their structural features. Addition of equimolar amounts of (*S*)-TRIP or TFA would then generate iminium salts **A**. At this stage, single-crystal X-ray diffraction studies could prove extremely useful to shed further light into the geometric and structural intricacies of these iminium ions in the solid state. Relating these structures to the *in situ* formed analogues in solution might finally allow us to gaze at "snapshots" of our transition state structures. Other acids [beside TFA and (*S*)-TRIP] might be tested—both to eventually facilitate crystallization of iminium ion **A** and for assessment of its counterion's influence on structure (Scheme 4.48).

The observed absolute configuration of 2,3-epoxycyclohexanone (**48a**) and 4-hydroxy-6-phenyl-2-hexanone (**96b**) (derived from the corresponding peroxyhemiketal **94b** through $P(OEt)_3$-reduction) was established by comparison of their optical rotation with literature values;[9, 36] absolute configurations of cyclic as well as acyclic epoxides **48** and **93**, and peroxyhemiketals **94** were assigned by analogy considering an uniform mechanistic path.

The absolute stereochemistry observed is the result of a *si(β)*-attack[29] in the [**13** · 2 TFA]-catalyzed epoxidation of cyclic enones **46** (n = 0–3) (*re(β)*-attack with catalyst [(*R*,*R*)-DPEN · (*S*)-TRIP]), and of a *re(β)*-attack in the corresponding reaction of acyclic enones **92** (Fig. 4.16a and b). Based on the assumption that the bias of the catalyst system is identical for both cyclic and acyclic substrates, the absolute stereochemistry observed for acyclic enones would be consistent with those substrates (or more precisely the corresponding *α*,*β*-unsaturated iminium ions generated with the amine salt catalyst) reacting preferentially in *s-trans* conformation.

[29] Due to a change in priority of the substituent attached to the *β*-carbon, compounds **46h-m** formally undergo attack from the *re(β)*-face in the presence of catalyst [**13** · 2 TFA], although in practice there is no change in facial selectivity.

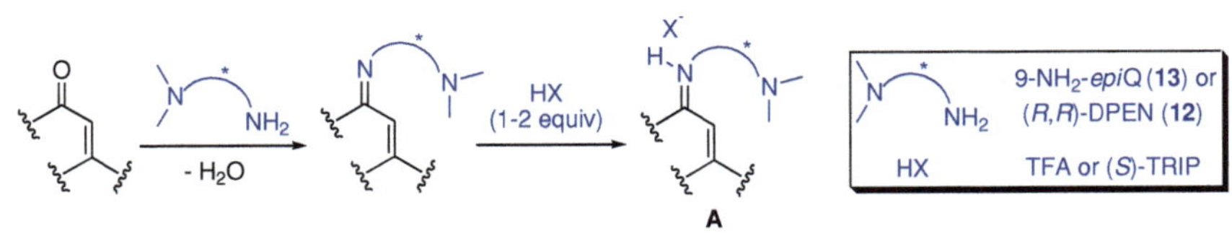

Scheme 4.48 Proposed preparation of α,β-unsaturated iminium ions **A** for X-ray structure analysis

Macrocyclic enones **46w-x** gave exclusively *trans*-configured epoxides when exposed to hydrogen peroxide in the presence of catalytic amounts of [**13** · 2 TXA]. These enone rings are large enough to allow *s-cis* and *s-trans* conformational interconversion and due to their conformationally flexibility, we assume them to behave similarly to the acyclic substrates—implying a *re*(β)-attack (Fig. 4.16c).

4.6 Preparation of Starting Materials

4.6.1 Preparation of Cyclic α,β-Unsaturated Ketones

The synthesis of six- to fifteen-membered cyclic α,β-unsaturated ketones was carried out following two general strategies: firstly, through the addition of Grignard reagents to 3-ethoxy-2-cyclohexenone (**135**) and 3-ethoxy-2-cycloheptenone (**136**) according to the method of Woods and co-workers (Table 4.40), [87, 88][30] and secondly, via Saegusa oxidation of the corresponding saturated cyclic ketones in case of six-, and eight- to fifteen-membered ring cyclic enones **80** and **46v-x** (Scheme 4.50) [89].

Whereas 3-ethoxy-2-cyclohex-2-enone (**135**) is commercially available, 3-ethoxy-2-cycloheptenone (**136**) was obtained from 1,3-cycloheptanedione and ethanol in the presence of a catalytic amount of *p*-toluenesulfonic acid at a Dean-Stark trap in 90% yield [90].

For the synthesis of 3-*tert*-butyl-2-cyclohexenone (**46j**) the addition of *tert*-butyl magnesium chloride to the vinylogous ester **135** proved to be less practical. The desired product was obtained in a non satisfying yield of <15% due to increased formation of by-products caused by using the sterically hindered Grignard reagent. However, *tert*-butyl-substituted enone **46j** could be readily accessed in 58% yield by a palladium-catalyzed allylic oxidation of 1-*tert*-butylcyclohexene developed by Corey and Yu (Scheme 4.49) [91].

Cyclic enones **80** and **46v-x** were prepared via *Saegusa* oxidation from the corresponding saturated cyclic ketones.[31] First, the cyclic ketones were converted to the trimethylsilyl enol ethers, which were then dissolved in dimethylsulfoxide

---

[30] Cyclic enones **46g,h** and **l** were prepared by Martin through the same route.

[31] Cyclic enones **46v-x** were prepared in collaboration with Wang.

Table 4.40 Synthesis of 3-substituted 2-cyclohexenones and 2-cycloheptenones **46** according to the method of Woods et al. [87, 88]

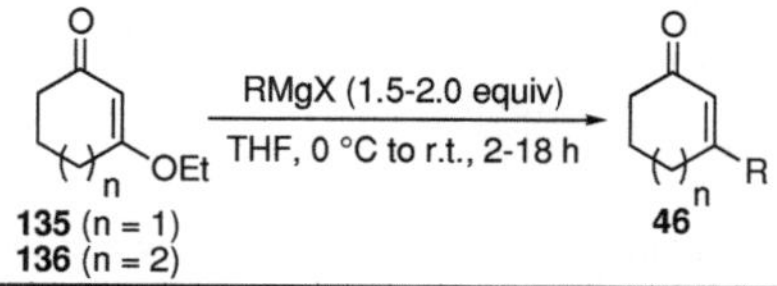

Entry	R	n	Product	Yield %[b]
1	*i*-Pr	1	**46i**	71
2	allyl	1	**46k**	47
3	$PhCH_2$	1	**46m**	76
4	Ph	1	**46n**	98
5	vinyl	1	**46o**	75
6	ethinyl	1	**46p**	77
7	Et	2	**46t**	45[a]
8	$PhCH_2$	2	**46u**	61[a]

[a] Overall yield from 1,3-cycloheptadione
[b] Yields of pure, isolated products

Scheme 4.49 Synthesis of 3-*tert*-butyl-2-cyclohexenone (**46j**) through palladium-catalyzed allylic oxidation

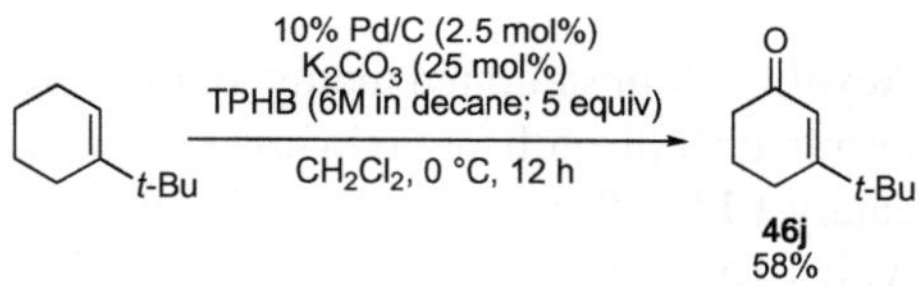

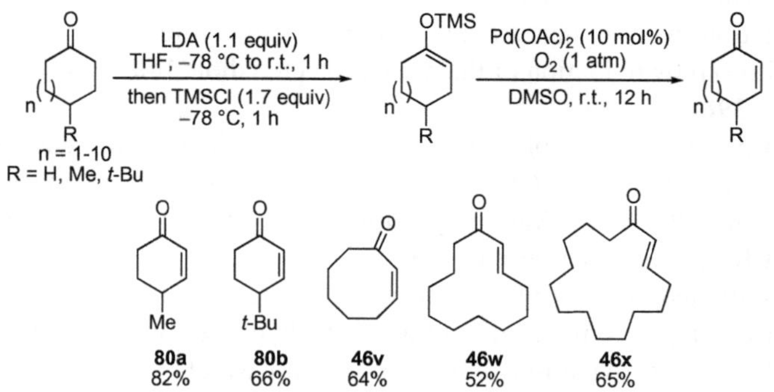

Scheme 4.50 Synthesis of cyclic enones **80** and **46v-x** through a *Saegusa* oxidation (*Larock* modification) [92]

and subjected to catalytic amounts of palladium acetate under an atmosphere of oxygen as stoichiometric oxidant (Scheme 4.50). Under those conditions (*Larock* modification) the use of catalytic palladium acetate is sufficient, whereas the original *Saegusa* oxidation requires stoichiometric palladium acetate and employs acetonitrile as the solvent [92].

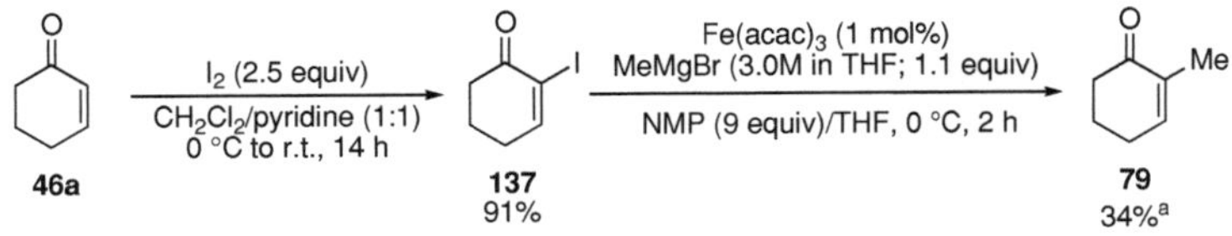

Scheme 4.51 Synthesis of 2-methyl-2-cyclohexenone (**79**) via α-iodination and iron-catalyzed cross coupling. ([a]Reduced yield of **79** due to its high volatility)

To further explore the substrate scope, we studied the asymmetric epoxidation of 2-methyl-2-cyclohexenone (**79**) as an example for an α-substituted cyclic α, β-unsaturated ketone. To this end, 2-methyl-2-cyclohexenone (**79**) was prepared in two steps from 2-cyclohexenone (**46a**) according to the reaction sequence depicted in Scheme 4.51 comprising the α-iodination of 2-cyclohexenone and subsequent iron-catalyzed cross coupling of 2-iodo-2-cyclohexenone (**137**) with methyl magnesium bromide [93].

4.6.2 Preparation of Acyclic α,β-Unsaturated Ketones

Acyclic α,β-unsaturated methylketones were prepared either by cross metathesis of a terminal olefin with methyl vinyl ketone in the presence of *Grubbs* second generation catalyst **138** (Table 4.41) [94, 95], or through a *Wittig* reaction of various aldehydes with 1-(triphenylphosphor-anylidene)-2-propanone (Table 4.42).

Most of the aldehydes were commercially available except for aldehydes **139h** and **139i**, which were prepared from the corresponding diols in two steps via monoprotection with either *tert*-butyl-dimethylsilyl chloride or dihydropyrane (DHP), followed by oxidation of the monoprotected diols **140** to the corresponding aldehydes **139h-i** (Scheme 4.52).

Comparing the results of the epoxidation and hydroperoxidation reactions conducted with enones **92** ($R^2 \neq$ Me) with those obtained in the reactions of methyl ketones **92** ($R^2 =$ Me) may allow us to evaluate the influence of the substituent R^2 on the reaction outcome.

α,β-Unsaturated ketones of the general structure **92** with $R^2 \neq$ Me were prepared in two steps from α,β-unsaturated aldehydes through Grignard addition followed by Swern oxidation or oxidation with MnO_2 of the resultant allylic alcohol (Scheme 4.53) [96].

Alternatively, an aldol-dehydration sequence provided concise access to such enones **92** ($R^2 \neq$ Me) [97]. This strategy was used for the synthesis of 2-methyl-4-hexen-3-one (**92o**) (Scheme 4.54).

To investigate the effect of the enone geometry on the asymmetric epoxidation reaction catalyzed by 9-amino(9-deoxy)*epi*quinine (**13**) TFA salt (cf. Scheme 4.20), we prepared (*Z*)-6-phenyl-3-hexen-2-one (*Z*)-**92b** in three steps from commercially available 4-phenyl-1-butyne according to a reaction sequence described by Heathcock et al. [97] (Scheme 4.55). Reaction of the lithiated alkyne with acetaldehyde

Table 4.41 Synthesis of acyclic α,β-unsaturated ketones **92** by cross metathesis

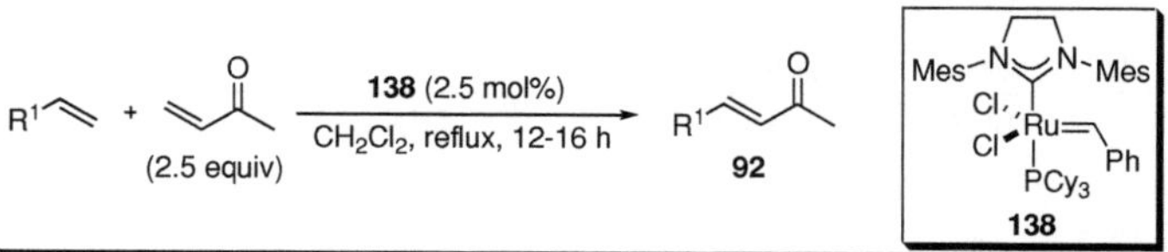

Entry	R^1	Enone	Yield [%][a]
1		**92g**	78
2	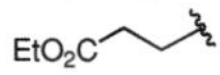	**92j**	80
3	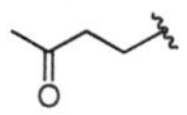	**92k**	99
4	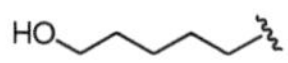	**92l**	94
5	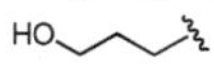	**92t**	99
6	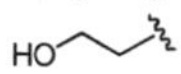	**92u**	64

[a] Yields of pure, isolated products

Table 4.42 Synthesis of various acyclic α,β-unsaturated ketones **92**, **101** and **105** through a *Wittig* reaction

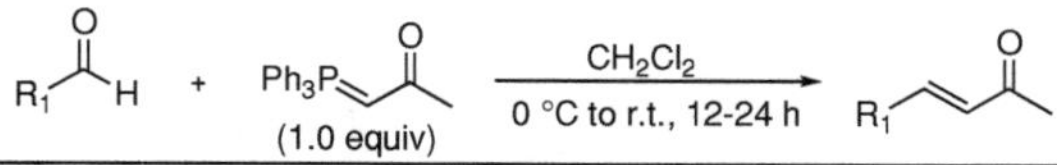

Entry	R^1	Enone	Yield [%][a]
1	$PhCH_2CH_2$	**92b**	75
2	$PhCH_2$	**92q**	26
3	*i*-Bu	**92d**	60
4	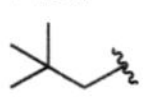	**92r**	56
5	Cy	**92e**	92
6	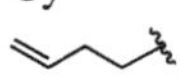	**92f**	88
7	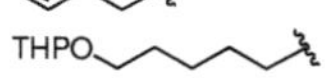	**92h**	85
8	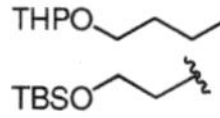	**92i**	72
9	CO_2Et	**101**	50
10	$BnOCH_2$	**105**	88

[a] Yields of pure, isolated products

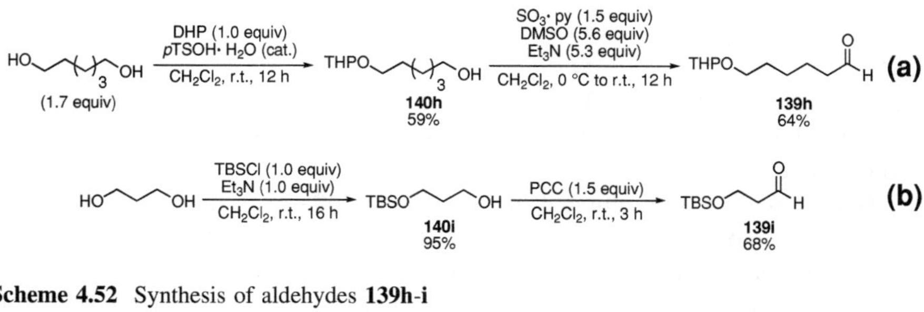

Scheme 4.52 Synthesis of aldehydes **139h-i**

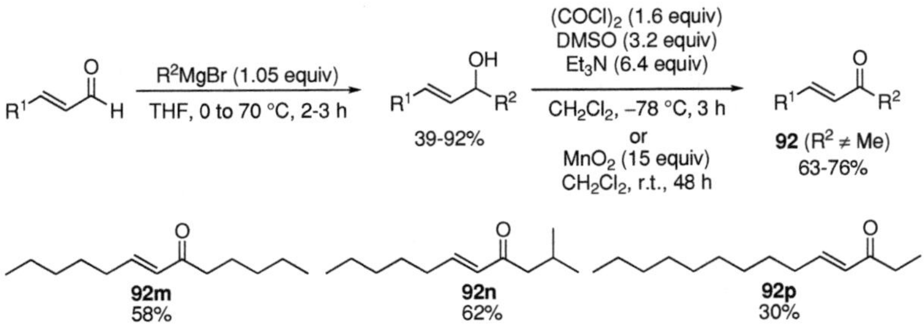

Scheme 4.53 Grignard addition-oxidation sequence: synthesis of enones **92m-n**, **p** (Enones **92m-n**, **p** were prepared by Wang.)

Scheme 4.54 Aldol-dehydration sequence: synthesis of 2-methyl-4-hexen-3-one (**92o**)

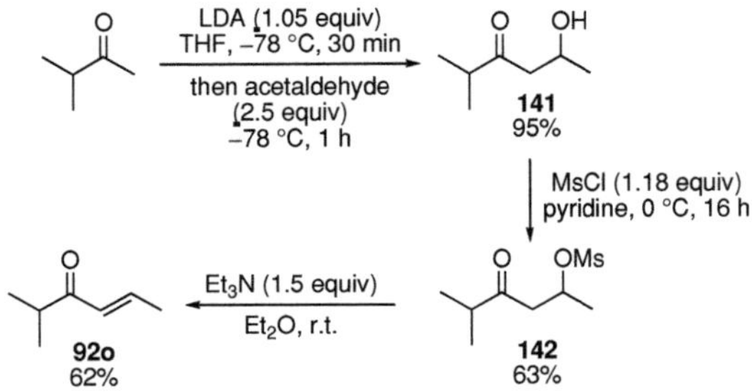

Scheme 4.55 Synthesis of (*Z*)-6-phenyl-3-hexen-2-one [(*Z*)-**92b**]

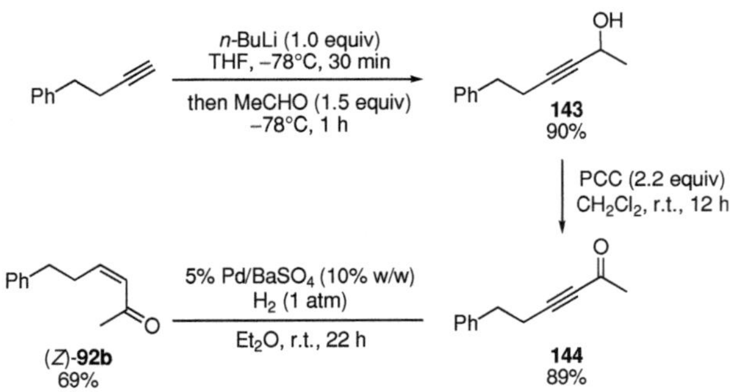

gave propargylic alcohol **143**, which was oxidized with PCC to the corresponding ketone **144**. Finally, *Lindlar* reduction of the alkyne gave the desired (*Z*)-enone in an overall yield of 55%.

Trisubstituted α,β-unsaturated methyl ketone **100**—both the (*E*)- and (*Z*)-isomer—were prepared from citral to study the effect of an additional substituent attached to the β-position and to see if those enones are suitable substrates in the asymmetric epoxidation/hydro-peroxidation reaction (see Sect. 4.2.2.5). Citral,

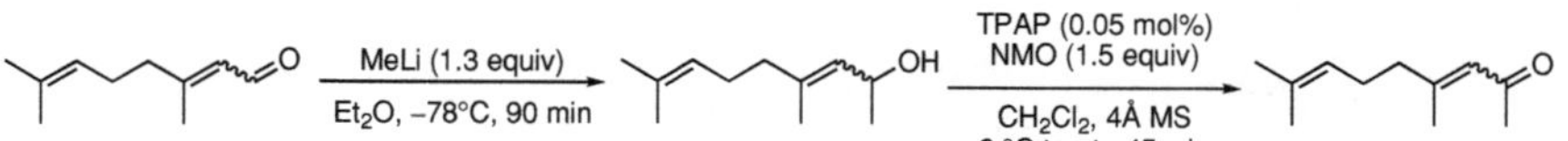

citral 64:34 *E*/*Z* **145** 90% **100** 89%

Scheme 4.56 Synthesis of 4,8-dimethylnona-3,7-dien-2-one (**100**)

which constitutes a mixture of geranial and neral in a 64:36 ratio, was converted to secondary alcohols **145** by the addition of methyl lithium. Subsequent *Ley* oxidation gave methyl ketones **100** as a mixture of (*E*)- and (*Z*)-isomers. Pure fractions of each isomer were readily obtained by column chromatography (Scheme 4.56) [98].

4.7 Catalyst Synthesis

4.7.1 Synthesis of Chiral Primary Amines Based on the Quinuclidine Scaffold

4.7.1.1 Synthesis of 9-Amino(9-deoxy) Cinchona Alkaloid Derivatives

9-Amino(9-deoxy)*epi*quinine (**13**) was first prepared by Brunner et al. [99] in 1995 from naturally abundant quinine via a Mitsunobu reaction with hydrazoic acid followed by an *in situ* Staudinger reduction of the azide intermediate. Whereas Brunner et al. [100, 101] mostly explored the catalytic potential of amides derived from 9-amino(9-deoxy)cinchona alkaloids in asymmetric transformations, 9-amino(9-deoxy)cinchona alkaloids themselves have only recently drawn considerable attention as primary amine catalysts in organocatalysis, both enamine and iminium ion catalysis (cf. Sect. 2.1.3.2) [102, 103].

In 2005, Sóos et al. [104] reported a modified synthesis of 9-amino (9-deoxy)cinchona alkaloids, in which hydrazoic acid was replaced by a less hazardous azide source—diphenylphosphoryl azide (DPPA). Starting from quinine (Q), 9-amino(9-deoxy)*epi*quinine (**13**) was obtained in 76% yield, and moreover, a range of 9-amino(9-deoxy)*epi*cinchona alkaloid derivatives were prepared from the parent 9-hydroxy-substituted compounds in good yields (42–77%) according to this operationally simple one pot-two step reaction sequence (Scheme 4.57).

In the case of 9-amino(9-deoxy)cinchona alkaloid derivatives **71** and **73**, the corresponding 9-hydroxy-substituted precursors, 6′-isopropoxycinchonidine (**146**) and *epi*quinine (**147**), respectively, were not commercially available. However, both compounds could be readily accessed from quinine in only two steps according to literature procedures (Scheme 4.58) [105, 106].

9-Amino(9-deoxy)*epi*cupreidine (**68**) bearing a free hydroxy group as potential hydrogen bonding donor point was prepared in only one step from 9-amino

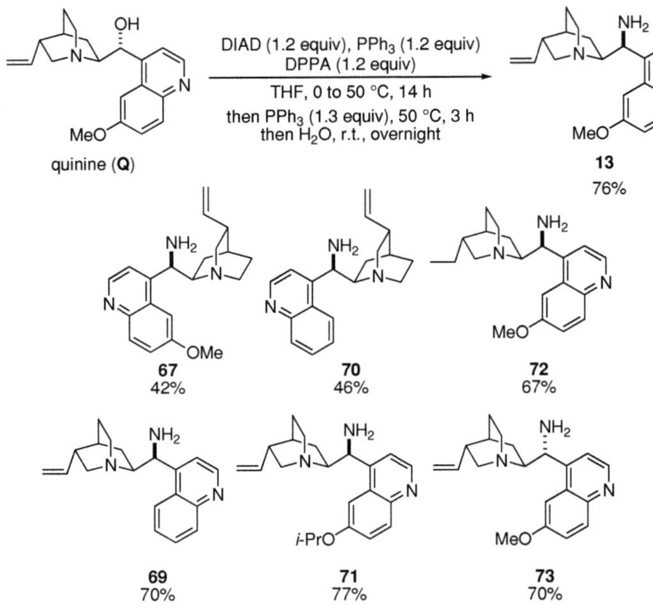

Scheme 4.57 Synthesis of 9-amino(9-deoxy)cinchona alkaloids and derivatives through a *Mitsunobu* reaction/*Staudinger* reduction sequence

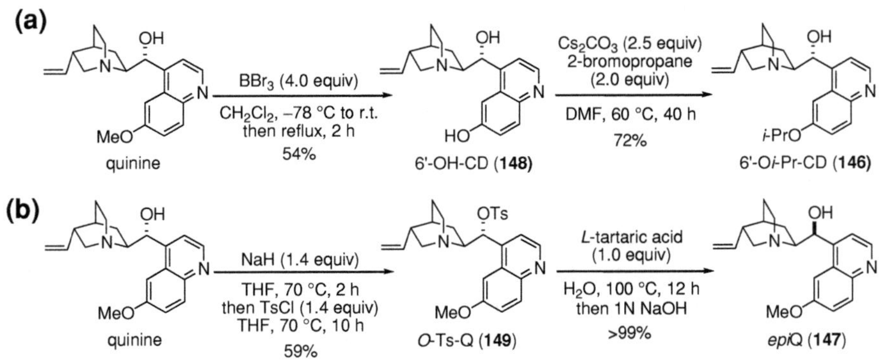

Scheme 4.58 **a** Synthesis of 6′-isopropoxycinchonidine (**146**). **b** Synthesis of *epi*quinine (**147**)

Scheme 4.59 Synthesis of 9-amino(9-deoxy) *epi*cupreidine (**68**)

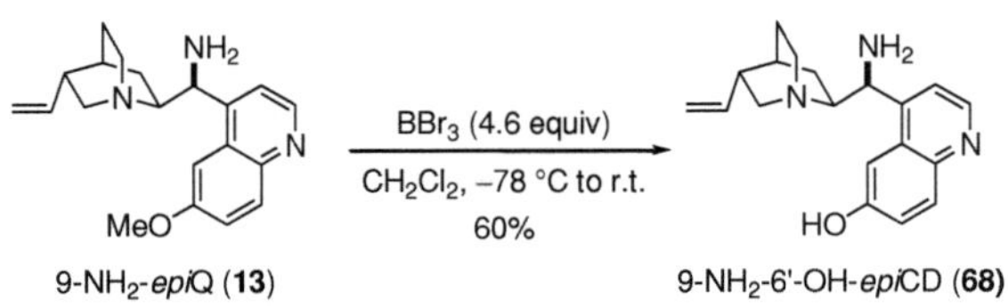

(9-deoxy)*epi*quinine (**13**) by borontribromide-mediated deprotection of the methoxy group (Scheme 4.59) [84].

4.7.1.2 *De-Novo* Design and Synthesis of α-(Aminobenzyl)Quinuclidine

α-(Aminobenzyl)quinuclidine (**86a**) can be regarded as truncated 9-amino (9-deoxy)cinchona alkaloids featuring a less hindered primary amine group due to the replacement of the quinoline moiety by a smaller phenyl group. Thus, we anticipated α-(aminobenzyl)quinuc-lidine (**86a**) in combination with a suitable acid co-catalyst to be considerably more effective in activating sterically demanding and less reactive enone substrates such as 2-cyclopentenone (**46r**) as iminium ions compared to, for instance, sterically congested 9-amino (9-deoxy)*epi*quinine (**13**) (cf. Sect. 4.1.2.5). We based our hypothesis on the observed high catalytic activity of salts of α-(aminomethyl)quinuclidine derivatives **63** and **64** featuring a primary amine function located at a primary carbon atom in the epoxidation of various cyclic enones including 2-cyclopentenone (**46r**) (cf. Table 4.18).

We planned to access compound **86** in two steps from quinuclidine. First, quinuclidine is lithiated at the α-position and then reacted with benzaldehyde to give α-(hydroxyl-benzyl)quinuclidine (**150**), which is in the second step converted into α-(amino-benzyl)quinuclidine (**86a**) by the reaction sequence described above (cf. Scheme 4.56) consisting of a Mitsunobu reaction to stereo-selectively introduce the azide and subsequent Staudinger reduction to afford the corresponding amine. Finally, classical resolution techniques or separation by chiral preparative HPLC would allow access to the pure enantiomers of the target compound **86a**.

The group of Kessar has demonstrated that α-deprotonation of quinuclidine with Schlosser-Lochmann base is possible upon activation of the quinuclidine nitrogen either as the *N*-oxide or by complexation with borontrifluoride [107, 108]. They also sought to develop this chemistry into an enantioselective process by performing the α-lithiation in the presence of (-)-sparteine, however attaining only moderate enantioselectivities (up to 70:30 *er*) [109].

Following their procedure, α-lithiation of the quinuclidine borontrifluoride complex and quenching of the α-lithiated species with benzaldehyde afforded α-(hydroxyl-benzyl)-quinuclidine (**150**) as an approximately 1:1 mixture of diastereomers,[32] which could be readily separated during the work-up procedure, and further purified by column chromatography. Recrystallization afforded both the *threo*- and the *erythro*-isomer of **150** in a yield of 26 and 31%, respectively (Scheme 4.60). Subsequently, both isomers were stereoselectively converted into amines **86a**, and the enantiomers of *erythro*-**86a** were separated by chiral HPLC.

[32] Despite several reruns, we were not able to reproduce the diastereoselectivity of >12:1 (*threo*/*erythro*) reported by the *Kessar* group.

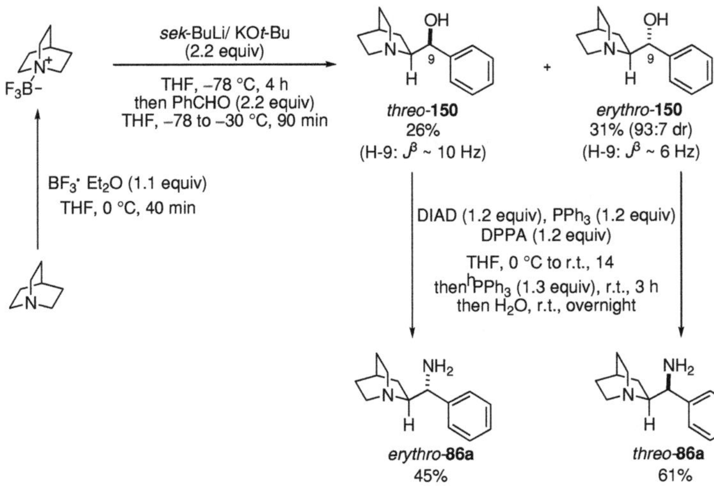

Scheme 4.60 Synthesis of racemic α-(aminobenzyl)quinuc-lidine (*threo*- and *erythro*-**86a**)

Despite numerous attempts, no separation was achieved until now in case of *threo*-**86a**, as it was the case at the alcohol stage (*erythro*-**150** and *threo*-**150**) [104].

4.7.2 Synthesis of Amino Acid-Derived Diamines

Recently, optically active diamines which are readily accessible from commercially available and naturally abundant amino acids in only a few steps, have emerged as highly efficient catalysts for a wide range of transformations proceeding via enamine or iminium ion catalysis [110]. Thus, we became interested to evaluate their catalytic potential in our epoxidation reaction, in particular in the asymmetric epoxidation of 2-cyclopentenone (**46r**) which had emerged as especially challenging substrate among the cyclic enone series. For this purpose, we prepared different diamines bearing (at least) one primary amine functionality through conventional synthetic routes, including peptide coupling chemistry. First, the amide bond is formed, which then gets reduced in the next step to obtain the desired diamines (Fig. 4.17).

For the synthesis of primary primary amine **87** and primary secondary amine **88**, the amide bond was formed by simple aminolysis of the respective amino acid ester, followed by reduction with lithium aluminium hydride (Scheme 4.61).

The synthesis of primary tertiary amines **89a-d** commenced with the peptide coupling between *N*-Boc-protected amino acids and dimethyl amine mediated by HBTU. After removal of the Boc group, reduction of the amides furnished the desired diamines in moderate to good yields (Scheme 4.62).

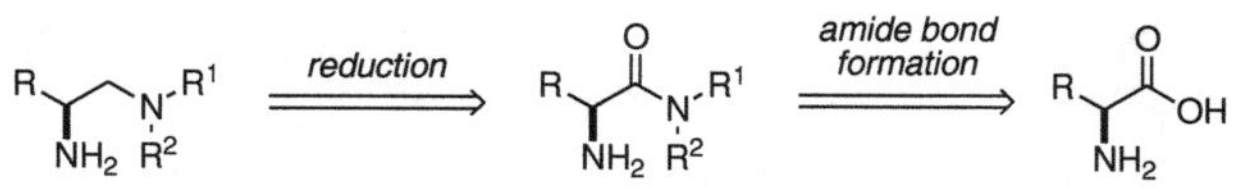

Fig. 4.17 General strategy for the synthesis of optically active diamines from the corresponding amino acids

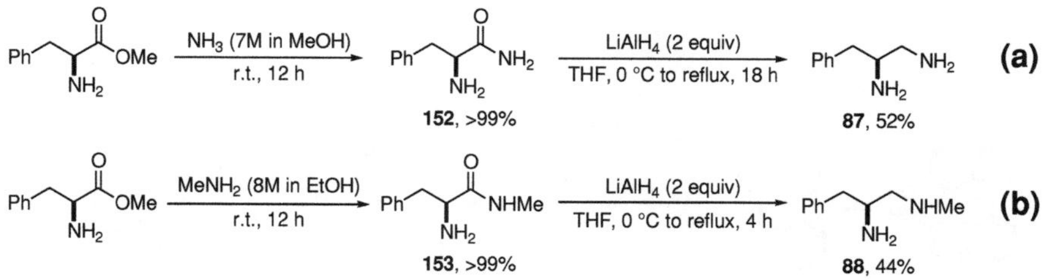

Scheme 4.61 **a** Synthesis of primary–primary diamine **87**. **b** Synthesis of primary–secondary diamine **88**

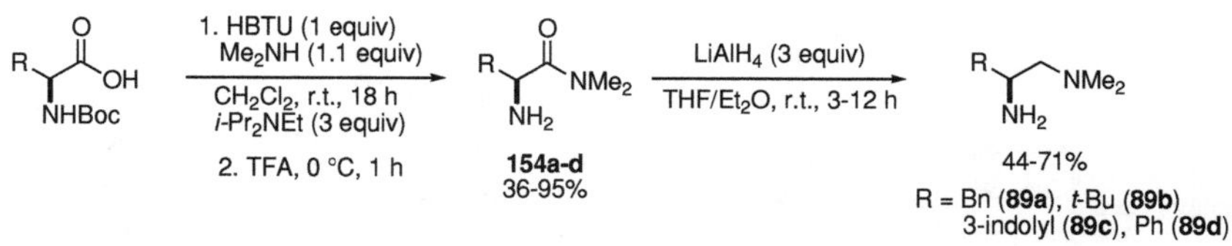

Scheme 4.62 Synthesis of primary tertiary diamine **89a-d**

All diamines **87–89** were tested as catalysts in the asymmetric epoxidation of 2-cyclopentenone, albeit with only moderate success (cf. Table 4.20).

4.7.3 Synthesis of BINOL-Derived Phosphoric Acids

At the outset of our studies, only chiral phosphoric acids derived from unsubstituted (*S*)- or (*R*)-BINOL were commercially available. All other chiral BINOL-derived phosphoric acids **7** were obtained via phosphorylation of appropriately 3, 3′-disubstituted BINOL derivatives **156**, which could be readily accessed from commercially available enantiomerically pure (*S*)- or (*R*)-BINOL in four steps involving cross coupling methods (Scheme 4.63).

Depending on the nature of the substituents R at the 3,3′-positions, two different cross-coupling strategies were applied: Most aromatic residues were successfully introduced via palladium-catalyzed Suzuki-Miyaura coupling of boronic diacid

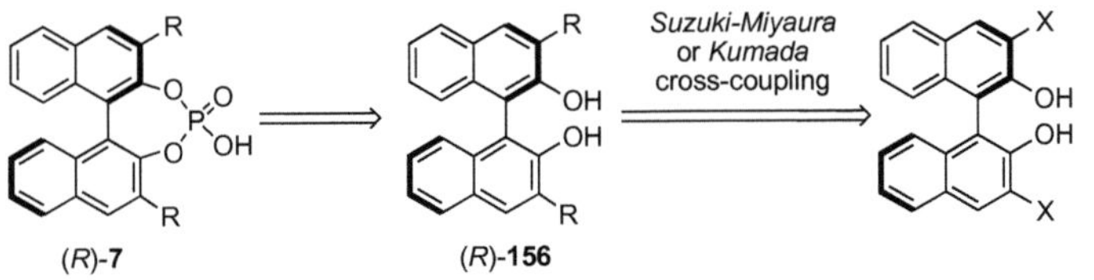

Scheme 4.63 Retrosynthetic approach to the preparation of 3,3′-disubstituted BINOL-derived phosphates **7**. (R = Ar; X = $B(OH)_2$, Br)

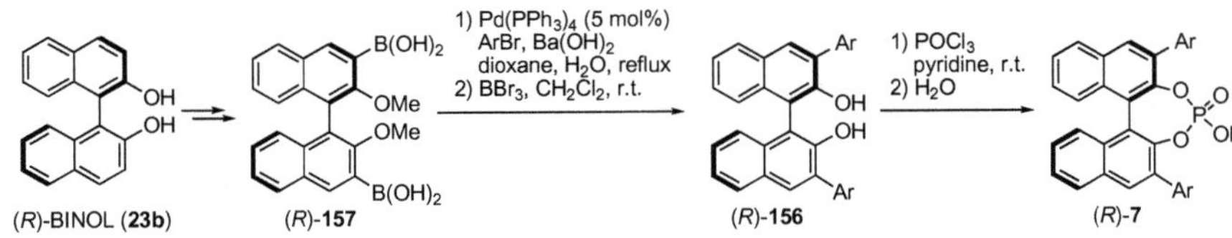

Scheme 4.64 General procedure for the synthesis of BINOL phosphates **7** via Suzuki–Miyaura cross-coupling

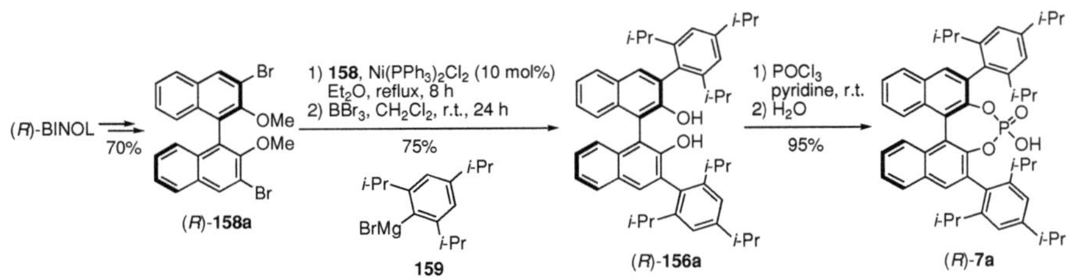

Scheme 4.65 Synthesis of TRIP (**7a**) via Kumada cross-coupling

157 and the respective aromatic halide according to the procedures of Jørgensen et al. and Wipf et al. [111, 112] (Scheme 4.64).

The introduction of sterically hindered substituents such as the 2,4,6-triisopropylphenyl group of the chiral phosphoric acid TRIP (**7a**) was accomplished via a nickel-catalyzed *Kumada* cross coupling of dibromide **158a** and the respective aryl magnesium bromide.[33] The pre-paration of the chiral phosphoric acid **TRIP** was developed and optimized by Seayad according to the procedures of Schrock et al. and Akiyama et al. [113–115] (Scheme 4.65).

---

[33] Suzuki–Miyaura reaction of boronic diacid **157** with 2,4,6-triisopropylphenyl bromide afforded poor yields of the dicoupling product.

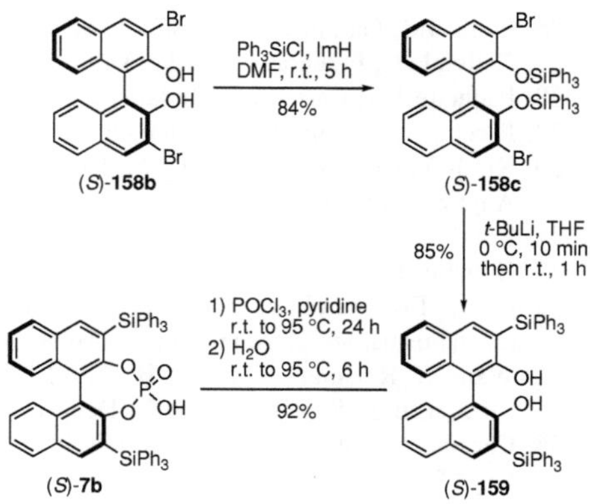

Scheme 4.66 Synthesis of bis(triphenylsilyl)-substituted phosphoric acid (*S*)-**7b** via -Osilyl to -Csilyl rearrangement

The synthesis of sterically hindered 3,3′-bis(triphenylsilyl)-1,1′-bi-2-naphthol hydrogen phosphate (**7b**) [116] relied on the remarkably facile 1,3-rearrangement of *O*-silylated dibromide **158c** to 3,3′-disilylated BINOL derivative **159** as the key step [117] (Scheme 4.66).

References

1. Wang X, List B (2008) Angew Chem Int Ed 47:1119
2. Martin NJA, List B (2006) J Am Chem Soc 128:13368
3. Kawara A, Taguchi T (1994) Tetrahedron Lett 35:8805
4. Yamaguchi M, Shiraishi T, Hirama M (1996) J Org Chem 61:3520
5. Yamaguchi M, Shiraishi T, Hirama M (1993) Angew Chem Int Ed 32:1176
6. Hanessian S, Pham V (2000) Org Lett 2:2975
7. Zhou J, Wakchaure V, Kraft P, List B (2008) Angew Chem Int Ed 47:7656
8. Mayer S, List B (2006) Angew Chem Int Ed 45:4193
9. Wynberg H, Marsman B (1980) J Org Chem 45:158
10. Martin NJA (2008) Ph.D. thesis
11. Kienle M, Argyrakis W, Baro A, Laschat S (2008) Tetrahedron Lett 49:1971
12. Alexakis A, Benhaim C (2002) Eur J Org Chem 2002:3221
13. Alexakis A, Bäckvall JE, Krause N, Pàmies O, Diéguez M (2008) Chem Rev 108:2796
14. Stork G, Danheiser RL (1973) J Org Chem 38:1775
15. Hoke ME, Brescia M-R, Bogaczyk S, DeShong P, King BW, Crimmins MT (2001) J Org Chem 67:327
16. Baldwin JE, Adlington RM, Mitchell MB (1993) J Chem Soc, Chem Commun (17):1332
17. Sucrow W, Rädecker G (1988) Chem Ber 121:219
18. Christian CF, Takeya T, Szymanski MJ, Singleton DA (2007) J Org Chem 72:6183
19. Ramachary DB, Ramakumar K, Kishor M (2005) Tetrahedron Lett 46:7037
20. Ramachary DB, Ramakumar K, Narayana VV (2008) Chem Eur J 14:9143
21. Yanagisawa A, Shinohara A, Takahashi H, Arai T (2007) Synlett 1:141
22. Hong L, Sun W, Liu C, Wang L, Wong K, Wang R (2009) Chem Eur J 15:11105
23. Yang Y-Q, Zhao G (2008) Chem Eur J 14:10888
24. Nazarov IN, Akhrem AA (1950) Zh Obshch Khim 20:2183
25. Payne G (1958) J Org Chem 23:310

26. Fraile JM, García JI, Mayoral JA, Sebti S, Tahirb R (2001) Green Chem 3:271
27. Cativiela C, Figueras F, Fraile JM, García JI, Mayoral JA (1995) Tetrahedron Lett 36:4125
28. Paquette LA, Carr RVC, Bellamy F (1978) J Am Chem Soc 100:6764
29. Rieche A, Schmitz E, Gründemann E (1960) Chem Ber 93:2443
30. Xie J-W, Chen W, Li R, Zeng M, Du W, Yue L, Chen Y-C, Wu Y, Zhu J, Deng J-G (2007) Angew Chem Int Ed 46:389
31. Erkkilä A, Pihko PM (2007) Eur J Org Chem 2007:4205
32. Hamann H-J, Bunge A, Liebscher J (2008) Chem Eur J 14:6849
33. Carey FA, Sundberg RJ (2007) Advanced organic chemistry. Part A: structure and mechanisms: structure and mechanisms, 5th edn. Springer, New York
34. Wang X, Reisinger CM, List B (2008) J Am Chem Soc 130:6070
35. Lu X, Liu Y, Sun B, Cindric B, Deng L (2008) J Am Chem Soc 130:8134
36. Carlone A, Bartoli G, Bosco M, Pesciaioli F, Ricci P, Sambri L, Melchiorre P (2007) Eur J Org Chem 33:5492
37. Barnier J-P, Morisson V, Blanco L (2001) Synth Commun 31:349
38. Kulinkovich OG, Astashko DA, Tyvorskii VI, Ilyina NA (2001) Synthesis 10:1453
39. Maulide N, Vanherck J-C, Gautier A, Marko IE (2007) Acc Chem Res 40:381
40. Bailey M, Staton I, Ashton PR, Markó IE, Ollis WD (1991) Tetrahedron: Asymmetry 2:495
41. Kayser RH, Pollack RM (1975) J Am Chem Soc 97:952
42. Schuurman RJW, Linden Ad, Grimbergen RPF, Nolte RJM, Scheeren HW (1996) Tetrahedron 52:8307
43. Aggarwal VK, Mereu A (1999) Chem Commun (22):2311
44. Bartoli G, Bosco M, Carlone A, Pesciaioli F, Sambri L, Melchiorre P (2007) Org Lett 9:1403
45. Xie J-W, Huang X, Fan L-P, Xu D-C, Li X-S, Su H, Wen Y-H (2009) Adv Synth Catal 351:3077
46. Fehr C, Galindo J (2005) Helv Chim Acta 88:3128
47. Dussault PH, Liu X (1999) Org Lett 1:1391
48. Kirihara M, Kakuda H, Ichinose M, Ochiai Y, Takizawa S, Mokuya A, Okubo K, Hatano A, Shiro M (2005) Tetrahedron 61:4831
49. Chung WK, Lam SK, Lo B, Liu LL, Wong W-T, Chiu P (2009) J Am Chem Soc 131:4556
50. Mukherjee S, Yang JW, Hoffmann S, List B (2007) Chem Rev 107:5471
51. Machajewski TD, Wong C-H (2000) Angew Chem Int Ed 39:1352
52. Geary LM, Hultin PG (2009) Tetrahedron: Asymmetry 20:131
53. Mahrwald RE (ed) (2004) Modern aldol reactions. Wiley-VCH, Weinheim
54. Guillena G, Nájera C, Ramón DJ (2007) Tetrahedron: Asymmetry 18:2249
55. Krattiger P, Kovasy R, Revell JD, Ivan S, Wennemers H (2005) Org Lett 7:1101
56. Tang Z, Jiang F, Cui X, Gong L-Z, Mi A-Q, Jiang Y-Z, Wu Y-D (2004) Proc Natl Acad Sci USA 101:5755
57. Tang Z, Jiang F, Yu L-T, Cui X, Gong L-Z, Mi A-Q, Jiang Y-Z, Wu Y-D (2003) J Am Chem Soc 125:5262
58. List B, Pojarliev P, Castello C (2001) Org Lett 3:573
59. Morisson V, Barnier J-P, Blanco L (1999) Tetrahedron Lett 40:4045
60. Griesbaum K, Kim WS (1992) J Org Chem 57:5574
61. Baumstark AL, Vasquez PC, Chen Y (1993) Heteroat Chem 4:175
62. Makarov SV (2001) Russ Chem Rev 70:885
63. Li DR, Murugan A, Falck JR (2007) J Am Chem Soc 130:46
64. Körber K, Risch P, Brückner R (2005) Synlett 19:2905
65. Corey EJ, Ensley HE (1973) J Org Chem 38:3187
66. Molander GA, Hahn G (1986) J Org Chem 51:2596
67. Miyashita M, Suzuki T, Yoshikoshi A (1987) Tetrahedron Lett 28:4293
68. Engman L, Stern D (2002) J Org Chem 59:5179
69. Jung ME, Allen DA (2008) Org Lett 10:2039
70. Miyashita M, Suzuki T, Hoshino M, Yoshikoshi A (1997) Tetrahedron 53:12469

71. Pidathala C, Hoang L, Vignola N, List B (2003) Angew Chem Int Ed 42:2785
72. Ghobril C, Sabot C, Mioskowski C, Baati R (2008) Eur J Org Chem (24):4104
73. McCullough KJ, Nojimab M (2001) Curr Org Synth 5:601
74. Dussault PH, Zope U (1995) Tetrahedron Lett 36:3655
75. Tokuyasu T, Ito T, Masuyama A, Nojima M (2000) Heterocycles 53:1293
76. Dussault PH, Lee RJ, Schultz JA, Suh YS (2000) Tetrahedron Lett 41:5457
77. Dussault PH, Lee IQ, Lee H-J, Lee RJ, Niu QJ, Schultz JA, Zope UR (2000) J Org Chem 65:8407
78. Dussault PH, Trullinger TK, Cho-Shultz S (2000) Tetrahedron 56:9213
79. Courtneidge JL, Bush M, Loh LS (1992) Tetrahedron 48:3835
80. Baba N, Oda J, Kawahara S, Hamada M (1989) Bull Inst Chem Res Kyoto Univ 67:121
81. Zajac WW, Walters TR, Darcy MG (1988) J Org Chem 53:5856
82. Baidya M, Horn M, Zipse H, Mayr H (2009) J Org Chem 74:7157
83. Aggarwal VK, Lopin C, Sandrinelli F (2003) J Am Chem Soc 125:7596
84. Chen W, Du W, Duan Y-Z, Wu Y, Yang S-Y, Chen Y-C (2007) Angew Chem Int Ed 46:7667
85. Kelly DR, Caroff E, Flood RW, Heal W, Roberts SM (2004) Chem Commun 2016
86. Kelly DR, Roberts SM (2006) Pept Sci 84:74
87. Woods GF, Tucker IW (1948) J Am Chem Soc 70:2174
88. Woods GF, Griswold PH, Armbrecht BH, Blumenthal DI, Plapinger R (1949) J Am Chem Soc 71:2028
89. Ito Y, Hirao T, Saegusa T (1978) J Org Chem 43:1011
90. Hadjiarapoglou L, Klein I, Spitzner D, de Meijere A (1996) Synthesis 525
91. Yu J-Q, Corey EJ (2002) Org Lett 4:2727
92. Larock RC, Hightower TR, Kraus GA, Hahn P, Zheng D (1995) Tetrahedron Lett 36:2423
93. Cahiez G, Avedissian H (1998) Synthesis 1199
94. Trnka TM, Grubbs RH (2000) Acc Chem Res 34:18
95. Chatterjee AK, Morgan JP, Scholl M, Grubbs RH (2000) J Am Chem Soc 122:3783
96. Tietze LF, Henrich M, Niklaus A, Buback M (1999) Chem Eur J 5:297
97. Oare DA, Henderson MA, Sanner MA, Heathcock CH (1990) J Org Chem 55:132
98. Usuda H, Kuramochi A, Kanai M, Shibasaki M (2004) Org Lett 6:4387
99. Brunner H, Bügler J, Nuber B (1995) Tetrahedron: Asymmetry 6:1699
100. Brunner H, Bügler J (1997) Bull Soc Chim Belg 106:77
101. Brunner H, Schmidt P (2000) Eur J Org Chem (11):2119
102. Zhang E, Fan C-A, Tu Y-Q, Zhang F-M, Song Y-L (2009) J Am Chem Soc 131:14626
103. McCooey SH, Connon SJ (2007) Org Lett 9:599
104. Vakulya B, Varga S, Csámpai A, Soós T (2005) Org Lett 7:1967
105. Berkessel A, Guixá M, Schmidt F, Neudörfl JM, Lex J (2007) Chem Eur J 13:4483
106. Franz MH, Röper S, Wartchow R, Hoffmann HMR (2004) J Org Chem 69:2983
107. Kessar SV, Singh P, Singh KN, Singh SK (1999) Chem Commun (19):1927
108. Kessar SV, Singh P, Vohra R, Kaur NP, Singh KN (1991) J Chem Soc, Chem Commun (8):568
109. Kessar SV, Singh P, Kaur A, Singh S (2003) ARKIVOC 3:120
110. Li J, Luo S, Cheng J-P (2009) J Org Chem 74:1747
111. Simonsen KB, Gothelf KV, Jorgensen KA (1998) J Org Chem 63:7536
112. Wipf P, Jung J-K (2000) J Org Chem 65:6319
113. Zhu SS, Cefalo DR, La DS, Jamieson JY, Davis WM, Hoveyda AH, Schrock RR (1999) J Am Chem Soc 121:8251
114. Seayad J, Seayad AM, List B (2006) J Am Chem Soc 128:1086
115. Akiyama T (2004) WO2004096753, PCT Int Appl
116. Storer RI, Carrera DE, Ni Y, MacMillan DWC (2005) J Am Chem Soc 128:84
117. Maruoka K, Itoh T, Araki Y, Shirasaka T, Yamamoto H (1988) Bull Chem Soc Jpn 61:2975

Chapter 5
Summary

Our laboratory has a long-term interest in the development of organocatalytic asymmetric epoxy-dation methods, since optically active epoxides are valuable chiral building blocks and versatile synthetic intermediates.

This Ph.D. thesis describes the successful development of organocatalytic asymmetric epoxidation and hydroperoxidation reactions of α,β-unsaturated ketones with aqueous hydrogen peroxide as the oxidant in the presence of catalytic amounts of a chiral primary amine salt. The focus was placed on challenging substrate classes such as simple cyclic enones and enolizable aliphatic acyclic enones. Indeed, and despite the wealth of enantioselective enone epoxidation methods known, these substrates have been systematically omitted and their enantioselective epoxidation is a challenge thus far unmet by the synthetic community.

At the outset, we concentrated our work on the asymmetric epoxidation of cyclic α,β-unsaturated ketones. In the course of our studies, we identified two powerful and complementary catalytic systems based on chiral primary amines for the desired transformation: the [(*R*,*R*)-DPEN · (*S*)-TRIP] salt and 9-amino (9-deoxy)*epi*quinine (9-NH_2-*epi*Q; **13**) as its trifluoroacetic acid salt [9-NH_2-*epi*Q · 2 TFA] (Fig. 5.1).

Using these chiral primary amine salts as catalysts at 10 mol% loadings, we achieved an efficient asymmetric epoxidation of variously substituted cyclic enones of different ring sizes, in good to high yields along with excellent enantioselectivities of up to >99.5:0.5 *er* in many cases (Scheme 5.1).

Mechanistically, we propose the reaction to proceed via activation of the α,β-unsaturated ketones as iminium ions (cf. Fig. 3.1). This assumption was supported by structure–activity relationships uncovered during our evaluation of modified catalyst candidates (cf. Sect. 4.5.1). Since primary amine catalysts equipped with a second basic amine functionality gave superior results in terms of enantioselectivity, we hypothesized that this additional binding site might bring the hydrogen peroxide into close proximity to the reactive center (via hydrogen bonding

Springer Theses, DOI: 10.1007/978-3-642-28118-1_5,

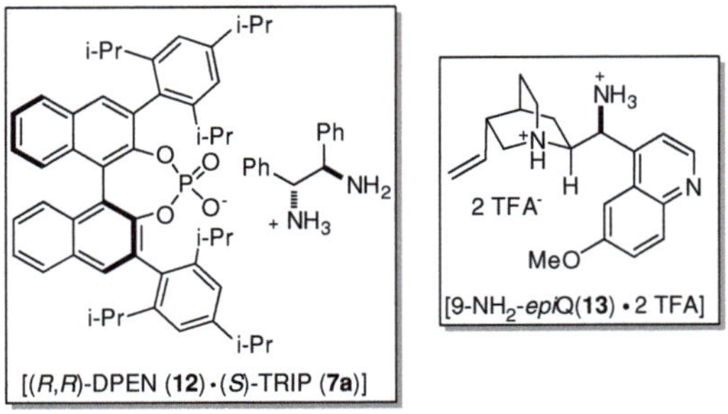

Fig. 5.1 Catalyst systems developed for the asymmetric epoxidation of cyclic α,β-enones

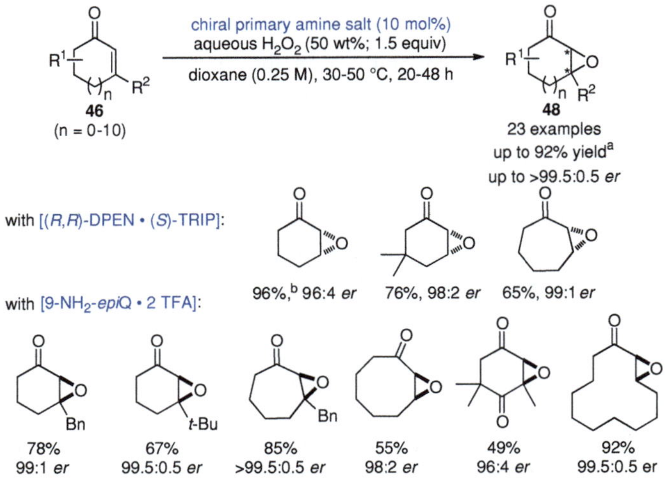

Scheme 5.1 Catalytic asymmetric epoxidation of cyclic α,β-unsaturated ketones. ([a]Yields of pure, isolated products; [b]Determined by GC with an internal standard method.)

interactions) and direct its attack to one of the enantiotopic faces of the double bond (Fig. 5.2).

In addition, our efforts were aimed at evaluating the synthetic potential of optically active cyclic α,β-epoxyketones **48**. We envisaged them to function as precursors to enantiomerically enriched cyclic aldol products through reductive epoxide opening. After screening different reduction methods, we identified conditions which provided cyclic aldol products **124** of different ring sizes in excellent yields. It should be noted that these aldols, which were obtained essentially without loss of optical purity, would be particularly difficult to access by any other synthetic means (Scheme 5.2).

We then became interested in expanding the scope of our catalytic asymmetric epoxidation reaction. Thus, the second part of this thesis concerned the work with acyclic α,β-unsaturated ketones. It could quickly be established that the

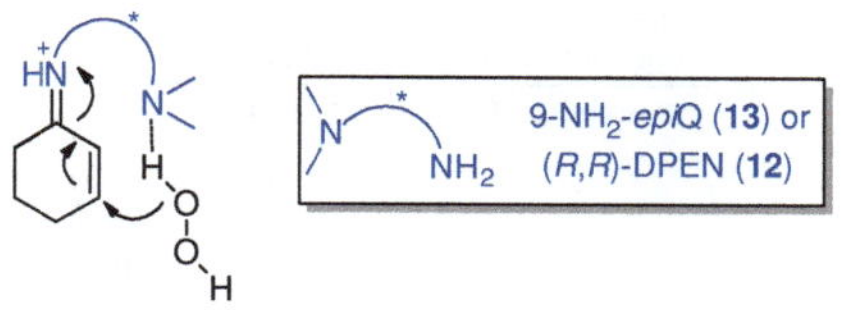

Fig. 5.2 Pre-transition state assembly invoking a directing effect of the bifunctional amine catalyst's second amino group

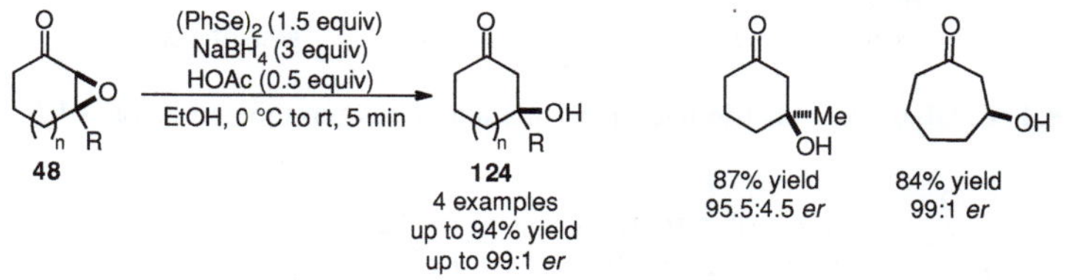

Scheme 5.2 Cyclic aldols through reductive epoxide cleavage

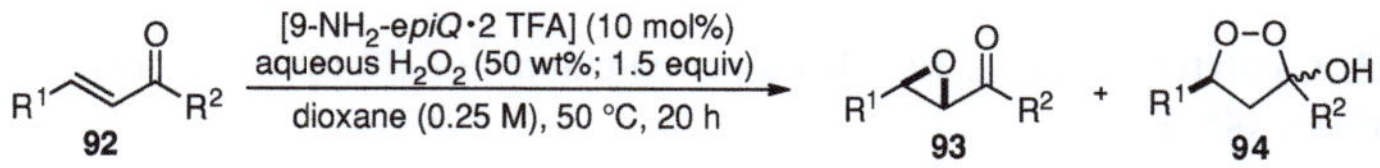

Scheme 5.3 [9-NH_2-*epi*Q · 2 TFA]-catalyzed reaction of acyclic α,β-enones with aqueous hydrogen peroxide

9-amino(9-deoxy)*epi*quinine (9-NH_2-*epi*Q; **13**) TFA salt was furthermore a viable catalytic system for the reaction of acyclic enones with aqueous hydrogen peroxide. In contrast to the reaction of cyclic substrates under identical reaction conditions, we found that acyclic enones **92** provide mixtures of *trans*-epoxides **93** along with unanticipated cyclic peroxyhemiketals of the general type **94** in varying ratios (Scheme 5.3).

Intriguingly, careful optimization of the catalyst composition and reaction conditions enabled us to direct the reaction either toward increasing peroxyhemiketal **94** or epoxide **93** formation. Thereby, we have established the first catalytic asymmetric hydro-peroxidation of α,β-unsaturated ketones, which delivers prized, synthetically and medicinally relevant 3-hydroxy-1,2-dioxolanes **94** (as ~1:1 mixtures of hemiketal isomers) in good yields along with high enantioselectivities of up to 98.5:1.5 *er* (Scheme 5.4). The reaction performed well within a wide range of aliphatic α,β-unsaturated ketones, whereas aromatic enones were not tolerated. Notably, the survival of a broad array of functional groups (featuring acid labile acetal and silyl ether functionalities) testifies the mildness of the adopted reaction conditions.

An exception among the cyclic enones constituted 2-cycloheptenone (**46s**) and the corresponding eight-membered ring enone (**46v**), which in analogous fashion

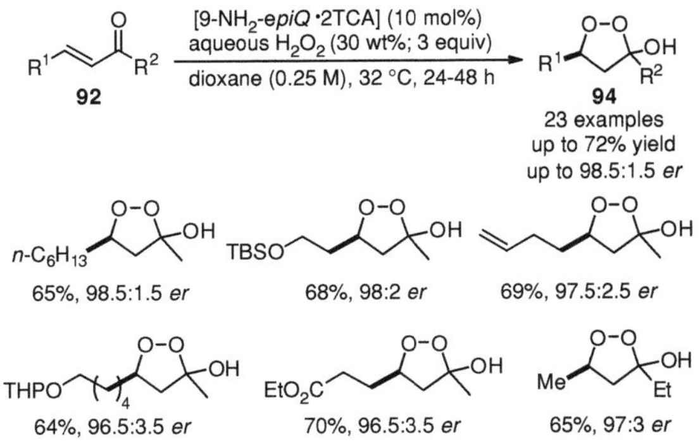

Scheme 5.4 Catalytic asymmetric hydroperoxidation of α,β-unsaturated ketones **92**

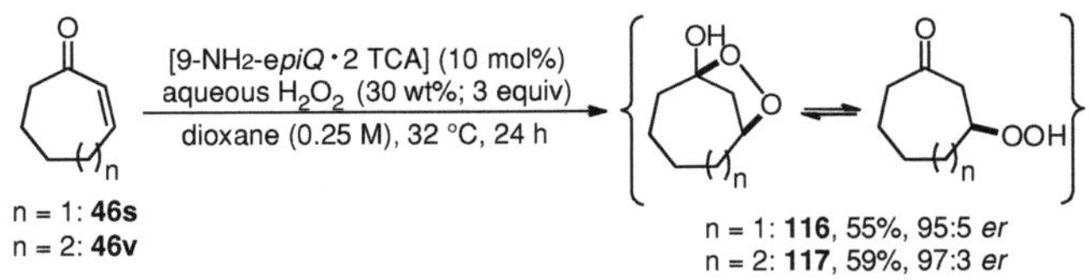

Scheme 5.5 Catalytic asymmetric hydroperoxidation of 2-cycloheptenone (**46s**) and 2-cyclooctenone (**48v**)

to the acyclic substrates afforded bicyclic peroxyhemiketals along with the expected epoxides (Scheme 5.5). Those bicyclic peroxyhemiketals, which have not been described previously, piqued our interest since their bicyclic peroxidic scaffold resembles potent antimalarial agents. After optimization, both compounds **116** and **117** were obtained in good yields and with high optical purity.

The versatility of cyclic peroxyhemiketals **94** (cf. Scheme 5.4) was illustrated by converting them into epoxides **93**, aldol products **96**, and 1,2-dioxolane **127** (Scheme 5.6).

Epoxides **93** were obtained via base-mediated rearrangement of crude hydroperoxidation products (cf. Sect. 4.2.2.2), whereas silane reduction of peroxyhemiketals **94** in the presence of a Brønsted acid afforded 1,2-dioxolane **127** (cf. Sect. 4.3.3). The reduction to aldol products **96** could be accomplished *in situ* by the addition of $P(OEt)_3$ directly to the reaction mixture once the hydroperoxidation of α,β-unsaturated ketones **92** was complete (cf. Sect. 4.3.1). Alternatively, the crude hydroperoxidation product could be subjected to catalytic hydrogenation. The hydroperoxidation-reduction sequence furnished aldol products which are formally derived from linear α-unbranched aldehydes with high enantioselectivities. This is particularly interesting in view of the fact that α-unbranched aldehydes still represent challenging substrates in proline-catalyzed aldol reactions.

Remarkably, aldols **96**, epoxides **93** and 1,2-dioxolane **127** could all be readily accessed in nearly enantiopure fashion, at will from a common peroxyhemiketal precursor.

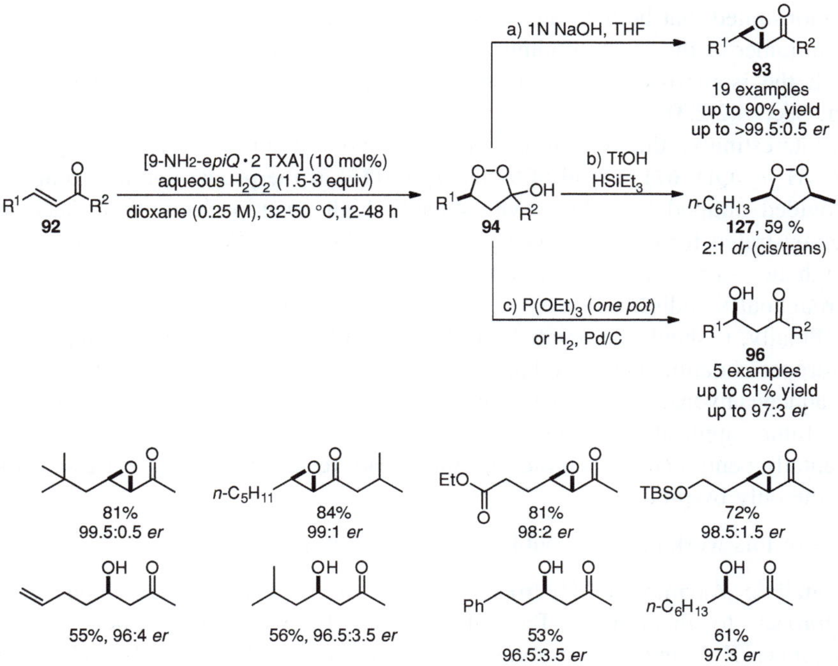

Scheme 5.6 Versatility of cyclic peroxyhemiketals **94**: syntheses of epoxides **93**, aldol products **96**, and 1,2-dioxolane **127**

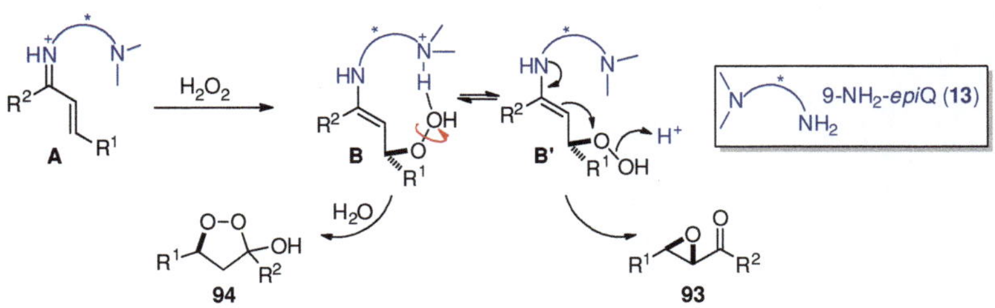

Scheme 5.7 Competing reaction pathways accounting for both the formation of peroxyhemiketals **94** and epoxides **93**

Our postulated mechanism invokes the formation of a peroxyenamine intermediate **B** resulting from the conjugate addition of hydrogen peroxide to the activated enone **A** (cf. Fig. 5.2). This intermediate can either undergo ring closure providing epoxides **93** or furnish peroxyhemiketals **94** upon hydrolysis (Scheme 5.7). Experimental support of this hypothesis was *inter alia* provided by monitoring the reaction by ESI–MS (cf. Sect. 4.5.2).

An intriguing feature of our asymmetric hydroperoxidation-epoxidation reaction of α,β-unsaturated ketones is the complete stereoconvergency. We have demonstrated that both *E*- and *Z*-isomers of a respective enone furnish the same enantiomer of the corresponding *trans*-epoxide in equally high enantioselectivity, with the isomerization most likely taking place via a dienamine intermediate (cf. Sect. 4.2.2.3).

Interestingly, the use of pseudoenantiomeric 9-amino(9-deoxy)*epi*quinidine (9-NH_2-*epi*QD; **67**) instead of 9-NH_2-*epi*Q (**13**) as the primary amine component provided antipodal products with equally high enantioselectivity as has been demonstrated for cyclic as well as acyclic substrates. We thus hold an efficient synthetic entry to either enantiomeric series of the compounds described throughout this dissertation.

Finally, it should be pointed out that a significant and general strength of the hydroperoxidation and epoxidation methods presented in this thesis lies in the use of aqueous hydrogen peroxide as the oxidant. Hydrogen peroxide meets all criteria for future applications potentially also on an industrial scale. It is an environmentally benign and economic oxidant: cheap, readily available, and gives water as the only by-product.

Part of this work has been published in scientific journals:

"Catalytic Asymmetric Hydroperoxidation of α,β-Unsaturated Ketones": An Approach to Enantiopure Peroxyhemiketals, Epoxides, and Aldols": C. M. Reisinger, X. Wang, B. List, *Angew. Chem., Int. Ed.* **2008**, *47*, 8112–8115; *Angew. Chem.*, **2008**, *120*, 8232–8235.

"Catalytic Asymmetric Epoxidation of Cyclic Enones": X. Wang, C. M. Reisinger, B. List, *J. Am. Chem. Soc.* **2008**, *130*, 6070–6071.

Chapter 6
Outlook

As can be inferred from the chemistry described in the previous chapters, there are exciting perspectives ahead for further development of our catalytic asymmetric epoxidation and hydroperoxidation reactions.

9-Amino(9-deoxy)*epi*quinine as its TFA salt ([9-NH_2-*epi*Q · 2 TFA]) proved to be a highly efficient and general iminium ion activator of cyclic as well as acyclic α,β-unsaturated ketone substrates, allowing the highly enantioselective conjugate addition of hydrogen peroxide. It might be desirable to explore the catalytic potential of such chiral primary amine salts in enantioselective epoxidation and hydroperoxidation reactions of other, sterically demanding α,β-unsaturated carbonyl compounds. In this context, future efforts might address the catalytic enantioselective epoxidation of α-substituted-α,β-unsaturated aldehydes in the presence of chiral primary amine salts such as [9-NH_2-*epi*Q · 2 TFA] and others [1, 2]. To date, there is no general and efficient method available for the direct epoxidation of this class of so-called α-branched-α,β-unsaturated aldehydes [3, 4]. Noteworthy, the corresponding optically active epoxyaldehydes constitute attractive chiral building blocks and pivotal intermediates in many natural product syntheses. Strikingly, current synthetic strategies typically rely on reaction sequences involving Sharpless asymmetric epoxidation of allylic alcohols followed by oxidation of the intermediate epoxyalcohols. A direct epoxidation of α-branched-α,β-unsaturated aldehydes would thus pave the way for the synthesis of such epoxyaldehydes with significantly improved redox economy [5].

Initial attempts toward the direct asymmetric epoxidation of α-substituted-α,β-enals afforded very promising results, and clearly warrant further optimization efforts. For instance, epoxyaldehyde **159** was formed in 37% yield along with good enantioselectivity of 93:7 *er* from 2-methyl-2-pentenal (**158**) in the presence of catalytic amounts of [9-NH_2-*epi*Q · 2 TFA] and hydrogen peroxide as the oxidant (Scheme 6.1).

Most of the work in this thesis focused on asymmetric epoxidations and hydroperoxidations of α,β-unsaturated ketones. Whereas aliphatic enones gave the

C. Reisinger, *Epoxidations and Hydroperoxidations of α,β-Unsaturated Ketones*, Springer Theses, DOI: 10.1007/978-3-642-28118-1_6,

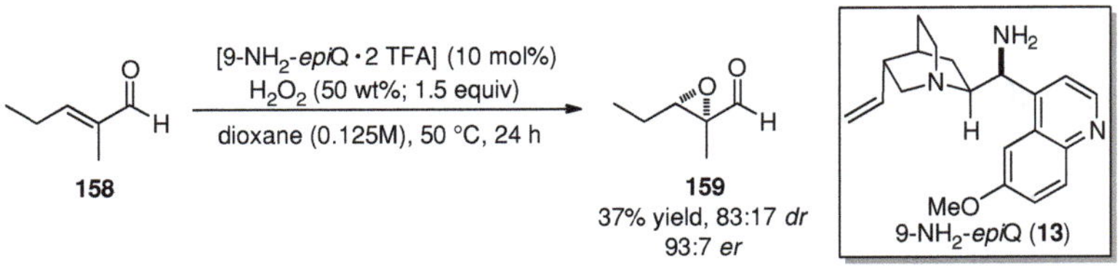

Scheme 6.1 Initial attempt toward the direct asymmetric epoxidation of α-substituted-α,β-unsaturated aldehydes

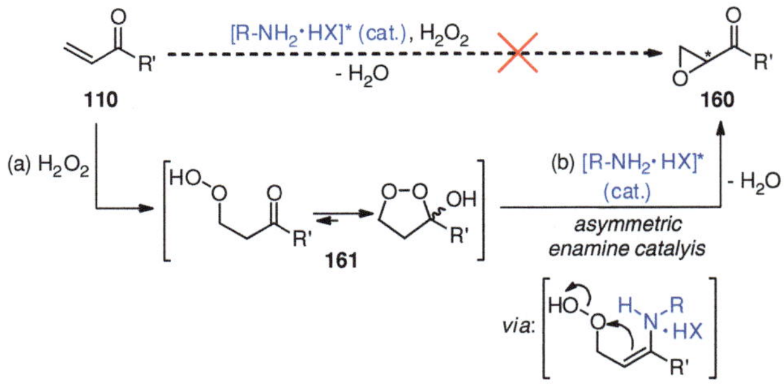

Scheme 6.2 Proposed two step approach to terminal α,β-epoxy ketones **160** starting from vinyl ketones **110**

corresponding α,β-epoxyketones with good yields and excellent enantioselectivities, aromatic enones were not tolerated. In addition, vinyl ketones **110** constituted the second major limitation regarding the scope of α,β-unsaturated ketones. These substrates proved to be essentially unreactive under the conditions of the catalytic reaction, an observation which was attributed to catalyst inhibition (cf. Sect. 4.2.2.4). Despite the synthetic importance of optically active terminal epoxides, to date no general and highly enantioselective method has been identified for their direct preparation; more so when terminal α,β-epoxyketones **160** are considered [6, 7]. Guided by this challenge, we envisage that the use of preformed achiral β-hydroperoxy ketones-cyclic peroxyhemiketals **161** might prove beneficial [8, 9], since they could allow us to circumvent difficulties associated with the conjugate addition of hydrogen peroxide in the presence of the primary amine catalyst [Scheme 6.2, step (a)]. Subsequent chiral primary amine or chiral primary amine salt-catalyzed transformation of such achiral β-hydroperoxy ketones via asymmetric enamine catalysis would give rise to the formation of the desired enantiomerically enriched epoxides [step (b)].

Preliminary experiments have demonstrated the viability of this approach. In the presence of [(*R*,*R*)-DPEN · (*S*)-TRIP] at a loading of 10 mol%, α,β-epoxyketone **160a** was formed in 50% yield and with encouraging enantioselectivity of 58.5:41.5 *er* from cyclic peroxyhemiketal **161a** (Scheme 6.3). Future work might be directed toward evaluating different catalyst motifs and optimizing the reaction conditions. In addition, the successful implementation of

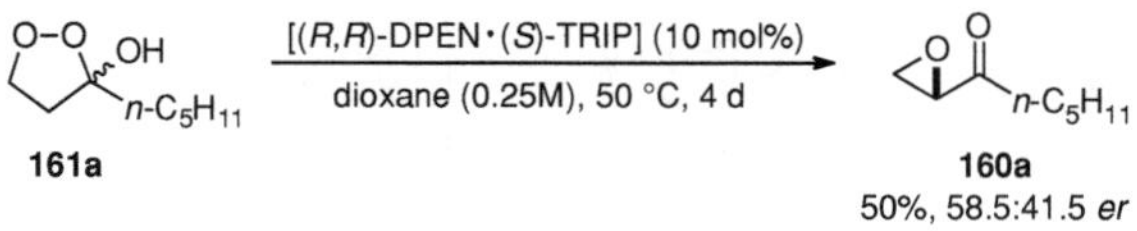

Scheme 6.3 [(*R*,*R*)-DPEN · (*S*)-TRIP]-catalyzed transformation of cyclic peroxyhemiketal **161a** into scalemic epoxyketone **160a**

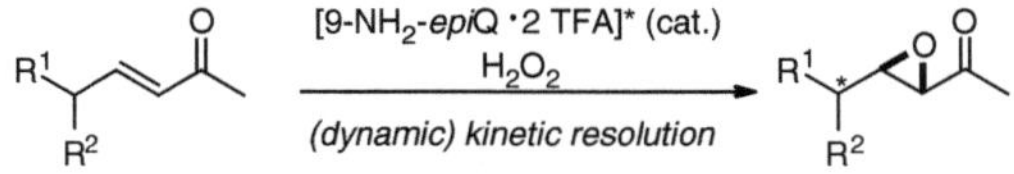

Scheme 6.4 Proposed (dynamic) kinetic resolution of α,β-unsaturated ketones via asymmetric epoxidation

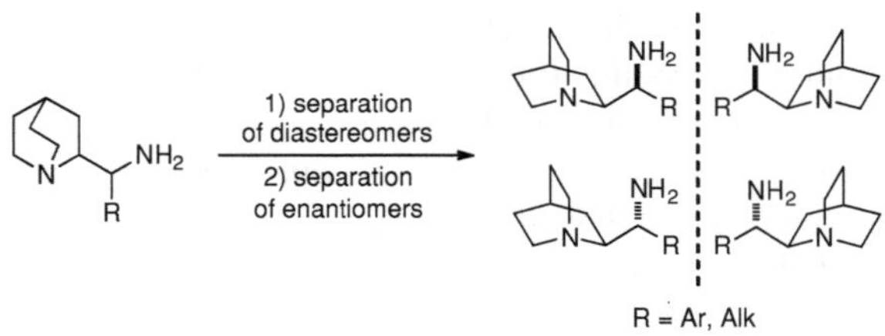

Scheme 6.5 Rapid library generation of *de-novo* designed chiral aminoquinuclidine derivatives

asymmetric enamine catalysis has earlier been illustrated by our group in the context of asymmetric epoxidations of both symmetrically β,β-disubstituted enals [10] and enones (cf. Sect. 4.2.2.5).

Another interesting expansion of these methodologies would be the coupling of our asymmetric epoxidation and hydroperoxidation of α,β-unsaturated ketones with an efficient kinetic or dynamic kinetic resolution process. Based on our previous experiments toward (dynamic) kinetic resolutions via asymmetric epoxidation within the bias of a cyclic system, future attempts might preferentially focus on **acyclic** racemic enone substrates according to the general strategy depicted in Scheme 6.4.

Progress has also been made toward the *de-novo* design of chiral primary amine catalysts of the general structure **86** based on the quinuclidine scaffold. Racemic synthesis of **86** according to the strategy introduced in Sect. 4.7.1.2, followed by the separation of the diastereomers and subsequent resolution would afford single enantiomers of both *threo-* and *erythro-*derivatives (Scheme 6.5). This promising direction of research might lead to the identification of new, potent primary amine catalysts with superior catalytic activity compared with parent Cinchona alkaloid-derived catalyst motifs. In particular, the *de-novo* approach may significantly facilitate catalyst optimization since one is not restricted to naturally abundant Cinchona alkaloids as starting materials. This further renders such compounds the

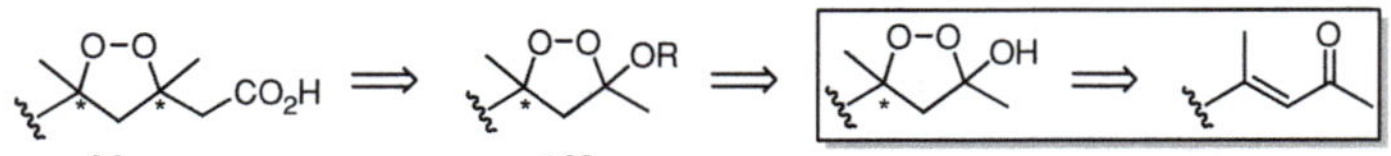

Fig. 6.1 Proposed retrosynthesis of 1,2-dioxolane-3-acetic acids from α,β-unsaturated ketones

Fig. 6.2 Natural products Novaxenicin B and Cespihypotin C

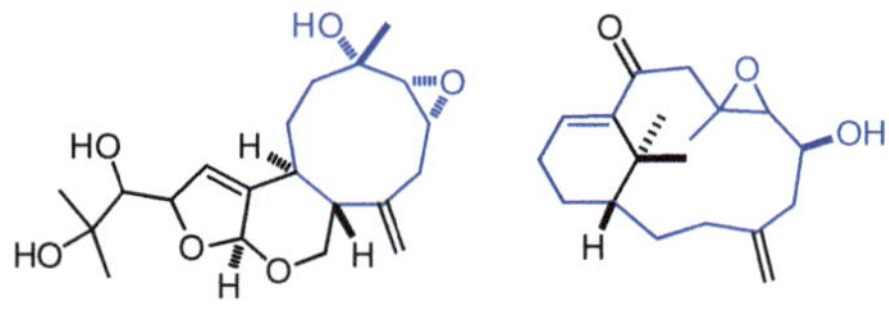

ideal platform to conduct systematic structure-selectivity studies [11]. Moreover, it nicely circumvents one of the most stringent, inherent limitations of the of Cinchona alkaloids in asymmetric catalysis: the inaccessibility of a truly enantiomeric form of these compounds.

Given the impressive results obtained with our epoxidation and hydroperoxidation processes, we believe that the methods described herein might be mature enough to be tested in the context of natural product synthesis. In particular, the optically active 3-hydroxy-1,2-dioxolanes made available in one step from α,β-unsaturated ketones through asymmetric hydroperoxidation could allow the development of a concise route to various members of the platinic acid natural product family. These naturally occurring peroxidic compounds display promising antitumor and antifungal activity [12]. Only one asymmetric synthesis of plakinic acid A (**36a**; cf. Fig. 2.6) has been reported to date. It is striking to note that this synthesis comprises 23 steps in a linear sequence with an overall yield of 5.7%, while requiring the use of ethereal hydrogen peroxide and semipreparative HPLC separation techniques [13]. Providing access to peroxyketals of the general type **162** in optically active form as described in Sect. 4.2.2.5 may pave the way to concise asymmetric syntheses of a broad range of 1,2-dioxolane acetic acid derivatives **36**. These should include naturally occurring representatives as well as synthetic analogues which might exhibit improved pharmacological activities, all prepared according to the unified retrosynthetic strategy depicted in Fig. 6.1 [12].

In recent years, macrocyclic natural products have gained increasing attention. They often display remarkable biological activities, and many of these compounds (or their derivatives) are used as drugs [14, 15]. Our intriguing discovery of an asymmetric epoxidation of macrocyclic enones proceeding under mild reaction conditions with extremely high stereocontrol might provide intriguing synthetic versatility. In particular, we find the possibility of effecting stereoselective, catalyst-controlled late-stage introduction of epoxide functionality onto a macrocyclic core very enticing. Such a possibility would clearly expand the toolbox of existing strategies toward the synthesis of macrocyclic natural products and their derivatives. One strength of our method is its stereoconvergency. Regardless of the geometry of a macrocyclic enone precursor (available *inter alia* through ring-closing metathesis or through Wittig and related olefination reactions), the (*E*)-configured macrocyclic epoxide would be generated exclusively. Moreover, pseudoenantiomeric amine catalysts (derived from quinine and quinidine) would allow concise access to diastereomeric products suitable for SAR studies with equally high enantioselectivity. Possible target molecules to illustrate the feasibility of a late-stage introduction of an epoxide moiety via chiral primary amine-catalyzed asymmetric epoxidation of enones might be among others Novaxenicin B [16] or Cespihypotin C (Fig. 6.2) [17].

References

1. Kano T, Tanaka Y, Osawa K, Yurino T, Maruoka K (2009) Chem Commun 15:1956
2. Galzerano P, Pesciaioli F, Mazzanti A, Bartoli G, Melchiorre P (2009) Angew Chem Int Ed 48:7892
3. Sparr C, Schweizer WB, Senn HM, Gilmour R (2009) Angew Chem Int Ed 48:3065
4. Erkkilä A, Pihko PM, Clarke M-R (2007) Adv Synth Catal 349:802
5. Burns NZ, Baran PS, Hoffmann RW (2009) Angew Chem Int Ed 48:2854
6. Pégorier L, Petit Y, Mambu A, Larchevêque M (1994) Synthesis 12:1403
7. Schaus SE, Brandes BD, Larrow JF, Tokunaga M, Hansen KB, Gould AE, Furrow ME, Jacobsen EN (2002) J Am Chem Soc 124:1307
8. Fraile JM, García JI, Mayoral JA, Sebti S, Tahirb R (2001) Green Chem 3:271
9. Cativiela C, Figueras F, Fraile JM, García JI, Mayoral JA (1995) Tetrahedron Lett 36:4125
10. Wang X, List B (2008) Angew Chem Int Ed 47:1119
11. Exner C, Pfaltz A, Studer M, Blaser H-U (2003) Adv Synth Catal 345:1253
12. Dussault PH, Liu X (1999) Org Lett 1:1391
13. Dai P, Trullinger TK, Liu X, Dussault PH (2006) J Org Chem 71:2283
14. Driggers EM, Hale SP, Lee J, Terrett NK (2008) Nat Rev Drug Discov 7:608
15. Wessjohann LA, Ruijter E, Garcia-Rivera D, Brandt W (2005) Mol Divers 9:171
16. Bishara A, Rudi A, Goldberg I, Benayahu Y, Kashman Y (2006) Tetrahedron 62:12092
17. Shen Y-C, Lin J-J, Wu Y-R, Chang J-Y, Duh C-Y, Lo KL (2006) Tetrahedron Lett 47:6651

Chapter 7
Experimental Part

7.1 General Experimental Conditions

Solvents and reagents

All solvents were purified by distillation before use following standard procedures. Absolute solvents were obtained by distillation over appropriate drying agent (*vide infra*) and then kept under an atmosphere of argon: diethyl ether, tetrahydrofuran, toluene, and *n*-hexane (sodium, benzophenone as indicator), chloroform, dichloromethane, triethylamine (calcium hydride), ethanol (magnesium). Absolute 1,4-dioxane, TBME, di(*n*-butyl)ether, DME, NMP, acetonitrile, and DMSO were purchased from Sigma-Aldrich and used as received. Other commercial reagents were obtained from various sources and used without further purification.

Inert gas atmosphere

Air and moisture-sensitive reactions were conducted under an argon atmosphere. Argon was obtained from *Air Liquide* with higher than 99.5% purity. All organocatalytic reactions within this thesis were carried out without exclusion of air and moisture.

Chromatographic methods

Reactions were mostly monitored by **thin layer chromatography** (**TLC**) using silica gel precoated glass plates (Merck, 0.25 mm thickness, silica gel 60F-254) or silica gel precoated aluminium foil plates (Macherey–Nagel MN, 0.20 mm thickness, Polygram SIL G/UV254). The spots were visualized with UV-light ($\lambda = 254$ nm) and/or by staining with anisaldehyde, phosphomolybdic acid, ninhydrin, dinitrophenylhydrazine, or potassium permanganate stains. Preparative scale TLC was conducted on Macherey–Nagel glass plates with a thickness of 0.25, 1, or 2 mm silica gel, respectively.

C. Reisinger, *Epoxidations and Hydroperoxidations of α,β-Unsaturated Ketones*,
Springer Theses, DOI: 10.1007/978-3-642-28118-1_7,

Flash column chromatography was performed using silica gel 60 (Merck, 60 Å, 230–400 mesh 0.040–0.063 mm) and separations were either conducted at slightly elevated pressure in a glass column or using the automated Sepacore Flash system from Büchi, consisting of fraction collector C-660, UV-photometer C-635, and pump module C-605.

Nuclear magnetic resonance spectroscopy (NMR)

Spectra were recorded on Bruker DPX 300 (^{1}H: 300 MHz, ^{13}C: 75 MHz), Bruker AV 400 (^{1}H: 400 MHz, ^{13}C: 100 MHz), and Bruker AV 500 (^{1}H: 500 MHz, ^{13}C: 125 MHz) spectrometers at room temperature (298 K). Chemical shifts for protons and carbons are reported in parts per million (ppm) relative to tetramethylsilane as internal standard or to the residual signal of the NMR solvents (*e.g.* CD_2Cl_2: δ H 5.32, δ C 53.8, $CDCl_3$: δ H 7.26, δ C 77.0, THF-d8: δ H 3.58, δ C 67.6). Chemical shifts for phosphorus are reported relative to H_3PO_4 as external standard. The ^{1}H NMR multiplicities are assigned as follows: singlet (s), doublet (d), triplet (t), q (quartet), quin (quintet), sext (sextet), sept (septet), m (multiplet), broad (br). The coupling constants (*J*) are reported in Hertz (Hz). The signals have been assigned using 1 and 2D experiments.

Gas chromatography (GC)

Gas chromatography (GC) was performed on HP 6890 and 5890 Series instruments (carrier gas: hydrogen) equipped with a split-mode capillary injection system and a flame ionization detector (FID).

The enantiomeric ratios of chiral molecules were determined using chiral columns containing the following chiral stationary phases:

BGB 176	2,3-dimethyl-6-*tert*-butyldimethyl-silyl-β-cyclo-dextrin (achiral component: *SE*-52 or *BGB*-15); 30 m × 0.25 × 0.25 mm
BGB 178	2,3-diethyl-6-*tert*-butyldimethyl-silyl-β-cyclo-dextrin (achiral component: *OV*-1701); 30 m × 0.25 × 0.25 mm
G-TA	trifluoroacetyl-γ-cyclodextrin; 30 m × 0.25 × 0.25 mm
Ivadex 1	dimethylpentyl-β-cyclodextrin (achiral component: *PS*086); 25 m × 0.25 × 0.25 mm
Ivadex 7	diethyl-*tert*-butyl-dimethyl-β-cyclodextrin; 25 m × 0.25 × 0.25 mm
Lipodex A	hexakis(2,3,6-tri-*O*-pentyl)-α-cyclodextrin; 25 m × 0.25 × 0.25 mm
Lipodex E	octakis(2,6-*O*-dipentyl-3-*O*-butyryl)-γ-cyclodextrin; 25 m × 0.25 ×0.25 mm

Lipodex G	octakis(2,3-*O*-dipentyl-6-*O*-methyl-γ-cyclodextrin; 25 m × 0.25 ×0.25 mm
Hydrodex-β-TBDAC	heptakis(2,3-*O*-diacetyl-6-*O*-*tert*-butyldimethyl-silyl)-β-cyclodextrin; 25 m × 0.25 × 0.25 mm.

GC-MS couplings were performed on an *Agilent Technology* GC 6890 Series and MSD 5973 (carrier gas: helium) with HP6890 Series Injector, employing an MN Optima®5 column (30 m × 0.25 × 0.25 mm). The mass spectra were recorded with an *Agilent Technology* 5973 Network MSD.

Mass spectrometry (MS)

Mass spectra were measured on a Finnigan MAT 8200 (70 eV) or MAT 8400 (70 eV) by electron ionization, chemical ionization, of fast atom/ion bombardment techniques. High resolution masses were determined on a Bruker APEX III FT-MS (7 T magnet). All masses are given in atomic units/elementary charge (*m/z*) and reported in percentage relative to the basic peak. The mechanistic studies (cf. Sect. 4.5.2) were performed by ESI-MS with a Finnigan Ultra Mass TSQ 7000.

Specific rotation ([α])

Optical rotations were measured on a *Perkin Elmer* 343 or *Rudolph Analytical Autopol IV* polarimeter using a 1 mL cell with a path length of 1 dm at the temperature and wavelength indicated, with "D" referring to the sodium D-line wavelength (589 nm). Concentrations are given in g/100 mL.

Determination of the optical purity

Enantiomeric ratios (*er*) were determined either by chiral GC or chiral HPLC analysis (as specified in the individual experiments) by comparing the samples with the appropriate racemic mixtures. The optical purity of peroxyhemiketals **94** was determined after converting it to the corresponding epoxide [with 1N NaOH (1 equiv) in Et_2O] or to the corresponding aldol-type product [with triethylphosphite (2 equiv) in Et_2O]. Racemic samples of epoxides were obtained by reaction of the enone with alkaline, aqueous hydrogen peroxide. These reactions were conducted in methanol in the presence of either *tert*-butylamine or NaOH as the catalytic base [1]. Racemic samples of aldol products were obtained through aldol reaction of the respective aldehyde with acetone catalyzed by either KOH or *rac*-proline. Racemic cyclic aldol products were obtained by reductive cleavage of the epoxide of racemic α,β-epoxyketones. The absolute configuration of 2,3-epoxycyclohexanone (**48a**) and 4-hydroxy-6-phenyl-2-hexanone (**96b**) [derived from the corresponding peroxyhemiketal **94b** through $P(OEt)_3$-reduction] was established by comparison of their optical rotation with literature values [2, 3]. All other absolute configurations were assigned by analogy.

7.2 Catalytic Asymmetric Epoxidation of Cyclic Enones

7.2.1 General Procedure

Conditions A:

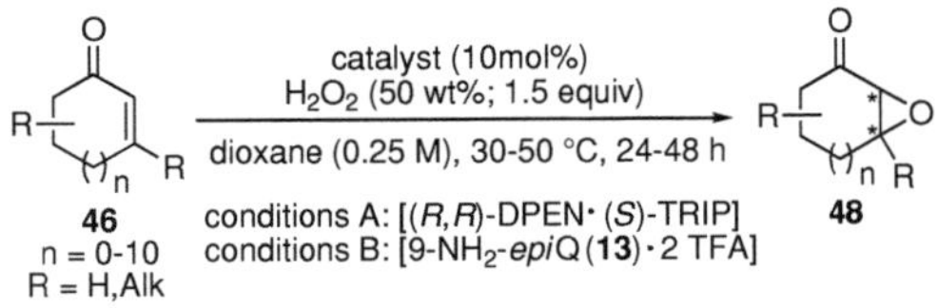

Conditions A: Reference [4] Catalyst salt [(*R*,*R*)-DPEN · (*S*)-TRIP] was prepared *in situ* by stirring (*R*,*R*)-DPEN ((*R*,*R*)-**12**; 10.6 mg, 0.05 mmol, 10 mol%) and (*S*)-TRIP [(*S*)-**7a**; 37.6 mg, 0.05 mmol, 10 mol%] in dioxane (2 mL) for 20 min at room temperature. Then, cyclic enone **46** (0.5 mmol, 1.0 equiv) was added, and 20 min later, aqueous hydrogen peroxide (50 wt%; 46 μL, 0.75 mmol, 1.5 equiv). After 12–72 h of stirring at 30–50 °C, the reaction mixture was extracted with Et_2O (3 × 10 mL). The combined organic phases were washed with brine, dried (Na_2SO_4), and filtered. For the highly volatile products **48a** and **48r**, the resulting solution was analyzed by GC for yield and *er* determination. Removal of the volatiles furnished the crude product, which was purified by flash column chromatography (silica gel, eluent: Et_2O-pentane) to afford pure cyclic α,β-epoxy ketone **48**.

Conditions B: Catalyst salt [**13** · 2 TFA] was prepared *in situ* by the addition of 9-amino(9-deoxy)*epi*quinine (**13**; 32.3 mg, 0.1 mmol, 10 mol%) to a solution of trifluoroacetic acid (TFA; 15.3 μL, 0.2 mmol, 20 mol%) in dioxane (4 mL). Then, cyclic enone **46** (1.0 mmol, 1.0 equiv) was added, and 20 min later, aqueous hydrogen peroxide (50 wt%; 92 μL, 1.5 mmol, 1.5 equiv). After 12–72 h of stirring at 30–50 °C, the reaction mixture was extracted with Et_2O (3 × 25 mL). The combined organic phases were washed with brine, dried (Na_2SO_4), and filtered. For the highly volatile products **48a** and **48r**, the resulting solution was analyzed by GC for yield and *er* determination. Removal of the volatiles furnished the crude product, which was purified by flash column chromatography (silica gel, eluent: Et_2O-pentane) to afford pure cyclic α,β-epoxy ketone **48**.

7.2.2 *Scope of Optically Active Cyclic α,β-Epoxyketones*

2,3-Epoxycyclohexanone (48a)

Conditions A:

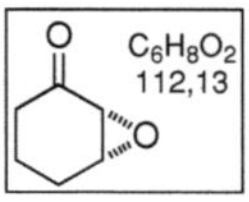

After 48 h at 35 °C, the conversion was determined to be 99% by GC using an Optima-5-Accent column (5 min at 40 °C, 5.0 °C/min until 100 °C, 25 °C/min until 250 °C, 1.0 min at 250 °C, 0.35 bar He; starting material: $\tau_R = 12.35$ min, product: $\tau_R = 15.66$ min). After purification by flash column chromatography (silica gel, 5–10% Et_2O in pentane) (2*R*,3*R*)-**48a** was obtained as a clear liquid [76 mg, 678 μmol, 68% (*reduced yield due to the high volatility of* **48a**); 96:4 *er*]. The enantiomeric ratio was determined by GC using a chiral BGB-176/BGB-15 column 30 m (80 °C, 1.2 °C/min until 105 °C, 20 °C/min until 220 °C, 10 min at 220 °C, 0.5 bar H_2); major enantiomer: $\tau_R = 16.62$ min, minor enantiomer: $\tau_R = 16.23$ min.

Optical rotation $[\alpha]_D^{23} = +101.2$ (c = 1.0, CH_2Cl_2, 96:4 *er* (2*R*,3*R*)), [Lit.:[2] (2*S*,3*S*)-**48a**, $[\alpha]_D^{rt} = -38.0$ (c = 0.82, CH_2Cl_2, 60:40 *er*)]

Conditions B:

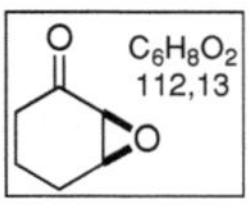

After 24 h at 30 °C, the conversion was determined to be 91% by GC using an Optima-5-Accent column (5 min at 40 °C, 5.0 °C/min until 100 °C, 25 °C/min until 250 °C, 1.0 min at 250 °C, 0.35 bar He; starting material: $\tau_R = 12.34$ min, product: $\tau_R = 15.65$ min). After purification by flash column chromatography (silica gel, 5–10% Et_2O in pentane) (2*S*,3*S*)-**48a** was obtained as a clear liquid [65 mg, 580 μmol, 58% (*reduced yield due to the high volatility of* **48a**); 97:3 *er*]. The enantiomeric ratio was determined by GC using a chiral BGB-176/BGB-15 column 30 m (80 °C, 1.2 °C/min until 105 °C, 20 °C/min until 220 °C, 10 min at 220 °C, 0.5 bar H_2); major enantiomer: $\tau_R = 16.22$ min, minor enantiomer: $\tau_R = 17.08$ min.

^{1}H NMR (500 MHz, CD_2Cl_2) δ 3.57–3.55 (m, 1H, $CH_2C_{epo}\boldsymbol{H}$), 3.15 (d, $J = 4.2$ Hz, 1H, $C_{epo}\boldsymbol{H}C(=O)$), 2.47 (dt, $J = 17.5$, 4.6 Hz, 1H, $C\boldsymbol{H}HC(=O)$), 2.26–2.19 (m, 1H, $C\boldsymbol{H}HC_{epo}H$), 2.07–2.00 (m, 1H, $CH\boldsymbol{H}C(=O)$), 1.94–1.84 (m, 2H, $CH\boldsymbol{H}C_{epo}H$ and $CH_2C\boldsymbol{H}HCH_2$), 1.69–1.60 (m, 1H, $CH_2CH\boldsymbol{H}CH_2$).

^{13}C NMR (100 MHz, CD_2Cl_2) δ 206.0 (*C*=O), 56.2 ($CH_2\boldsymbol{C}H_{epo}$), 55.4 ($\boldsymbol{C}H_{epo}C(=O)$), 36.7 ($\boldsymbol{C}H_2C(=O)$), 23.2 ($\boldsymbol{C}H_2CH_{epo}$), 17.3 ($CH_2\boldsymbol{C}H_2CH_2$).

The analytical data were identical in all respects to those of the commercially available 2,3-epoxycyclohexanone (**48a**; Sigma-Aldrich).

(2*R*,3*R*)-4,4-Dimethyl-2,3-epoxycyclohexanone ((2*R*,3*R*)-**48b**)

Conditions A:

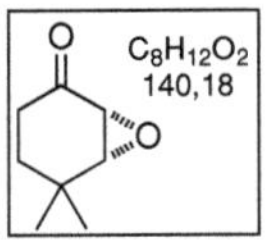

The title compound was isolated after 48 h at 35 °C and purification by flash column chromatography (silica gel, 2% Et_2O in pentane) as a clear liquid (56 mg, 400 µmol, 80%; 97:3 *er*). The enantiomeric ratio was determined by GC using a chiral Hydrodex-β-TBDAc column 25 m (80 °C, 1.5 °C/min until 170 °C, 20 °C/min until 220 °C, 10 min at 220 °C, 0.6 bar H_2); major enantiomer: $\tau_R = 25.75$ min, minor enantiomer: $\tau_R = 21.91$ min.

^{1}H NMR (500 MHz, $CDCl_3$) δ 3.20 (d, $J = 3.9$ Hz, 1H, $C_{epo}\boldsymbol{H}C(=O)$), 3.15 (dd, $J = 4.0$, 1.3 Hz, 1H, $CMe_2C_{epo}\boldsymbol{H}$), 2.37 (ddd, $J = 18.9$, 6.3, 3.1 Hz, 1H, $C\boldsymbol{H}HC(=O)$), 2.20–2.13 (m, 1H, $CH\boldsymbol{H}C(=O)$), 1.88 (ddd, $J = 13.6$, 11.8, 6.4 Hz, 1H, $C\boldsymbol{H}HCqMe_2$), 1.31 (dddd, $J = 13.6$, 7.1, 3.0, 1.2 Hz, 1H, $CH\boldsymbol{H}CqMe_2$), 1.19 (s, 3H, $C\boldsymbol{H}_3$), 1.04 (s, 3H, $C\boldsymbol{H}_3$).

^{13}C NMR (75 MHz, CD_2Cl_2) δ 205.9 ($\boldsymbol{C}$=O), 64.1 ($CMe_2\boldsymbol{C}H_{epo}$), 55.9 ($\boldsymbol{C}H_{epo}C(=O)$), 33.1 ($\boldsymbol{C}H_2C(=O)$), 30.7 ($\boldsymbol{C}qMe_2$), 29.7 ($\boldsymbol{C}H_2CqMe_2$), 27.4 ($\boldsymbol{C}H_3$), 22.8 ($\boldsymbol{C}H_3$).

MS (EI) *m*/*z* (%) 140 [M^+] (14), 124 (2), 111 (20), 97 (27), 85 (58), 69 (100), 55 (73), 43 (29), 41 (95), 39 (32), 29 (31).

HRMS calcd for $C_8H_{12}O_2$ [M^+] 140.0839, found 140.0837.

(2*R*,3*R*)-5,5-Dimethyl-2,3-epoxycyclohexanone ((2*R*,3*R*)-**48c**)

Conditions A:

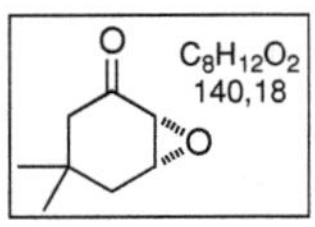

The title compound was isolated after 48 h at 35 °C and purification by flash column chromatography (silica gel, 3% Et_2O in pentane) as a clear liquid (53 mg, 757 µmol, 76%; 98:2 *er*). The enantiomeric ratio was determined by GC using a chiral Hydrodex-β-TBDAc column 25 m (80 °C, 1.2 °C/min until 120 °C, 20 °C/min until 220 °C, 10 min at 220 °C, 0.5 bar H_2); major enantiomer: $\tau_R = 29.57$ min, minor enantiomer: $\tau_R = 27.55$ min.

^{1}H NMR (300 MHz, CD_2Cl_2) δ 3.48 (t, $J = 4.1$ Hz, 1H, $CH_2C_{epo}\boldsymbol{H}$), 3.14 (d, $J = 3.8$ Hz, 1H, $C_{epo}\boldsymbol{H}C(=O)$), 2.61 (d, $J = 13.2$ Hz, 1H, $C\boldsymbol{H}HC(=O)$), 2.00 (d, $J = 15.4$ Hz, 1H, $C\boldsymbol{H}HCH_{epo}$), 1.86–1.74 (m, 2H, $CH\boldsymbol{H}C(=O)$ and $CH\boldsymbol{H}CH_{epo}$), 1.00 (s, 3H, $C\boldsymbol{H}_3$), 0.90 (s, 3H, $C\boldsymbol{H}_3$).

^{13}C NMR (75 MHz, CD_2Cl_2) δ 207.5 (***C***=O), 57.4 (***C***H_{epo}C(=O)), 55.0 (CH_2***C***H_{epo}), 49.0 (***C***H_2C(=O)), 37.5 (***C***H_2CH_{epo}), 37.4 (***C***qMe_2), 31.0 (***C***H_3), 28.1 (***C***H_3).

GC-MS (EI-DE) *m/z* (%) 140 [M^+] (20), 125 (1), 112 (4), 97 (20), 83 (100), 79 (8), 69 (28), 55 (90), 53 (13), 43 (19), 41 (73), 39 (34), 27 (20).

HRMS calcd for $C_8H_{12}O_2$ [M^+] 140.0839, found 140.0837.

(2*R*,3*R*)-3,5,5-Trimethyl-2,3-epoxycyclohexanone ((2*R*,3*R*)-**48d**)

Conditions A:

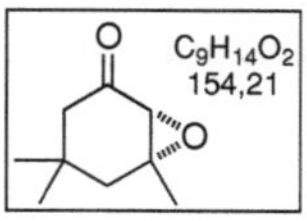

The title compound was isolated after 48 h at 50 °C and purification by flash column chromatography (silica gel, 1.5% Et_2O in pentane) as a clear liquid (48 mg, 623 μmol, 62%; 96:4 *er*). The enantiomeric ratio was determined by GC using a chiral Hydrodex-β-TBDAc column 25 m (60 °C, 1.0 °C/min until 90 °C, 20 °C/min until 220 °C, 10 min at 220 °C, 0.5 bar H_2); major enantiomer: τ_R = 27.80 min, minor enantiomer: τ_R = 27.24 min.

^{1}H NMR (500 MHz, $CDCl_3$) δ 3.01 (s, 1H, C_{epo}***H***), 2.58 (d, *J* = 13.5 Hz, 1H, C***H***HC(=O)), 2.04 (d, *J* = 14.7 Hz, 1H, Cq_{epo}C***H***H), 1.77 (ddd, *J* = 13.4, 2.0, 1.0 Hz, 1H, CH***H***C(=O)), 1.70 (dd, *J* = 14.8, 2.2 Hz, 1H, Cq_{epo}CH***H***), 1.38 (s, 3H, Cq_{epo}C***H***$_3$), 0.98 (s, 3H, C***H***$_3$), 0.87 (s, 3H, C***H***$_3$).

^{13}C NMR (75 MHz, $CDCl_3$) δ 207.8 (***C***=O), 64.2 (***C***q_{epo}), 61.4 (***C***H_{epo}), 48.0 (***C***H_2C(=O)), 42.7 (***C***H_2Cq_{epo}), 36.1 (***C***qMe_2), 30.8 (Cq(***C***H_3)$_2$), 27.8 (Cq(***C***H_3)$_2$), 24.0 (Cq_{epo}***C***H_3).

MS (EI-DE) *m/z* (%) 154 [M^+] (30), 139 (33), 126 (17), 111 (12), 97 (25), 83 (100), 69 (50), 55 (39), 53 (9), 43 (40), 41 (65), 29 (23).

HRMS calcd for $C_9H_{14}O_2$ [M^+] 154.0992, found 154.0994.

(2*S*,3*S*)-2,3-Epoxy-3-methylcyclohexanone ((2*S*,3*S*)-**48f**)

Conditions B:

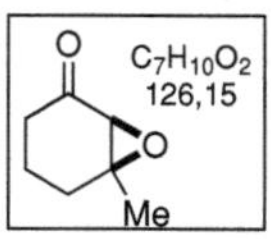

The title compound was isolated after 24 h at 30 °C and purification by flash column chromatography (silica gel, 2–10% Et_2O in pentane) as a clear oil (88 mg, 698 μmol, 70% [*reduced yield due to the high volatility of* **48f**); 98:2 *er*]. The enantiomeric ratio was determined by GC using a chiral BGB-178/OV-1701 column 30 m (80 °C, 1.2 °C/min until 105 °C, 20 °C/min until 220 °C, 10 min at 220 °C, 0.5 bar H_2); major enantiomer: τ_R = 16.79 min, minor enantiomer: τ_R = 19.18 min.

^{1}H NMR (400 MHz, CD_2Cl_2) δ 3.01 (s, 1H, $C_{epo}\boldsymbol{H}$), 2.42 (dt, J = 17.6, 4.1 Hz, 1H, $C\boldsymbol{H}HC(=O)$), 2.13–1.84 (m, 4H, $CH\boldsymbol{H}C(=O)$, $C\boldsymbol{H}_2Cq_{epo}$, and $CH_2C\boldsymbol{H}HCH_2$), 1.66–1.58 (m, 1H, $CH_2CH\boldsymbol{H}CH_2$), 1.42 (s, 3H, $C\boldsymbol{H}_3$).

^{13}C NMR (75 MHz, CD_2Cl_2) δ 206.7 ($\boldsymbol{C}$=O), 62.6 ($\boldsymbol{C}q_{epo}$), 62.3 ($\boldsymbol{C}H_{epo}$), 36.1 ($\boldsymbol{C}H_2C(=O)$), 28.7 ($\boldsymbol{C}H_2Cq_{epo}$), 22.3 ($\boldsymbol{C}H_3$), 17.5 ($CH_2\boldsymbol{C}H_2CH_2$).

MS (EI-DE) *m/z* (%) 126 [M$^+$] (78), 111 (8), 97 (46), 83 (35), 81 (21), 79 (5), 71 (79), 69 (26), 67 (8), 58 (3), 55 (61), 53 (12), 43 (89), 41 (100), 39 (49), 31 (2), 27 (36).

HRMS (EI-FE) calcd for $C_7H_{10}O_2$ [M$^+$] 126.0680, found 126.0681.

(2*S*,3*S*)-3-Ethyl-2,3-epoxycyclohexanone ((2*S*,3*S*)-**48g**)

Conditions B:

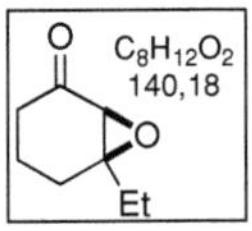

The title compound was isolated after 30 h at 50 °C and purification by flash column chromatography (silica gel, 5–10% Et_2O in pentane) as a clear oil (102 mg, 728 μmol, 73%; 98.5:1.5 *er*). The enantiomeric ratio was determined by GC using a chiral BGB-178/OV-1701 column 30 m (80 °C, 1.2 °C/min until 115 °C, 20 °C/min until 220 °C, 10 min at 220 °C, 0.5 bar H_2); major enantiomer: τ_R = 23.73 min, minor enantiomer: τ_R = 25.43 min.

^{1}H NMR (400 MHz, CD_2Cl_2) δ 3.03 (s, 1H, $C_{epo}\boldsymbol{H}$), 2.45 (dt, J = 17.4, 4.3 Hz, 1H, $C\boldsymbol{H}HC(=O)$), 2.12–1.95 (m, 2H, $CH\boldsymbol{H}C(=O)$ and $C\boldsymbol{H}HCq_{epo}$), 1.94–1.83 (m, 2H, $CH\boldsymbol{H}Cq_{epo}$ and $CH_2C\boldsymbol{H}HCH_2$), 1.78–1.61 (m, 3H, $C\boldsymbol{H}_2CH_3$ and $CH_2CH\boldsymbol{H}CH_2$), 0.96 (t, J = 7.5 Hz, 3H, $C\boldsymbol{H}_3$).

^{13}C NMR (100 MHz, CD_2Cl_2) δ 206.9 ($\boldsymbol{C}$=O), 66.3 ($\boldsymbol{C}q_{epo}$), 61.1 ($\boldsymbol{C}H_{epo}$), 36.4 ($\boldsymbol{C}H_2C(=O)$), 29.2 ($\boldsymbol{C}H_2CH_3$), 26.5 ($\boldsymbol{C}H_2Cq_{epo}$), 17.8 ($CH_2\boldsymbol{C}H_2CH_2$), 8.8 ($\boldsymbol{C}H_3$).

MS (EI-DE) *m/z* (%) 140 [M$^+$] (90), 125 (5), 111 (49), 107 (1), 97 (55), 95 (24), 85 (79), 79 (12), 67 (24), 57 (30), 55 (84), 43 (38), 41 (100), 39 (58), 29 (92), 27 (77).

HRMS (EI-FE) calcd for $C_8H_{12}O_2$ [M$^+$] 140.0836, found 140.0837.

(2*S*,3*R*)-2,3-Epoxy-3-isobutylcyclohexanone ((2*S*,3*R*)-**48h**)

Conditions B:

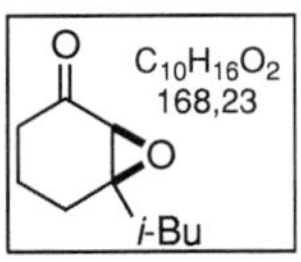

The title compound was isolated after 48 h at 50 °C and purification by flash column chromatography (silica gel, 5–10% Et_2O in pentane) as a clear oil (123 mg, 731 μmol, 73%; 98:2 *er*). The enantiomeric ratio was determined by GC using a chiral BGB-178/OV-1701 column 30 m (80 °C, 0.8 °C/min until 120 °C, 20 °C/min until 220 °C, 10 min at 220 °C, 0.5 bar H_2); major enantiomer: τ_R = 43.77 min, minor enantiomer: τ_R = 45.20 min.

^{1}H NMR (400 MHz, CD_2Cl_2) δ 2.98 (s, 1H, $C_{epo}\boldsymbol{H}$), 2.45 (dt, $J = 17.4$, 4.3 Hz, 1H, C(=O)C***H***H), 2.11–1.81 (m, 5H, C(=O)CH***H***, $C\boldsymbol{H}_2Cq_{epo}$, C***H***$Me_2$, and $CH_2C\boldsymbol{H}HCH_2$), 1.67–1.60 (m, 2H, $CH_2CH\boldsymbol{H}CH_2$ and C***H***H*i*-Pr), 1.40 (dd, $J = 13.8$, 8.1 Hz, 1H, CH***H****i*-Pr), 0.96 (d, $J = 6.6$ Hz, 3H, $CH(C\boldsymbol{H}_3)_2$), 0.91 (d, $J = 6.6$ Hz, 3H, $CH(C\boldsymbol{H}_3)_2$).

^{13}C NMR (75 MHz, CD_2Cl_2) δ 206.9 (***C***=O), 64.8 ($\boldsymbol{C}q_{epo}$), 61.8 ($\boldsymbol{C}H_{epo}$), 45.4 ($\boldsymbol{C}H_2$*i*-Pr), 36.4 (C(=O)$\boldsymbol{C}H_2$), 26.5 ($\boldsymbol{C}H_2Cq_{epo}$), 25.4 (***C***$HMe_2$), 23.2 ($CH(\boldsymbol{C}H_3)_2$), 22.7 ($CH(\boldsymbol{C}H_3)_2$), 17.7 ($CH_2\boldsymbol{C}H_2CH_2$).

MS (EI-DE) *m/z* (%) 168 [M^+] (45), 153 (7), 139 (7), 126 (36), 112 (31), 79 (32), 67 (43), 55 (56), 41 (100), 39 (43), 27 (54).

HRMS (EI-FE) calcd for $C_{10}H_{16}O_2$ [M^+] 168.1149, found 168.1150.

(2*S*,3*R*)-2,3-Epoxy-3-isopropylcyclohexanone ((2*S*,3*R*)-**48i**)

Conditions B:

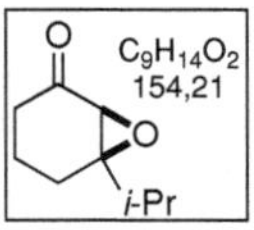

The title compound was isolated after 48 h at 50 °C and purification by flash column chromatography (silica gel, 5–10% Et_2O in pentane) as a clear oil (121 mg, 785 μmol, 79%; 99:1 *er*). The enantiomeric ratio was determined by GC using a chiral BGB-178/OV-1701 column 30 m (80 °C, 1.2 °C/min until 120 °C, 20 °C/min until 220 °C, 10 min at 220 °C, 0.5 bar H_2); major enantiomer: $\tau_R = 27.79$ min, minor enantiomer: $\tau_R = 29.24$ min.

^{1}H NMR (400 MHz, CD_2Cl_2) δ 3.03 (s, 1H, $C_{epo}\boldsymbol{H}$), 2.46 (dt, $J = 17.3$, 4.7 Hz, 1H, C***H***HC(=O)), 2.11–1.99 (m, 2H, CH***H***C(=O) and C***H***HCq_{epo}), 1.95–1.83 (m, 2H, $CH\boldsymbol{H}Cq_{epo}$ and $CH_2C\boldsymbol{H}HCH_2$), 1.70–1.60 (m, 2H, $CH_2CH\boldsymbol{H}CH_2$ and C***H***Me_2), 1.03 (d, $J = 6.8$ Hz, 3H, $CH(C\boldsymbol{H}_3)_2$), 0.97 (d, $J = 7.0$ Hz, 3H, $CH(C\boldsymbol{H}_3)_2$).

^{13}C NMR (100 MHz, CD_2Cl_2) δ 207.0 (***C***=O), 69.3 ($\boldsymbol{C}q_{epo}$), 61.1 ($\boldsymbol{C}H_{epo}$), 36.6 ($\boldsymbol{C}H_2$C(=O)), 34.5 (***C***HMe_2), 23.3 ($\boldsymbol{C}H_2Cq_{epo}$), 18.1 ($CH_2CH_2CH_2$), 18.0 ($CH(\boldsymbol{C}H_3)_2$), 17.9 ($CH(\boldsymbol{C}H_3)_2$).

MS (EI-DE) *m/z* (%) 154 [M^+] (61), 139 (4), 125 (24), 111 (78), 99 (34), 81 (41), 71 (6), 69 (30), 55 (93), 53 (18), 43 (84), 41 (100), 29 (32), 27 (56).

HRMS (EI-FE) calcd for $C_9H_{14}O_2$ [M^+] 154.0993, found 154.0994.

(2*S*,3*R*)-3-*tert*-Butyl-2,3-epoxycyclohexanone ((2*S*,3*R*)-**48j**)

Conditions B:

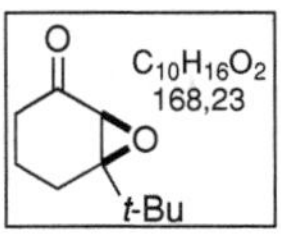

The title compound was isolated after 48 h at 50 °C and purification by flash column chromatography (silica gel, 5–10% Et_2O in pentane) as a clear oil (57 mg,

339 μmol, 68%; 99.5:0.5 *er*). The enantiomeric ratio was determined by GC using a chiral Hydrodex-β-TBDAc column 25 m (100 °C, 1.2 °C/min until 125 °C, 18 °C/min until 220 °C, 10 min at 320 °C, 0.5 bar H_2); major enantiomer: τ_R = 13.22 min, minor enantiomer: τ_R = 14.25 min.

^{1}H NMR (500 MHz, CD_2Cl_2) δ 3.21 (s, 1H, C_{epo}***H***), 2.46 (dt, *J* = 17.7, 4.7 Hz, 1H, C***H***HC(=O)), 2.19–2.13 (m, 1H, C***H***$_2$), 2.07–1.99 (m, 1H, C***H***$_2$), 1.95–1.86 (m, 2H, C***H***$_2$), 1.67–1.60 (m, 1H, C***H***$_2$), 0.97 (s, 9H, C***H***$_3$).

^{13}C NMR (125 MHz, CD_2Cl_2) δ 207.3 (***C***=O), 70.5 (***C***q_{epo}), 59.2 (***C***$_{epo}$H), 36.1 (***C***H$_2$C(=O)), 34.0 (***C***qMe$_3$), 25.4 (3C, ***C***H$_3$), 23.2 (Cq$_{epo}$***C***H$_2$), 17.8 (CH$_2$***C***H$_2$CH$_2$).

MS (EI-DE) *m*/*z* (%) 168 [M$^+$] (22), 153 (5), 139 (14), 125 (39), 113 (15), 96 (87), 83 (24), 69 (61), 55 (100), 41 (57), 29 (26).

HRMS (EI-FE) calcd for $C_{10}H_{16}O_2$ [M$^+$] 168.1150, found 168.1150.

(2*S*,3*R*)-2,3-Epoxy-3-allylcyclohexanone ((2*S*,3*R*)-**48k**)

Conditions B:

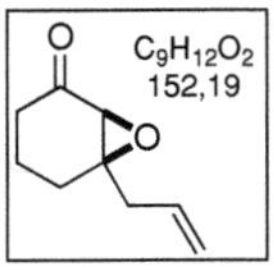

The title compound was isolated after 48 h at 50 °C and purification by flash column chromatography (silica gel, 2–15% Et_2O in pentane) as a clear oil (34.8 mg, 229 μmol, 23%; 97.5:2.5 *er*). The enantiomeric ratio was determined by GC using a chiral Hydrodex-β-TBDAc column 25 m (21 min at 125 °C, 14 °C/min until 230 °C, 10 min at 230 °C, 0.6 bar H_2); major enantiomer: τ_R = 24.39 min, minor enantiomer: τ_R = 25.96 min.

^{1}H NMR (500 MHz, CD_2Cl_2) δ 5.77 (ddt, *J* = 17.0, 10.1, 7.1 Hz, 1H, C***H***=CH$_2$), 5.17–5.13 (m, 2H, CH=C***H***$_2$), 3.06 (s, 1H, C$_{epo}$***H***), 2.48–2.38 (m, 3H, C***H***$_2$CH=CH$_2$ and C***H***HC(=O)), 2.13–2.09 (m, 1H, C***H***$_{2,\ cycl.}$), 2.05–1.98 (m, 1H, C***H***$_{2,\ cycl.}$), 1.93–1.85 (m, 2H, C***H***$_{2,\ cycl.}$), 1.67–1.61 (m, 1H, C***H***$_{2,\ cycl.}$).

^{13}C NMR (125 MHz, CD_2Cl_2) δ 206.5 (***C***=O), 132.3 (***C***H=CH$_2$), 119.1 (CH=***C***H$_2$), 64.7 (***C***q$_{epo}$), 60.6 (***C***H$_{epo}$), 40.6 (***C***H$_2$CH=CH$_2$), 36.3 (***C***H$_2$C(=O)), 26.7 (***C***H$_2$Cq$_{epo}$), 17.7 (CH$_2$***C***H$_2$CH$_2$).

MS (EI-DE) *m*/*z* (%) 152 [M$^+$] (8), 137 (4), 123 (60), 109 (13), 97 (41), 91 (11), 79 (53), 69 (29), 67 (61), 55 (85), 53 (23), 41 (100), 39 (55), 29 (18), 27 (31).

HRMS (EI-FE) calcd for $C_9H_{12}O_2$ [M$^+$] 152.0836, found 152.0837.

(*E*)-3-(1-propenyl)cyclohex-2-enone (76)

The title compound (70.2 mg, 516 μmol, 52%) was obtained as a side product in the epoxidation reaction of 3-allylcyclohex-2-enone (**46k**) under the **conditions B**.

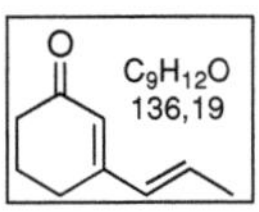

^{1}H NMR (500 MHz, CD_2Cl_2) δ 6.30–6.22 (m, 2H, Cq=C***H*** and CH=C***H***CH_3), 5.81–5.80 (m, 1H, C***H***=CHCH_3), 2.45 (t, J = 5.9 Hz, C$\boldsymbol{H}_2$C(=O)), 2.34 (t, J = 6.6 Hz, C$\boldsymbol{H}_2$Cq=CH), 2.00 (quint, J = 6.4 Hz, CH_2C$\boldsymbol{H}_2$$CH_2$), 1.87 (d, J = 5.6 Hz, C$\boldsymbol{H}_3$).

^{13}C NMR (125 MHz, CD_2Cl_2) δ 202.2 (***C***=O), 157.9 (***C***q=CH), 134.2 (***C***H=CH), 133.0 (CH=***C***H), 126.4 (Cq=***C***H), 38.2 (***C***H_2C(=O)), 25.4 (***C***H_2Cq=CH), 23.0 (CH_2***C***$H_2$$CH_2$), 19.2 (***C***$H_3$).

(2*S*,3*R*)-2,3-Epoxy-3-phenethylcyclohexanone ((2*S*,3*R*)-**48l**)

Conditions B:

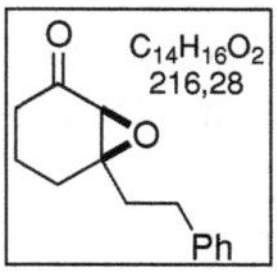

The title compound was isolated after 48 h at 50 °C and purification by flash column chromatography (silica gel, 1–10% Et_2O in pentane) as a clear oil (182 mg, 843 µmol, 84%; 98.5:1.5 *er*). The enantiomeric ratio was determined by GC using a chiral BGB-176/BGB-15 column 30 m (100 °C, 1.2 °C/min until 180 °C, 20 °C/min until 220 °C, 10 min at 220 °C, 0.5 bar H_2); major enantiomer: τ_R = 56.99 min, minor enantiomer: τ_R = 58.46 min.

^{1}H NMR (400 MHz, CD_2Cl_2) δ 7.31–7.27 (m, 2H, C$\boldsymbol{H}_{ar}$), 7.23–7.18 (m, 3H, C$\boldsymbol{H}_{ar}$), 3.01 (s, 1H, C_{epo}***H***), 2.74 (t, J = 8.3 Hz, 2H, C$\boldsymbol{H}_2$Ph), 2.46 (dt, J = 17.0, 4.7 Hz, 1H, C(=O)C***H***H), 2.20–2.12 (m, 1H, C***H***HCq_{epo}), 2.08–1.87 (m, 5H, C(=O)CH***H***, CH***H***Cq_{epo}, CH_2C***H***HCH_2, and C$\boldsymbol{H}_2$$CH_2$Ph), 1.71–1.62 (m, 1H, CH_2CH***H***CH_2).

^{13}C NMR (100 MHz, CD_2Cl_2) δ 206.5 (***C***=O), 141.5 (***C***q_{ar}), 128.8 (2C, ***C***H_{ar}), 128.6 (2C, ***C***H_{ar}), 126.5 (***C***$H_{ar,\ p}$), 65.1 (***C***q_{epo}), 61.4 (***C***H_{epo}), 38.2 (***C***$H_2$$CH_2$Ph), 36.3 (C(=O)***C***$H_2$), 31.2 (***C***$H_2$Ph), 26.9 (***C***$H_2$$Cq_{epo}$), 17.8 ($CH_2$***C***$H_2$$CH_2$).

MS (EI-DE) *m*/*z* (%) 216 [M^+] (2), 198 (2), 187 (1), 169 (3), 154 (1), 143 (11), 128 (4), 104 (100), 97 (13), 91 (69), 79 (13), 69 (5), 65 (17), 55 (12), 41 (31).

HRMS (EI-DE) calcd for $C_{14}H_{16}O_2$ [M^+] 216.1149, found 216.1150.

3-Benzyl-2,3-epoxycyclohexanone (48m)

Conditions B:

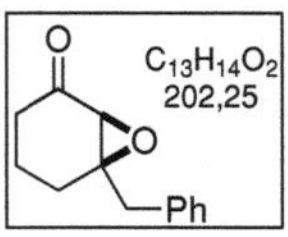

The title compound (2*S*,3*R*)-**48m** was isolated after 48 h at 50 °C and purification by flash column chromatography (silica gel, 2–10% Et_2O in pentane) as a clear oil (158 mg, 781 µmol, 78%; 99:1 *er*). The enantiomeric ratio was determined by GC using a chiral Hydrodex-β-TBDAc column 25 m (100 °C, 1.2 °C/min until

175 °C, 20 °C/min until 220 °C, 10 min at 220 °C, 0.4 bar H_2); major enantiomer: τ_R = 45.74 min, minor enantiomer: τ_R = 46.65 min.

Conditions B:

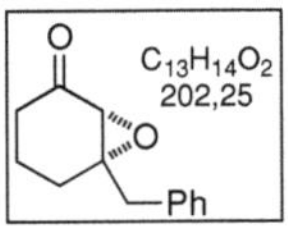

9-amino(9-deoxy)*epi*quinidine (**67**) was used (instead of **13**). The title compound (2*R*,3*S*)-**48a** was isolated after 48 h at 50 °C and purification by flash column chromatography (silica gel, 2–10% Et_2O in pentane) as a clear oil (156 mg, 771 μmol, 77%; 98.5:1.5 *er*). The enantiomeric ratio was determined by GC using a chiral Hydrodex-*β*-TBDAc column 25 m (100 °C, 1.2 °C/min until 175 °C, 20 °C/min until 220 °C, 10 min at 220 °C, 0.4 bar H_2); major enantiomer: τ_R = 46.37 min, minor enantiomer: τ_R = 45.88 min.

^{1}H NMR (400 MHz, CD_2Cl_2) δ 7.34–7.21 (m, 5H, $C_6\boldsymbol{H}_5$), 3.04 (s, 1H, $C_{epo}\boldsymbol{H}$), 2.98 (s, 2H, $C\boldsymbol{H}_2$Ph), 2.44 (dt, *J* = 17.7, 4.4 Hz, 1H, C(=O)C***H***H), 2.10–1.95 (m, 2H, C(=O)CH***H*** and C***H***HCq_{epo}), 1.89–1.80 (m, 2H, CH***H***Cq_{epo} and CH_2C***H***HCH), 1.65–1.56 (m, 1H, CH_2CH***H***CH_2).

^{13}C NMR (100 MHz, CD_2Cl_2) δ 206.4 (***C***=O), 136.2 (***C***q_{ar}), 130.0 (2C, ***C***H_{ar}), 128.8 (2C, ***C***H_{ar}), 127.3 (***C***$H_{ar,\ p}$), 65.4 (***C***q_{epo}), 60.8 (***C***H_{epo}), 42.4 (***C***H_2Ph), 36.4 (C(=O)***C***H_2), 26.7 (***C***H_2Cq_{epo}), 17.6 (CH_2***C***H_2CH_2).

MS (EI-DE) *m*/*z* (%) 202 [M^+] (60), 184 (5), 173 (53), 156 (4), 145 (12), 129 (46), 117 (21), 91 (100), 78 (11), 65 (28), 55 (28), 51 (15), 39 (36).

HRMS (EI-DE) calcd for $C_{13}H_{14}O_2$ [M^+] 202.0991, found 202.0994.

(2*R*,3*S*)-2,6,6-Trimethyl-2,3-epoxy-1,4-cyclohexanedione ((2*R*,3*S*)-**48q**)

Conditions B:

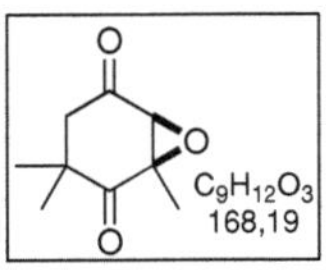

20 mol% catalytic salt [**13** · 2 TFA] was used. The title compound was isolated after 48 h at 50 °C and purification by flash column chromatography (silica gel, 2.5% Et_2O in pentane) as a clear oil (82 mg, 488 μmol, 49%; 96:4 *er*). The enantiomeric ratio was determined by GC using a chiral BGB-178/OV-1701 column 30 m (80 °C, 1.2 °C/min until 130 °C, 20 °C/min until 220 °C, 10 min at 220 °C, 0.5 bar H_2); major enantiomer: τ_R = 19.27 min, minor enantiomer: τ_R = 22.67 min.

^{1}H NMR (300 MHz, CD_2Cl_2) δ 3.49 (d, *J* = 1.1 Hz, 1H, $C_{epo}\boldsymbol{H}$), 3.15 (d, *J* = 13.6 Hz, 1H, C***H***H), 2.15 (dd, *J* = 13.5, 1.2 Hz, 1H, CH***H***), 1.55 (s, 3H, $C\boldsymbol{H}_3$), 1.28 (s, 3H, $C\boldsymbol{H}_3$), 1.08 (s, 3H, $C\boldsymbol{H}_3$).

^{13}C NMR (75 MHz, $CDCl_3$) δ 205.6 (***C***=O), 204.1 (***C***=O), 64.8 (***C***q_{epo}), 62.8 (***C***H_{epo}), 47.0 (***C***H_2), 45.5 (***C***qMe_2), 26.9 (***C***H_3), 26.1 (***C***H_3), 15.9 (Cq_{epo}***C***H_3).

MS (EI) *m/z* (%) 168 [M^+] (59), 153 (27), 125 (65), 85 (77), 83 (21), 69 (34), 56 (100), 43 (91), 41 (95), 39 (41), 27 (34).

HRMS calcd for $C_9H_{12}O_3$ [M^+] 168.0785, found 168.0786.

(2*R*,3*R*)-2,3-Epoxycyclopentanone ((2*R*,3*R*)-**48r**)

Conditions A:

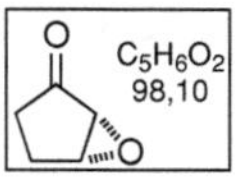

After 48 h at 50 °C, the conversion was determined to be 33% by GC and the enantiomeric ratio was determined to be 90:10 *er* by chiral GC using a Lipodex E column 25 m (60 °C, 1.2 °C/min until 90 °C, 18 °C/min until 220 °C, 10 min at 220 °C, 0.5 bar H_2); major enantiomer: τ_R = 14.49 min, minor enantiomer: τ_R = 13.33 min.

^{1}H NMR (300 MHz, CD_2Cl_2) δ 3.89–3.88 (m, 1H, $CH_2C\boldsymbol{H}_{epo}$), 3.26 (d, J = 2.6 Hz, 1H, $C\boldsymbol{H}_{epo}C(=O)$), 2.37–2.17 (m, 2H, $C\boldsymbol{H}H$), 2.09–1.97 (m, 2H, $CH\boldsymbol{H}$).

^{13}C NMR (75 MHz, CD_2Cl_2) δ 210.3 ($\boldsymbol{C}$=O), 58.3 ($\boldsymbol{C}H_{epo}C(=O)$), 55.2 ($CH_2\boldsymbol{C}H_{epo}$), 30.9 ($\boldsymbol{C}H_2C(=O)$), 23.5 ($\boldsymbol{C}H_2CH_{epo}$).

GC-MS (GC-EI) *m/z* (%) 98 [M^+] (41), 82 (13), 69 (25), 55 (22), 53 (6), 42 (100), 39 (37), 31 (1), 27 (18).

HRMS (EI-FE) calcd for $C_5H_6O_2$ [M^+] 98.0367, found 98.0368.

2,3-Epoxycycloheptanone (48s)

Conditions A:

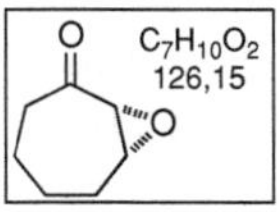

The title compound (2*R*,3*R*)-**48s** was isolated after 20 h at 30 °C and purification by flash column chromatography (silica gel, 3–15% Et_2O in pentane) as a clear liquid (41 mg, 325 μmol, 65%; 99:1 *er*). The enantiomeric ratio was determined by GC using a chiral Hydrodex-β-TBDAc column 25 m (80 °C, 1.5 °C/min until 120 °C, 20 °C/min until 220 °C, 10 min at 220 °C, 0.6 bar H_2); major enantiomer: τ_R = 23.46 min, minor enantiomer: τ_R = 20.62 min.

Conditions B:

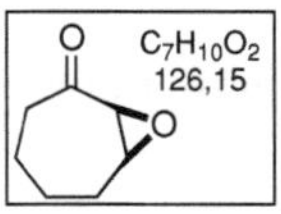

The title compound (2*S*,3*S*)-**48s** was isolated after 24 h at 50 °C and purification by flash column chromatography (silica gel, 3–15% Et_2O in pentane) as a clear liquid (78 mg, 621 μmol, 62%; >99.5:0.5 *er*). The enantiomeric ratio was determined by GC using a chiral Hydrodex-β-TBDAc column 25 m (80 °C, 1.5 °C/min until 120 °C, 20 °C/min until 220 °C, 10 min at 220 °C, 0.6 bar H_2); major enantiomer: τ_R = 20.66 min, minor enantiomer: τ_R = 24.29 min.

^{1}H NMR (300 MHz, CD_2Cl_2) δ 3.37–3.35 (m, 2H, C***H***$_{epo}$), 2.62 (ddd, J = 13.6, 11.3, 3.8 Hz, 1H, C(=O)C***H***H), 2.48–2.40 (m, 1H, C***H***HCH$_{epo}$), 2.30–2.23 (m, 1H, C(=O)CH***H***), 1.87–1.63 (m, 4H, C(=O)CH$_2$C***H***H, CH***H***CH$_{epo}$, and C***H***$_2$CH$_2$CH$_{epo}$), 1.06–0.91 (m, 1H, C(=O)CH$_2$CH***H***).

^{13}C NMR (75 MHz, CD_2Cl_2) δ 210.1 (***C***=O), 59.5 (C(=O)***C***H$_{epo}$), 55.1 (CH$_2$***C***H$_{epo}$), 40.6 (C(=O)***C***H$_2$), 27.5 (***C***H$_2$CH$_{epo}$), 23.6 (C(=O)CH$_2$***C***H$_2$), 23.1 (***C***H$_2$CH$_2$CH$_{epo}$).

GC-MS (GC-EI) *m/z* (%) 126 [M$^+$] (16), 110 (12), 97 (33), 83 (24), 81 (35), 79 (14), 70 (58), 68 (32), 55 (79), 41 (100), 39 (52), 27 (37).

HRMS (EI-FE) calcd for $C_7H_{10}O_2$ [M$^+$] 126.0679, found 126.0681.

(2*S*,3*S*)-2,3-Epoxy-3-ethylcycloheptanone ((2*S*,3*S*)-**48t**)

Conditions B:

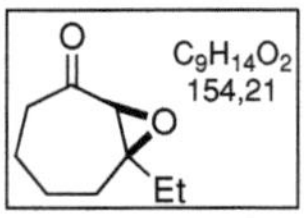

The title compound was isolated after 20 h at 50 °C and purification by flash column chromatography (silica gel, 5% Et_2O in pentane) as a clear oil (127 mg, 824 μmol, 82%; >99.5:0.5 *er*). The enantiomeric ratio was determined by GC using a chiral BGB-176/BGB-15 column 30 m (80 °C, 1.2 °C/min until 220 °C, 10 min at 220 °C, 0.5 bar H_2); major enantiomer: τ_R = 18.25 min, minor enantiomer: τ_R = 19.50 min.

^{1}H NMR (400 MHz, CD_2Cl_2) δ 3.20 (d, J = 1.3 Hz, 1H, C$_{epo}$***H***), 2.67 (ddd, J = 13.5, 11.0, 4.1 Hz, 1H, C(=O)C***H***H), 2.24–2.18 (m, 1H, C(=O)CH***H***), 2.18–2.12 (m, 1H, C***H***HCq$_{epo}$), 1.87–1.66 (m, 5H, CH***H***Cq$_{epo}$, C(=O)CH$_2$C***H***H, C***H***$_2$CH$_2$Cq$_{epo}$, and C***H***HCH$_3$), 1.59–1.49 (m, 1H, CH***H***CH$_3$), 1.13–1.02 (m, 1H, C(=O)CH$_2$CH***H***), 0.93 (t, J = 7.5 Hz, 3H, C***H***$_3$).

^{13}C NMR (75 MHz, CD_2Cl_2) δ 210.7 (***C***=O), 65.0 (***C***H$_{epo}$), 64.9 (***C***q$_{epo}$), 41.1 (C(=O)***C***H$_2$), 31.9 (***C***H$_2$CH$_3$), 31.4 (***C***H$_2$Cq$_{epo}$), 25.3 (C(=O)CH$_2$***C***H$_2$), 23.9 (***C***H$_2$CH$_2$Cq$_{epo}$), 9.1 (***C***H$_3$).

GC-MS (GC-EI) *m/z* (%) 154 [M$^+$] (1), 138 (1), 125 (11), 109 (12), 98 (68), 93 (8), 83 (29), 67 (35), 55 (100), 53 (16), 41 (65), 29 (57), 27 (23).

HRMS (EI-FE) calcd for $C_9H_{14}O_2$ [M$^+$] 154.0995, found 154.0994.

(2*S*,3*R*)-3-Benzyl-2,3-epoxycycloheptanone ((2*S*,3*R*)-**48u**)

Conditions B:

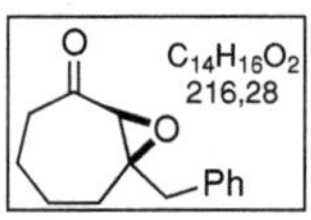

The title compound was isolated after 20 h at 50 °C and purification by flash column chromatography (silica gel, 5–10% Et_2O in pentane) as a clear oil (184 mg, 851 μmol, 85%; >99.5:0.5 *er*). The enantiomeric ratio was determined by HPLC using a chiral Chiralpak IA column (2% *i*-PrOH in heptane, 0.5 mL/min); major enantiomer: $\tau_R = 13.74$ min, minor enantiomer: $\tau_R = 14.83$ min.

^{1}H NMR (400 MHz, CD_2Cl_2) δ 7.33–7.22 (m, 5H, $C_6\boldsymbol{H}_5$), 3.20 (d, *J* = 1.2 Hz, 1H, $C_{epo}\boldsymbol{H}$), 2.94 (dd, *J* = 18.8, 14.3 Hz, 2H, $C\boldsymbol{H}_2$Ph), 2.67 (ddd, *J* = 13.4, 11.0, 4.0 Hz, 1H, C(=O)C***H***H), 2.25–2.16 (m, 2H, C(=O)CH***H*** and C***H***HCq_{epo}), 1.86–1.63 (m, 4H, CH***H***Cq_{epo}, $C\boldsymbol{H}_2CH_2Cq_{epo}$, and C(=O)$CH_2$C***H***H), 1.11–1.00 (m, 1H, C(=O)CH_2CH***H***).

^{13}C NMR (75 MHz, CD_2Cl_2) δ 210.3 (***C***=O), 136.5 (***C***q_{ar}), 130.1 (2C, ***C***H_{ar}), 128.8 (2C, ***C***H_{ar}), 127.2 (***C***$H_{ar,\ p}$), 64.6 (***C***H_{epo}), 64.1 (***C***q_{epo}), 44.6 (***C***H_2Ph), 41.1 (C(=O)***C***H_2), 31.7 (***C***H_2Cq_{epo}), 25.2 (C(=O)CH_2***C***H_2), 23.9 (***C***$H_2CH_2Cq_{epo}$).

MS (EI-DE) *m/z* (%) 216 [M^+] (22), 198 (3), 187 (19), 169 (6), 159 (3), 143 (7), 129 (12), 118 (30), 104 (4), 97 (20), 91 (100), 78 (8), 65 (18), 55 (11), 41 (23).

HRMS (EI-DE) calcd for $C_{14}H_{16}O_2$ [M^+] 216.1152, found 216.1150.

(2*S*,3*S*)-2,3-Epoxycyclooctanone ((2*S*,3*S*)-**48v**)

Conditions B:

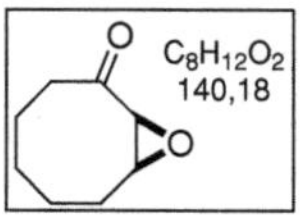

After 24 h at 50 °C and base treatment of the crude product in THF, purification by flash column chromatography (silica gel, 10–40% Et_2O in pentane) provided the title compound as a colorless solid (77 mg, 550 μmol, 55%; 98:2 *er*). The enantiomeric ratio was determined by GC using a chiral BGB-176/SE-52 30 m (80 °C, 2 °C/min until 135 °C, 18 °C/min until 220 °C, 10 min at 220 °C, 0.6 bar H_2); major enantiomer: $\tau_R = 23.43$ min, minor enantiomer: $\tau_R = 21.95$ min.

^{1}H NMR (500 MHz, CD_2Cl_2) δ 3.70 (d, *J* = 5.4 Hz, 1H, C***H***$_{epo}$C(=O)), 3.22 (ddd, *J* = 9.8, 5.3, 3.7 Hz, 1H, CH_2C***H***$_{epo}$), 2.66 (ddd, *J* = 13.2, 7.7, 4.2 Hz, 1H, C***H***HC(=O)), 2.30 (ddd, *J* = 13.6, 10.1, 3.6 Hz, 1H, CH***H***C(=O)), 2.19–2.14 (m, 1H, C***H***HCH_{epo}), 1.94–1.87 (m, 1H, C***H***HCH_2C(=O)), 1.77–1.70 (m, 1H, C***H***H$(CH_2)_2CH_{epo}$), 1.69–1.62 (m, 1H, CH***H***CH_2C(=O)), 1.60–1.51 (m, 2H, $C\boldsymbol{H}_2CH_2CH_{epo}$), 1.49–1.40 (m, 1H, CH***H***$(CH_2)_2CH_{epo}$), 0.99–0.91 (m, 1H, CH***H***CH_{epo}).

^{13}C NMR (125 MHz, CD_2Cl_2) δ 207.3 (***C***=O), 58.8 (***C***H_{epo}C(=O)), 55.7 (CH_2***C***H_{epo}), 43.0 (***C***H_2C(=O)), 27.2 (2C, ***C***H_2CH_{epo} and ***C***$H_2(CH_2)_2CH_{epo}$), 24.7 (***C***H_2CH_2C(=O)), 24.6 (***C***$H_2CH_2CH_{epo}$).

MS (EI-DE) *m/z* (%) 140 [M^+] (18), 111 (5), 97 (16), 83 (37), 79 (17), 70 (30), 57 (27), 55 (100), 53 (11), 41 (84), 39 (43), 27 (45).

HRMS (EI-DE) calcd for $C_8H_{12}O_2$ [M^+] 140.0836, found 140.0837.

(*E*)-(2*S*,3*R*)-2,3-Epoxycyclododecanone ((2*S*,3*R*)-**48w**)

Conditions B:

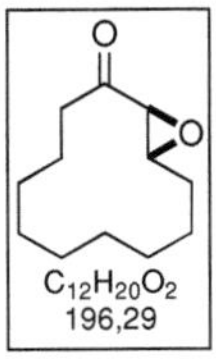

The title compound was isolated after 20 h at 50 °C and purification by flash column chromatography (silica gel, 10–20% Et_2O in pentane) as a white solid (45 mg, 229 μmol, 92%; 99.5:0.5 *er*). The enantiomeric ratio was determined by GC using a chiral Hydrodex-*β*-TBDAc column 25 m (100 °C, 1.2 °C/min until 170 °C, 20 °C/min until 220 °C, 10 min at 320 °C, 0.6 bar H_2); major enantiomer: $\tau_R = 54.36$ min, minor enantiomer: $\tau_R = 53.35$ min.

^{1}H NMR (500 MHz, CD_2Cl_2) δ 3.51 (d, $J = 1.6$ Hz, 1H, C_{epo}***H***C(=O)), 2.93 (td, $J = 9.5$, 2.2 Hz, 1H, CH_2C_{epo}***H***), 2.91–2.87 (m, 1H, C***H***HC(=O)), 2.32–2.67 (m, 1H, CH***H***C(=O)), 2.23–2.17 (m, 1H, C***H***HC_{epo}H), 1.84–1.69 (m, 3H, –(C$\boldsymbol{H}_2$)$_n$–), 1.54–1.35 (m, 9H, –(C$\boldsymbol{H}_2$)$_n$–), 1.29–1.24 (m, 1H, –(C$\boldsymbol{H}_2$)$_n$–), 1.20–1.12 (m, 2H, –(C$\boldsymbol{H}_2$)$_n$– and CH***H***C_{epo}H).

^{13}C NMR (125 MHz, CD_2Cl_2) δ 206.9 (***C***=O), 61.2 ($CH_2\boldsymbol{C}_{epo}$H), 59.2 ($\boldsymbol{C}_{epo}$HC(=O)), 41.4 (***C***H_2C(=O)), 32.2 (***C***H_2C_{epo}H), 26.9 (***C***H_2), 26.5 (***C***H_2), 25.7 (***C***H_2), 25.5 (***C***H_2), 24.4 (***C***H_2), 24.0 (***C***H_2), 23.2 (***C***H_2).

MS (EI) *m*/*z* (%) 196 [M^+] (8), 178 (1), 168 (1), 149 (2), 139 (4), 135 (6), 121 (14), 111 (29), 107 (8), 98 (54), 84 (31), 67 (36), 55 (100), 41 (92), 29 (39).

HRMS (EI-DE) calcd for $C_{12}H_{20}O_2$ [M^+] 196.1465, found 196.1463.

(*E*)-2,3-Epoxycyclopentadecanone (**48x**)

Conditions B:

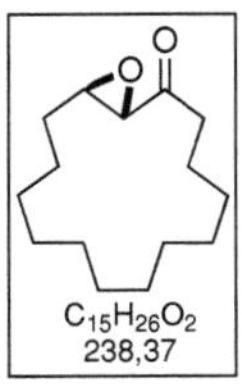

The title compound (2*S*,3*R*)-**48x** was isolated after 20 h at 50 °C and purification by flash column chromatography (silica gel, 5–15% Et_2O in pentane) as a white solid (52 mg, 218 μmol, 87%; 99.5:0.5 *er*). The enantiomeric ratio was determined by GC using a chiral Hydrodex-*β*-TBDAc column 25 m (100 °C, 1.2 °C/min until 185 °C, 18 °C/min until 220 °C, 10 min at 320 °C, 0.6 bar H_2); major enantiomer: $\tau_R = 65.00$ min, minor enantiomer: $\tau_R = 63.68$ min.

Conditions B:

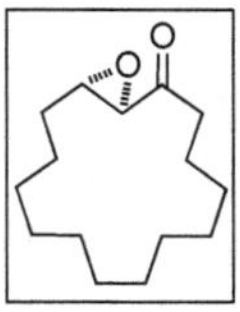

9-amino(9-deoxy)*epi*quinidine (**67**) was used (instead of **13**). The title compound (2*R*,3*S*)-**48x** was isolated after 20 h at 50 °C and purification by flash column chromatography (silica gel, 5–15% Et_2O in pentane) as a white solid (51 mg, 215 μmol, 86%; 99.5:0.5 *er*). The enantiomeric ratio was determined by GC using a chiral Hydrodex-*β*-TBDAc column 25 m (100 °C, 1.2 °C/min until 190 °C, 18 °C/min until 230 °C, 5 min at 320 °C, 0.6 bar H_2); major enantiomer: τ_R = 64.72 min, minor enantiomer: τ_R = 65.60 min.

^{1}H NMR (500 MHz, CD_2Cl_2) δ 3.27 (d, J = 1.9 Hz, 1H, C_{epo}***H***C(=O)), 3.01 (td, J = 8.6, 2.7 Hz, 1H, CH_2C_{epo}***H***), 2.47–2.39 (m, 2H, C***H***$_2$C(=O)), 2.06–2.00 (m, 1H, C***H***HC_{epo}H), 1.73–1.63 (m, 2H, –(C***H***$_2$)$_n$–), 1.53–1.50 (m, 2H, –(C***H***$_2$)$_n$–), 1.39–1.21 (m, 17H, –(C***H***$_2$)$_n$– and C_{epo}HCH***H***).

^{13}C NMR (125 MHz, CD_2Cl_2) δ 207.0 (***C***=O), 60.2 (***C***$_{epo}$HC(=O)), 59.2 (CH_2***C***$_{epo}$H), 37.6 (***C***H_2C(=O)), 31.0 (CH_2***C***$_{epo}$H), 27.5 (***C***H_2), 27.1 (***C***H_2), 26.9 (***C***H_2), 26.6 (***C***H_2), 26.4 (***C***H_2), 26.4 (***C***H_2), 26.2 (***C***H_2), 25.7 (***C***H_2), 24.9 (***C***H_2), 22.6 (***C***H_2).

MS (EI) *m/z* (%) 238 [M^+] (19), 220 (1), 209 (1), 195 (3), 177 (3), 163 (3), 149 (3), 121 (12), 111 (29), 98 (48), 81 (38), 67 (39), 55 (100), 41 (86), 29 (34).

HRMS (EI-DE) calcd for $C_{15}H_{26}O_2$ [M^+] 238.1931, found 238.1933.

(2*S*,3*S*)-2,3-Epoxy-4-methylcyclohexanone ((2*S*,3*S*)-**81a**)

Conditions B:

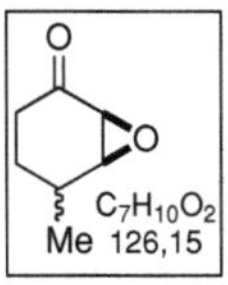

After 24 h at 30 °C, the conversion was determined to be 94% [54:46 *dr* (*trans*/*cis*)] by GC/MS. The enantiomeric ratio was determined by GC using a chiral BGB-178/OV-1701 column 30 m (80 °C, 0.8 °C/min until 105 °C, 20 °C/min until 220 °C, 10 min at 220 °C, 0.5 bar H_2). *trans*-**81a**: 93.5:6.5 *er* (major enantiomer: τ_R = 27.25 min, minor enantiomer: τ_R = 29.13 min); *cis*-**81a**: 99:1 *er* (minor enantiomer: τ_R = 28.64 min, minor enantiomer: τ_R = 29.65 min).

trans-**81a**: **^{1}H NMR** (500 MHz, CD_2Cl_2) δ 3.37 (dd, J = 3.7, 2.2 Hz, 1H, CHC***H***$_{epo}$), 3.15 (d, J = 4.0 Hz, 1H, C***H***$_{epo}$C(=O)), 2.46–2.40 (m, 2H, C***H***Me and C***H***HC(=O)), 2.18–2.04 (m, 2H, CH***H***C(=O) and C***H***HCH), 1.48–1.41 (m, 1H, CH***H***CH), 1.08 (d, J = 7.2 Hz, 3H, C***H***$_3$).

^{13}C NMR (125 MHz, CD_2Cl_2) δ 206.5 (***C***=O), 66.0 (CH***C***H$_{epo}$), 61.7 (***C***H$_{epo}$ C(=O)), 33.3 (***C***H_2C(=O)), 27.9 (***C***HMe), 25.6 (***C***H_2CH), 15.9 (***C***H_3).

GC-MS (GC-EI) *m/z* (%) 126 [M^+].

HRMS (EI-FE) calcd for $C_7H_{10}O_2$ [M^+] 126.0680, found 126.0681.

cis-**81a**: ^{1}**H NMR** (500 MHz, CD_2Cl_2) δ 3.38 (d, J = 3.9 Hz, 1H, CHC$\boldsymbol{H}_{epo}$), 3.16 (d, J = 3.8 Hz, 1H, C$\boldsymbol{H}_{epo}$C(=O)), 2.42 (ddd, J = 18.6, 5.0, 3.5 Hz, 1H, C***H***HC(=O)), 2.19–2.12 (m, 1H, C***H***Me), 2.07 (ddd, J = 18.6, 11.8, 6.9 Hz, 1H, CH***H***C(=O)), 1.67–1.57 (m, 2H, C$\boldsymbol{H}_2$CH), 1.22 (d, J = 6.9 Hz, 3H, C$\boldsymbol{H}_3$).

^{13}**C NMR** (125 MHz, CD_2Cl_2) δ 205.6 (***C***=O), 60.1 (CH***C***H_{epo}), 55.9 (***C***H_{epo}C(=O)), 36.5 (***C***H_2C(=O)), 29.3 (***C***HMe), 24.1 (***C***H_2CH), 18.8 (***C***H_3).

GC-MS (GC-EI) *m/z* (%) 126 [M^+].

HRMS (EI-FE) calcd for $C_7H_{10}O_2$ [M^+] 126.0680, found 126.0681.

(2*S*,3*S*)-2,3-Epoxy-4-*tert*-butylcyclohexanone ((2*S*,3*S*)-**81b**)

Conditions B:

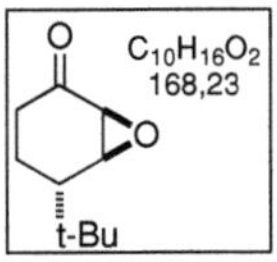

After 96 h at 50 °C, the conversion was determined to be 56% by GC/MS with a yield of epoxide **81b** of 48% [92:8 *dr* (*trans*/*cis*)]. The enantiomeric ratio was determined by GC using a chiral Hydrodex-β-TBDAc column 25 m (100 °C, 1.2 °C/min until 220 °C, 10 min at 220 °C, 0.5 bar H_2). *trans*-**81b**[1]**:** 79.5:20.5 *er* (major enantiomer: τ_R = 26.00 min, minor enantiomer: τ_R = 29.01 min) [5–7].

trans-**81b**: ^{1}**H NMR** (400 MHz, $CDCl_3$) δ 3.49–3.47 (m, 1H, CHC$\boldsymbol{H}_{epo}$), 3.15 (d, J = 3.8 Hz, 1H, C$\boldsymbol{H}_{epo}$C(=O)), 2.70–2.62 (m, 1H, C***H***HC(=O)) 2.15–2.09 (m, 1H,CH***H***C(=O)), 2.02–1.94 (m, 2H, C***H****t*-Bu and C***H***HCH), 1.74–1.64 (m, 1H, CH***H***CH), 0.99 (s, 9H, C$\boldsymbol{H}_3$).

^{13}**C NMR** (100 MHz, $CDCl_3$) δ 208.3 (***C***=O), 61.5 (CH***C***H_{epo}), 55.2 (***C***H_{epo} C(=O)), 43.5 (***C***H*t*-Bu), 34.7 (***C***H_2C(=O)), 32.6 (***C***qMe_3), 27.6 (3C, ***C***H_3), 26.1 (***C***H_2CH).

MS (EI-DE) *m/z* (%) 168 [M^+] (8), 153 (1), 139 (4), 125 (2), 112 (21), 107 (4), 97 (8), 83 (24), 70 (32), 67 (11), 57 (100), 41 (43), 39 (16), 29 (19).

HRMS (EI-FE) calcd for $C_{10}H_{16}O_2$ [M^+] 168.1152, found 168.1150.

***trans*-(2*S*,3*S*,5*S*)-2,3-Epoxy-3-methyl-5-phenylcyclohexanone** (*trans*-(2*S*,3*S*,5*S*)-**84**)

Conditions B:

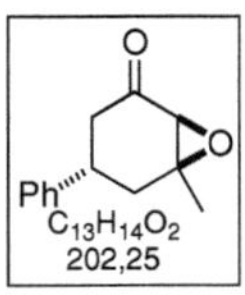

[1] Relative configuration was assigned on the basis of NOE correlations in collaboration with the NMR department of the Max-Planck-Institut für Kohlenforschung.

After 48 h at 50 °C, the conversion was determined to be 83% by GC/MS with a yield of epoxide **84** of 75% [97:3 *dr* (*trans*/*cis*)]. The enantiomeric ratio was determined to be 98.5:1.5 by chiral GC using a Hydrodex-β-TBDAc column 25 m (80 °C, 1.2 °C/min until 180 °C, 20 °C/min until 220 °C, 10 min at 220 °C, 0.5 bar H_2); major enantiomer: $\tau_R = 67.61$ min, minor enantiomer: $\tau_R = 69.91$ min).

^{1}H NMR (500 MHz, CD_2Cl_2) δ 7.33 (t, $J = 7.7$ Hz, 2H, C$\boldsymbol{H}_{Ph,\ m}$), 7.24–7.20 (m, 3H, C$\boldsymbol{H}_{Ph,\ o,\ p}$), 3.34 (tt, $J = 12.3$, 4.5 Hz, 1H, C***H***Ph), 3.14 (s, 1H, C$\boldsymbol{H}_{epo}$), 2.63 (dd, $J = 18.7$, 4.9 Hz, 1H, C***H***HC(=O)), 2.34 (dd, $J = 14.3$, 3.9 Hz, 1H, C***H***HCqMe), 2.23 (dd, $J = 18.6$, 12.7 Hz, 1H, CH***H***C(=O)), 2.11 (dd, $J = 15.0$, 12.4 Hz, 1H, CH***H***CqMe), 1.49 (s, 3H, C$\boldsymbol{H}_3$).

^{13}C NMR (125 MHz, CD_2Cl_2) δ 205.3 (***C***=O), 143.8 (***C***q_{Ph}), 129.0 (2C, ***C***$H_{Ph,\ m}$), 127.3 (2C, ***C***$H_{Ph,\ o}$), 127.1 (***C***$H_{Ph,\ p}$), 61.5 (***C***H_{epo}), 61.4 (***C***q_{epo}), 44.0 (***C***H_2C(=O)), 36.8 (***C***H_2Cq_{epo}), 34.3 (***C***HPh), 21.9 (***C***H_3).

MS (EI-DE) *m*/*z* (%) 202 [M^+] (23), 184 (17), 174 (68), 159 (84), 145 (47), 131 (100), 115 (49), 103 (84), 91 (74), 85 (33), 77 (62), 69 (50), 65 (22), 51 (31), 43 (54), 27 (16).

HRMS (EI-FE) calcd for $C_{13}H_{14}O_2$ [M^+] 202.0993, found 202.0994.

(2*S*,3*S*)-2,3-Epoxy-3-phenethyl-cyclopentanone ((2*S*,3*S*)-**91**)

Conditions A:

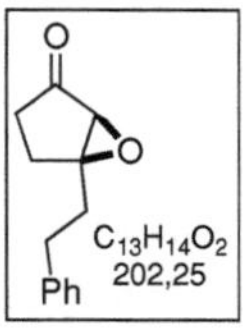

Amine **64** was used [instead of (*R*,*R*)-DPEN (**12**)]. The title compound was isolated after 48 h at 50 °C and purification by flash column chromatography (silica gel, 15–20% Et_2O in pentane) as a clear oil (66 mg, 326 µmol, 65%; 85.5:14.5 *er*). The enantiomeric ratio was determined by GC using a chiral Lipodex E column 25 m (80 °C, 1.2 °C/min until 180 °C, 18 °C/min until 220 °C, 10 min at 320 °C, 0.5 bar H_2); major enantiomer: $\tau_R = 68.62$ min, minor enantiomer: $\tau_R = 67.99$ min.

^{1}H NMR (500 MHz, CD_2Cl_2) δ 7.31–7.27 (m, 2H, C$\boldsymbol{H}_{Ph,\ m}$), 7.21–7.19 (m, 3H, C$\boldsymbol{H}_{Ph,\ o,\ p}$), 3.10 (s, 1H, C_{epo}***H***), 2.81–2.71 (m, 2H, PhC$\boldsymbol{H}_2$$CH_2$), 2.37–2.27 (m, 2H, CH_2C***H***HC(=O) and C***H***HCH_2C(=O)), 2.23–2.12 (m, 2H, PhCH_2C$\boldsymbol{H}_2$), 2.10–2.03 (m, 1H, CH_2CH***H***C(=O)), 2.02–1.97 (m, 1H, CH***H***CH_2C(=O)).

^{13}C NMR (125 MHz, CD_2Cl_2) δ 210.5 (***C***=O), 141.1 (***C***q_{Ph}), 128.5 (2C, ***C***$H_{Ph,\ m}$), 128.3 (2C, ***C***$H_{Ph,\ o}$), 126.2 (***C***$H_{ar,\ p}$), 68.5 (***C***q_{epo}), 60.4 (***C***$_{epo}$H), 33.6 (PhCH_2***C***H_2), 32.5 (***C***H_2C(=O)), 31.1 (Ph***C***H_2), 26.0 (***C***H_2CH_2C(=O)).

MS (EI-DE) *m*/*z* (%) 202 [M^+] (7), 173 (1), 155 (1), 142 (2), 130 (28), 104 (11), 91 (100), 83 (19), 77 (6), 65 (12), 55 (11), 39 (7), 29 (5).

HRMS (EI-FE) calcd for $C_{13}H_{14}O_2$ [M^+] 202.0992, found 202.0994.

7.3 Catalytic Asymmetric Hydroperoxidation and Epoxidation of Acyclic Enones

7.3.1 Catalytic Asymmetric Hydroperoxidation of Acyclic Enones

7.3.1.1 General Procedure

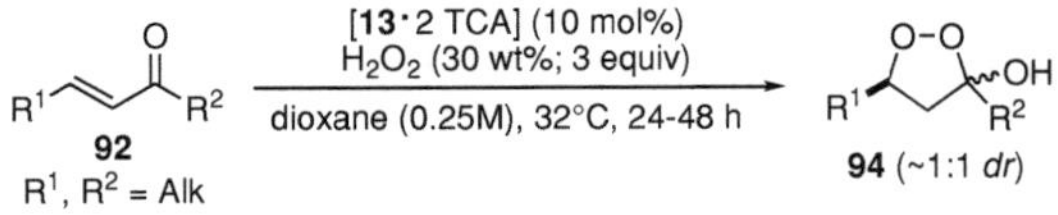

Reference [8] Catalyst salt [**13** · 2 TCA] was prepared *in situ* by the addition of 9-amino(9-deoxy)*epi*quinine (**13**; 32.3 mg, 0.1 mmol, 10 mol%) to a solution of trichloroacetic acid (32.6 mg, 0.2 mmol, 20 mol%) in dioxane (4 mL). Then, enone **92** (1.0 mmol, 1.0 equiv) was added, and 20 min later, aqueous hydrogen peroxide (30 wt%, 304 μL, 3.0 mmol, 3 equiv). After 20–48 h of stirring at 32 °C, the reaction mixture was extracted with Et_2O (3 × 25 mL) and the combined organic phases were washed with brine, dried (Na_2SO_4), filtered, and concentrated *in vacuo*. The crude product was subjected to flash column chromatography (silica gel, eluent: Et_2O-pentane) to afford the corresponding pure peroxyhemiketal **94**. The optical purity was determined after converting the peroxyhemiketal to the corresponding epoxide [with 1 N NaOH (1 equiv) in Et_2O] or to the corresponding aldol-type product [with triethylphosphite (2 equiv) in Et_2O].

7.3.1.2 Scope of Optically Active 3-Hydroxy-1,2-dioxolanes

(5*R*)-5-Hexyl-3-methyl-1,2-dioxolan-3-ol ((5*R*)-**94a**):

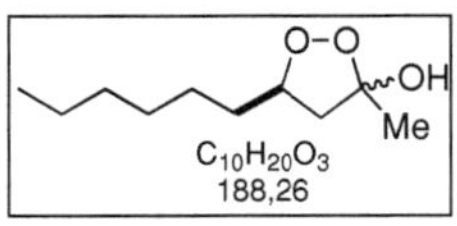

Peroxyhemiketal **94a** was prepared according to the general procedure. The reaction mixture was stirred for 36 h at 32 °C. Purification of the crude product by flash column chromatography (silica gel, 5–10% Et_2O in pentane) provided the title compound **94a** as a colorless oil (123 mg, 653 μmol, 65%; 98.5:1.5 *er*). The enantiomeric ratio was determined after converting peroxyhemiketal **94a** into the corresponding epoxide **93a**. The enantiomers were analyzed by GC using a chiral BGB-176/SE-52 column 29.5 m (80 °C, 1.2 °C/min until 115 °C, 18 °C/min until 220 °C, 10 min at 220 °C, 0.6 bar H_2); major enantiomer:

τ_R = 24.41 min, minor enantiomer: τ_R = 26.10 min. *Characterized as mixture of hemiketal epimers* (1:1 *dr*).

^{1}H NMR (500 MHz, CD_2Cl_2) δ 4.41–4.36 (m, 1H), 4.31–4.26 (m, 1H), 2.96 (br s, 2H), 2.77–2.72 (m, 1H), 2.69–2.65 (m, 1H), 2.26–2.18 (m, 2H), 1.72–1.55 (m, 4H), 1.52 (s, 3H), 1.51 (s, 3H), 1.47–1.35 (m, 2H), 1.34–1.24 (m, 14H), 0.89–0.87 (m, 6H).

^{13}C NMR (125 MHz, CD_2Cl_2) δ 105.9, 105.0, 82.7, 81.3, 53.1, 52.8, 35.2, 32.6, 32.3, 32.2, 29.8, 29.7, 26.7, 26.5, 23.4, 23.1, 23.1, 22.9, 14.4, 14.4.

MS (EI-DE) *m/z* (%) 188 [M^+] (1), 155 (56), 137 (3), 113 (8), 95 (7), 81 (5), 71 (63), 55 (14), 43 (100), 29 (12).

HRMS (EI-FE) calcd for $C_{10}H_{20}O_3$ [M^+] 188.1413, found 188.1412.

(5*R*)-3-Methyl-5-phenethyl-1,2-dioxolan-3-ol ((5*R*)-**94b**):

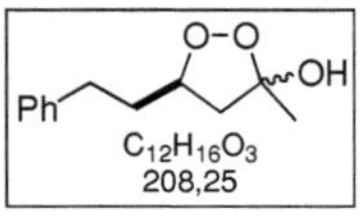

Peroxyhemiketal **94a** was prepared according to the general procedure. The reaction mixture was stirred for 36 h at 32 °C. Purification of the crude product by flash column chromatography (silica gel, 5–40% Et_2O in pentane) provided the title compound **94a** as a colorless oil (141 mg, 677 μmol, 68%; 97:3 *er*). The enantiomeric ratio was determined after converting peroxyhemiketal **94b** into the corresponding epoxide **93b**. The enantiomers were analyzed by GC using a chiral BGB-176/BGB-15 column 30 m (80 °C, 1.2 °C/min until 145 °C, 20 °C/min until 220 °C, 10 min at 220 °C, 0.6 bar H_2); major enantiomer: τ_R = 46.17 min, minor enantiomer: τ_R = 47.42 min. *Characterized as mixture of hemiketal epimers* (1:1 *dr*).

^{1}H NMR (500 MHz, CD_2Cl_2) δ 7.30–7.27 (m, 4H), 7.21–7.18 (m, 6H), 4.43–4.37 (m, 1H), 4.34–4.28 (m, 1H), 2.94–2.93 (m, 2H), 2.80–2.61 (m, 6H), 2.30–2.22 (m, 2H), 2.06–1.97 (m, 2H), 1.95–1.88 (m, 1H), 1.77–1.70 (m, 1H), 1.54 (s, 3H), 1.51 (s, 3H).

^{13}C NMR (125 MHz, CD_2Cl_2) δ 142.0, 141.8, 129.0, 129.0, 128.9, 128.9, 126.6, 126.5, 106.0, 105.0, 81.8, 80.5, 52.9, 52.8, 37.0, 34.5, 32.9, 32.7, 23.4, 22.8.

MS (EI-DE) *m/z* (%) 208 [M^+] (trace), 190 (3), 174 (10), 159 (2), 148 (2), 131 (15), 117 (21), 104 (93), 91 (100), 87 (7), 77 (15), 65 (14), 58 (4), 51 (9), 43 (77), 39 (6).

HRMS (CI-FE, *i*-butane) calcd for $C_{12}H_{17}O_3$ [M + H]$^+$ 209.1175, found 209.1178.

(5*R*)-3-Ethyl-5-methyl-1,2-dioxolan-3-ol ((5*R*)-**94c**):

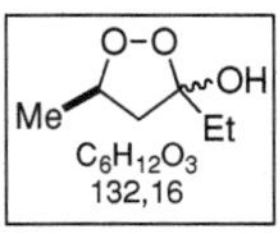

The title compound was isolated after 24 h at 32 °C and purification by flash column chromatography (silica gel, 10–30% Et_2O in pentane) as a colorless oil

(74 mg, 560 μmol, 56%; 97:3 *er*). The enantiomeric ratio was determined after converting dioxolane **94c** to the corresponding epoxide **93c**. The enantiomers were analyzed by GC using a chiral BGB-178/BGB-15 column 30 m (60 °C, 1.0 °C/min until 80 °C, 18 °C/min until 220 °C, 10 min at 220 °C, 0.4 bar H_2); major enantiomer: $\tau_R = 14.73$ min, minor enantiomer: $\tau_R = 15.98$ min. *Characterized as a mixture of hemiketal epimers* (1:1 *dr*).

^{1}H NMR (400 MHz, CD_2Cl_2) δ 4.60 (app. sext, $J = 6.2$ Hz, 1H, C***H***), 4.40–4.32 (m, 1H, C***H***), 3.12 (br s, 2H, O***H***), 2.71 (dd overlapped, $J = 12.8$, 7.5 Hz, 1H, $C\boldsymbol{H}H_{cycl.}$), 2.69 (dd overlapped, $J = 12.6$, 7.1 Hz, 1H, $C\boldsymbol{H}H_{cycl.}$), 2.17 (dd overlapped, $J = 12.6$, 5.8 Hz, 1H, $CH\boldsymbol{H}_{cycl.}$), 2.13 (dd overlapped, $J = 12.8$, 8.5 Hz, 1H, $CH\boldsymbol{H}_{cycl.}$), 1.84–1.71 (m, 4H, $C\boldsymbol{H}_2CH_3$), 1.30 (d, $J = 6.3$ Hz, 3H, $C\boldsymbol{H}_3CH$), 1.26 (d, $J = 6.1$ Hz, 3H, $C\boldsymbol{H}_3CH$), 1.02 (t overlapped, $J = 7.5$ Hz, 3H, $CH_2C\boldsymbol{H}_3$), 1.00 (t overlapped, $J = 7.6$ Hz, 3H, $CH_2C\boldsymbol{H}_3$).

^{13}C NMR (100 MHz, CD_2Cl_2) 108.5 (*C*q_{OH}), 107.3 (*C*q_{OH}), 78.4 (*C*H), 76.9 (*C*H), 52.3 ($\mathit{C}H_{2,\ cycl.}$), 52.0 ($\mathit{C}H_{2,\ cycl.}$), 29.9 ($\mathit{C}H_2CH_3$), 29.7 ($\mathit{C}H_2CH_3$), 20.3 ($\mathit{C}H_3CH$), 17.0 ($\mathit{C}H_3CH$), 9.2 ($CH_2\mathit{C}H_3$), 8.9 ($CH_2\mathit{C}H_3$).

MS (EI-DE) *m/z* (%) 132 [M$^+$] (3), 115 (1), 99 (100), 91 (1), 87 (3), 85 (27), 81 (4), 75 (3), 71 (24), 69 (3), 61 (25), 57 (59), 43 (60), 31 (9), 29 (40), 27 (14).

HRMS (EI-FE) calcd for $C_6H_{12}O_3$ [M$^+$] 132.0785, found 132.0786.

(5*R*)-5-Isobutyl-3-methyl-1,2-dioxolan-3-ol ((5*R*)-**94d**):

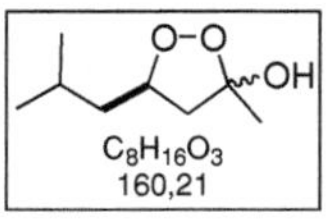

Peroxyhemiketal **94a** was prepared according to the general procedure. The reaction mixture was stirred for 36 h at 32 °C. Purification of the crude product by flash column chromatography (silica gel, 5–30% Et_2O in pentane) provided the title compound **94a** as a colorless oil (98 mg, 612 μmol, 61%; 97.5:2.5 *er*). The enantiomeric ratio was determined after converting peroxyhemiketal **94d** into the corresponding epoxide **93d**. The enantiomers were analyzed by GC using a chiral BGB-176/BGB-15 column 30 m (60 °C, 0.8 °C/min until 80 °C, 20 °C/min until 220 °C, 10 min at 220 °C, 0.6 bar H_2); major enantiomer: $\tau_R = 15.10$ min, minor enantiomer: $\tau_R = 17.20$ min. *Characterized as mixture of hemiketal epimers* (1:1 *dr*).

^{1}H NMR (500 MHz, CD_2Cl_2) δ 4.53–4.47 (m, 1H), 4.39–4.33 (m, 1H), 3.03–2.98 (m, 2H), 2.79–2.75 (m, 1H), 2.71–2.67 (m, 1H), 2.25–2.21 (m, 1H), 2.20–2.17 (m, 1H), 1.73–1.56 (m, 4H), 1.53 (s, 3H), 1.51 (s, 3H), 1.50–1.44 (m, 1H), 1.30–1.25 (m, 1H), 0.93–0.90 (m, 12H).

^{13}C NMR (125 MHz, CD_2Cl_2) δ 105.9, 104.9, 81.3, 79.7, 53.7, 53.2, 44.1, 41.7, 26.6, 26.1, 23.5, 23.3, 23.2, 22.9, 22.8, 22.5.

MS (EI-DE) *m/z* (%) 160 [M$^+$] (2), 135 (0.09), 127 (39), 109 (10), 95 (3), 85 (12), 77 (1), 71 (96), 69 (14), 57 (23), 43 (100), 41 (25), 29 (18).

HRMS (EI-FE) calcd for $C_8H_{16}O_3$ [M]$^+$ 160.1098, found 160.1099.

(5*S*)-5-Cyclohexyl-3-methyl-1,2-dioxolan-3-ol ((5*S*)-**94e**):

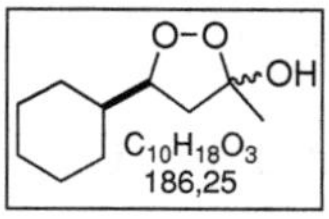

Peroxyhemiketal **94e** was prepared according to the general procedure. The reaction mixture was stirred for 48 h at 32 °C. Purification of the crude product by flash column chromatography (silica gel, 5–30% Et_2O in pentane) provided the title compound **94e** as a colorless oil (101 mg, 542 µmol, 54%; 98:2 *er*). The enantiomeric ratio was determined after converting peroxyhemiketal **94e** into the corresponding epoxide **93e**. The enantiomers were analyzed by GC using a chiral BGB-176/SE-52 column 29.5 m (80 °C, 1.2 °C/min until 130 °C, 20 °C/min until 220 °C, 10 min at 220 °C, 0.5 bar H_2); major enantiomer: $\tau_R = 32.73$ min, minor enantiomer: $\tau_R = 33.99$ min. *Characterized as mixture of hemiketal epimers* (1:1 *dr*).

^{1}H NMR (500 MHz, CD_2Cl_2) δ 4.11–4.06 (m, 1H), 4.05–4.00 (m, 1H), 3.00 (br s, 2H), 2.68–2.64 (m, 1H), 2.60–2.56 (m, 1H), 2.34–2.28 (m, 2H), 1.91–1.86 (m, 1H), 1.83–1.79 (m, 1H), 1.75–1.53 (m, 10H), 1.52 (s, 3H), 1.50 (s, 3H), 1.31–1.12 (m, 6H), 1.08–0.90 (m, 4H).

^{13}C NMR (125 MHz, CD_2Cl_2) δ 105.8, 105.0, 86.7, 85.2, 51.1, 50.3, 42.6, 41.2, 30.3, 29.9, 29.7, 29.3, 26.8, 26.7, 26.4, 26.3, 26.1, 26.0, 23.4, 22.6.

MS (EI-DE) *m*/*z* (%) 186 [M^+] (2), 153 (50), 135 (14), 125 (1), 111 (8), 95 (16), 83 (71), 71 (60), 67 (20), 55 (100), 43 (95), 29 (21).

HRMS (CI-FE, NH_3) calcd for $C_{10}H_{22}NO_3$ $[M + NH_4]^+$ 204.1598, found 204.1600.

(5*R*)-5-(3-Butenyl)-3-methyl-1,2-dioxolan-3-ol ((5*R*)-**94f**):

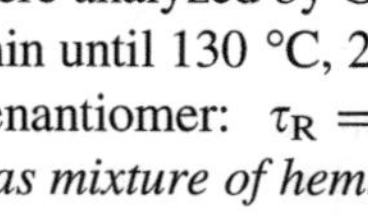

Peroxyhemiketal **94f** was prepared according to the general procedure. The reaction mixture was stirred for 36 h at 32 °C. Purification of the crude product by flash column chromatography (silica gel, 5–30% Et_2O in pentane) provided the title compound **94f** as a colorless oil (109 mg, 689 µmol, 69%; 97.5:2.5 *er*). The enantiomeric ratio was determined after converting peroxyhemiketal **94f** into the corresponding aldol-type product **96f**. The enantiomers were analyzed by HPLC using a chiral Chiralpak IA column (10% *i*PrOH/heptane, 0.5 mL/min); major enantiomer: $\tau_R = 13.19$ min, minor enantiomer: $\tau_R = 12.41$ min. *Characterized as mixture of hemiketal epimers* (1:1 *dr*).

^{1}H NMR (500 MHz, CD_2Cl_2) δ 5.86–5.77 (m, 2H), 5.07–5.02 (m, 2H), 5.00–4.96 (m, 2H), 4.44–4.39 (m, 1H), 4.34–4.28 (m, 1H), 3.10 (br s, 2H), 2.80–2.76 (m, 1H), 2.69–2.65 (m, 1H), 2.29–2.25 (m, 1H), 2.24–2.21 (m, 1H), 2.19–2.05 (m, 4H), 1.83–1.75 (m, 2H), 1.73–1.54 (m, 2H), 1.53 (s, 3H), 1.51 (s, 3H).

^{13}C NMR (125 MHz, CD_2Cl_2) δ 138.3, 138.1, 115.5, 115.4, 105.9, 105.0, 81.9, 80.6, 52.8, 52.7, 34.4, 31.9, 30.9, 30.7, 23.4, 22.8.

MS (EI-DE) *m*/*z* (%) 140 (trace), 125 (10), 107 (2), 97 (2), 83 (10), 71 (18), 55 (28), 43 (100), 41 (13), 29 (14).

HRMS (CI-FE, *i*-butane) calcd for $C_8H_{15}O_3$ $[M + H]^+$ 159.1020, found 159.1021.

(5*R*)-5-(3-Bromopropyl)-3-methyl-1,2-dioxolan-3-ol ((5*R*)-**94g**):

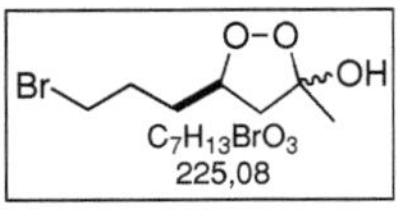

Peroxyhemiketal **94g** was prepared according to the general procedure. The reaction mixture was stirred for 24 h at 32 °C. Purification of the crude product by flash column chromatography (silica gel, 25–40% Et_2O in pentane) provided the title compound **94g** as a colorless oil (81 mg, 360 μmol, 72%; 97:3 *er*). The enantiomeric ratio was determined after converting dioxolane **94g** into the corresponding epoxide **93g**. The enantiomers were analyzed by GC using a chiral BGB-176/BGB-15 column 30 m (100 °C, 1.2 °C/min until 135 °C, 18 °C/min until 220 °C, 5 min at 320 °C, 0.5 bar H_2); major enantiomer: $\tau_R = 19.07$ min, minor enantiomer: $\tau_R = 20.64$ min. *Contains traces of* (*E*)-*7-chloro*-3,4-*epoxy-heptan-2-one. Characterized as mixture of hemiketal epimers* (1:1 *dr*).

^{1}H NMR (400 MHz, THF-d8) δ 5.36 (s, 1H, O***H***), 5.33 (s, 1H, O***H***), 4.38–4.32 (m, 1H, C***H***), 4.25–4.18 (m, 1H, C***H***), 3.51–3.40 (m, 4H, C***H***$_2$Br), 2.68–2.64 (m, 1H, C***H***$_2$COH), 2.61–2.55 (m, 1H, C***H***$_2$COH), 2.22–2.17 (m, 1H, C***H***$_2$COH), 2.16–2.10 (m, 1H, C***H***$_2$COH), 2.00–1.80 (m, 4H, C***H***$_2$), 1.79–1.68 (m, 3H, C***H***$_2$), 1.63–1.54 (m, 1H, C***H***$_2$), 1.41 (s, 3H, C***H***$_3$), 1.39 (s, 3H, C***H***$_3$).

^{13}C NMR (100 MHz, THF-d8): δ 105.9 (***C***$_q$), 104.9 (***C***$_q$), 81.9 (***C***H), 80.7 (***C***H), 53.6 (CO***C***H$_2$CO), 53.5 (CO***C***H$_2$CO), 34.4 (***C***H$_2$), 34.4 (***C***H$_2$, 2C), 31.9 (***C***H$_2$), 31.0 (***C***H$_2$), 30.9 (***C***H$_2$), 24.6 (***C***H$_3$), 23.6 (***C***H$_3$).

(5*R*)-3-Methyl-5-(5-(tetrahydro-2*H*-pyran-2-yloxy)pentyl)-1,2-dioxolan-3-ol ((5*R*)-**94h**):

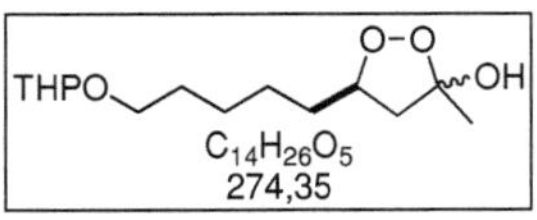

Peroxyhemiketal **94h** was prepared according to the general procedure. The reaction mixture was stirred for 36 h at 32 °C. Purification of the crude product by flash column chromategraphy (silica gel, 30–40% Et_2O in pentane) provided the title compound **94h** (87 mg, 318 μmol, 64%; 96.5:3.5 *er*) as a colorless oil. The enantiomeric ratio was determined after converting peroxyhemiketal **94h** into the corresponding THP-deprotected epoxide **99**. The enantiomers were analyzed by GC using a chiral Ivadex 1 column 25 m (80 °C, 1.0 °C/min until 155 °C, 18 °C/min until 220 °C, 10 min at 320 °C, 0.6 bar H_2); major enantiomer: $\tau_R = 60.79$ min, minor enantiomer: $\tau_R = 61.52$ min. *Characterized as mixture of diastereomers.*

^{1}H NMR (400 MHz, THF-d8) δ 5.30 (s, 2H, O*H*), 5.26 (s, 2H, O*H*), 4.53 (t, $J = 3.2$ Hz, 4H, C$\boldsymbol{H}_{THP}$), 4.34–4.28 (m, 2H, C*H*), 4.21–4.14 (m, 2H, C*H*), 3.80–3.74 (m, 4H, OC*H*H$_{THP}$), 3.67 (dt, $J = 9.5$, 6.6 Hz, 4H, THPOC*H*H), 3.44–3.38 (m, 4H, OCH*H*$_{THP}$), 3.30 (dt, $J = 9.5$, 6.4 Hz, 4H, THPOCH*H*), 2.65–2.53 (m, 4H, C$\boldsymbol{H}_2$Cq$_{OH}$), 2.18–2.07 (m, 4H, C$\boldsymbol{H}_2$Cq$_{OH}$), 1.84–1.76 (m, 4H, C$\boldsymbol{H}_2$), 1.65–1.42 (m, 40H, C$\boldsymbol{H}_2$), 1.40 (s, 12H, C$\boldsymbol{H}_3$), 1.38 (s, 12H, C$\boldsymbol{H}_3$).

^{13}C NMR (100 MHz, THF-d8) δ 105.8 (2C, *C*q$_{OH}$), 104.9 (2C, *C*q$_{OH}$), 99.3 (4C, O*C*HO), 82.7 (2C, CH$_2$*C*HCH$_2$), 81.3 (2C, CH$_2$*C*HCH$_2$), 68.0 (4C, THPO*C*H$_2$), 62.3 (4C, O*C*H$_{2,\ THP}$), 54.0 (2C, *C*H$_2$Cq$_{OH}$), 53.6 (2C, *C*H$_2$Cq$_{OH}$), 35.8 (2C, *C*H$_2$), 33.3 (2C, *C*H$_2$), 31.9 (4C, *C*H$_2$), 31.0 (2C, *C*H$_2$), 30.9 (2C, *C*H$_2$), 27.6 (2C, *C*H$_2$), 27.5 (2C, *C*H$_2$), 27.5 (2C, *C*H$_2$), 27.3 (2C, *C*H$_2$), 26.9 (4C, *C*H$_2$), 24.4 (2C, *C*H$_3$), 23.8 (2C, *C*H$_3$), 20.5 (4C, *C*H$_2$).

MS (EI-DE) *m*/*z* (%) 274 [M$^+$] (trace), 185 (2), 173 (1), 155 (1), 139 (1), 115 (1), 99 (6), 85 (100), 81 (9), 69 (9), 55 (13), 43 (38).

HRMS (ESI+) calcd for $C_{14}H_{26}O_5Na$ [(M + Na)$^+$] 297.1670, found 297.1672.

(5*R*)-5-(2-(*tert*-Butyldimethylsilyloxy)ethyl)-3-methyl-1,2-dioxolan-3-ol ((5*R*)-**94i**):

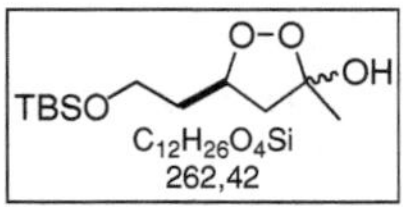

Peroxyhemiketal **94i** was prepared according to the general procedure. The reaction mixture was stirred for 36 h at 32 °C. Purification of the crude product by flash column chromatography (silica gel, 15–20% Et_2O in pentane) provided the title compound as a colorless oil (89 mg, 340 µmol, 68%; 98:2 *er*). The enantiomeric ratio was determined after converting peroxyhemiketal **94i** into the corresponding epoxide **93i**. The enantiomers were analyzed by GC using a chiral BGB-176/BGB-15 column 30 m (80 °C, 1.2 °C/min until 150 °C, 18 °C/min until 220 °C, 5 min at 320 °C, 0.5 bar H_2); major enantiomer: $\tau_R = 34.84$ min, minor enantiomer: $\tau_R = 35.96$ min. *Characterized as mixture of hemiketal epimers* (1:1 *dr*).

^{1}H NMR (400 MHz, THF-d8) δ 5.34 (s, 1H, O*H*), 5.28 (s, 1H, O*H*), 4.52–4.46 (m, 1H, C*H*), 4.35–4.28 (m, 1H, C*H*), 3.85–3.64 (m, 4H, TBSOC$\boldsymbol{H}_2$), 2.69–2.55 (m, 2H, C$\boldsymbol{H}_2$CqOH), 2.25–2.15 (m, 2H, C$\boldsymbol{H}_2$CqOH), 1.92–1.75 (m, 3H, TBSOCH$_2$C$\boldsymbol{H}_2$), 1.64–1.55 (m, 1H, TBSOCH$_2$C$\boldsymbol{H}_2$), 1.41 (s, 3H, Cq$_{OH}$C$\boldsymbol{H}_3$), 1.39 (s, 3H, Cq$_{OH}$C$\boldsymbol{H}_3$), 0.90 (s, 18H, Cq(C$\boldsymbol{H}_3$)$_3$), 0.04 (s, 12H, Si(C$\boldsymbol{H}_3$)$_2$).

^{13}C NMR (100 MHz, THF-d8) δ 105.8 (*C*q$_{OH}$), 104.9 (*C*q$_{OH}$), 79.9 (*C*H), 78.3 (*C*H), 61.3 (TBSO*C*H$_2$), 61.0 (TBSO*C*H$_2$), 53.8 (*C*H$_2$CqOH), 53.6 (*C*H$_2$CqOH), 38.9 (TBSOCH$_2$*C*H$_2$), 36.7 (TBSOCH$_2$*C*H$_2$), 26.6 (6C, Cq(*C*H$_3$)$_3$), 24.3 (CqOH*C*H$_3$), 23.7 (CqOH*C*H$_3$), 19.2 (2C, *C*qMe$_3$), 5.02 (2C, Si(*C*H$_3$)$_2$), 5.05 (2C, Si(*C*H$_3$)$_2$).

MS (EI-DE) *m/z* (%) 205 (1), 187 (1), 173 (1), 157 (2), 145 (8), 131 (32), 115 (22), 101 (26), 89 (9), 75 (100), 59 (12), 43 (24), 29 (4).

HRMS (ESI+) calcd for $C_{12}H_{26}O_4SiNa$ [$(M + Na)^+$] 285.1491, found 285.1493.

(5*R*)-5-(3-Ethoxy-3-oxopropyl)-3-methyl-1,2-dioxolan-3-ol ((5*R*)-**94j**):

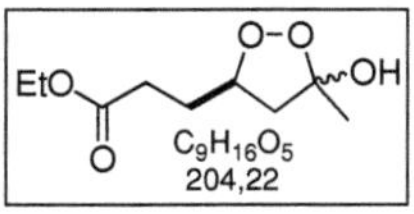

Peroxyhemiketal **94j** was prepared according to the general procedure. The reaction mixture was stirred for 24 h at 32 °C. Purification of the crude product by flash column chromatography (silica gel, 25–50% Et_2O in pentane) provided the title compound **94j** (71 mg, 348 μmol, 70%; 96.5:3.5 *er*) as a colorless oil. The enantiomeric ratio was determined after converting peroxyhemiketal **94j** into the corresponding epoxide **93j**. The enantiomers were analyzed by GC using a chiral BGB-176/SE-52 column 30 m (80 °C, 1.2 °C/min until 130 °C, 18 °C/min until 220 °C, 5 min at 320 °C, 0.5 bar H_2); major enantiomer: $\tau_R = 36.32$ min, minor enantiomer: $\tau_R = 37.59$ min. *Characterized as mixture of hemiketal epimers* (1:1 *dr*).

^{1}H NMR (400 MHz, CD_2Cl_2) δ 4.47–4.41 (m, 1H, C***H***), 4.38–4.31 (m, 1H, C***H***), 4.11 (q overlapped, $J = 7.2$ Hz, 2H, C$\boldsymbol{H}_2$CH$_3$), 4.10 (q overlapped, $J = 7.1$ Hz, 2H, C$\boldsymbol{H}_2$CH$_3$), 2.83–2.78 (m, 1H, C$\boldsymbol{H}_2$Cq$_{OH}$), 2.71–2.51 (m, 3H, C$\boldsymbol{H}_2$C$_{OH}$ and O***H***), 2.45–2.22 (m, 6H, C$\boldsymbol{H}_2$), 2.05–1.90 (m, 3H, C$\boldsymbol{H}_2$), 1.81–1.73 (m, 1H, C$\boldsymbol{H}_2$), 1.53 (s, 3H, C$\boldsymbol{H}_3$), 1.51 (s, 3H, C$\boldsymbol{H}_3$), 1.24 (t overlapped, $J = 7.2$ Hz, 3H, CH$_2$C$\boldsymbol{H}_3$), 1.23 (t overlapped, $J = 7.3$ Hz, 3H, CH$_2$C$\boldsymbol{H}_3$).

^{13}C NMR (100 MHz, CD_2Cl_2) δ 172.9 (***C*** =O), 172.8 (***C***=O), 105.5 (***C***q$_{OH}$), 104.5 (***C***q$_{OH}$), 80.8 (***C***H), 79.8 (***C***H), 60.7 (***C***H$_2$CH$_3$), 60.5 (***C***H$_2$CH$_3$), 52.3 (***C***H$_{2,\ cycl.}$), 51.9 (***C***H$_{2,\ cycl.}$), 30.7 (***C***H$_2$), 30.6 (***C***H$_2$), 29.9 (***C***H$_2$), 27.2 (***C***H$_2$), 22.9 (***C***H$_3$), 22.3 (***C***H$_3$), 14.1 (CH$_2$***C***H$_3$), 14.0 (CH$_2$***C***H$_3$).

MS (EI) *m/z* (%) 172 (3), 141 (2), 126 (11), 115 (5), 98 (28), 85 (13), 73 (12), 55 (13), 43 (100), 29 (37).

HRMS (ESI+) calcd for $C_9H_{16}O_5Na$ [$(M + Na)^+$] 227.0890, found 227.0890.

(5*R*)-5-(3-Oxobutyl)-3-methyl-1,2-dioxolan-3-ol ((5*R*)-**94k**):

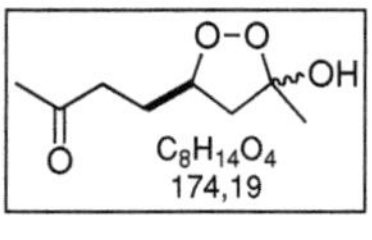

Peroxyhemiketal **94k** was prepared according to the general procedure. Catalyst [**13** · 2 TFA] (10 mol%) was used together with 50 wt% aqueous hydrogen peroxide (1.5 equiv) as oxidant. The reaction mixture was stirred for 24 h at 32 °C. Purification of the crude product by flash column chromatography (silica gel, 30–50% Et_2O in pentane) provided the title compound **94k** (26 mg, 151 μmol, 30%; 96:4 *er*) as a colorless oil. The enantiomeric ratio was determined after converting peroxyhemiketal **94k** into the corresponding epoxide **93k** The enantiomers were analyzed by GC using a chiral BGB-176/SE-52 column 30 m (80 °C, 1.2 °C/min until 120 °C, 20 °C/min until 220 °C, 10 min at 220 °C, 0.5 bar H_2); major enantiomer: $\tau_R = 28.87$ min, minor enantiomer: $\tau_R = 30.28$ min. *Characterized as a mixture of three hemiketal isomers* (~1:1:0.8).

^{1}H NMR (500 MHz, THF-d8) 3-*hydroxy*-1,2-*dioxolane* [*mixture of hemiketal epimers* (~1:1 *dr*)]:

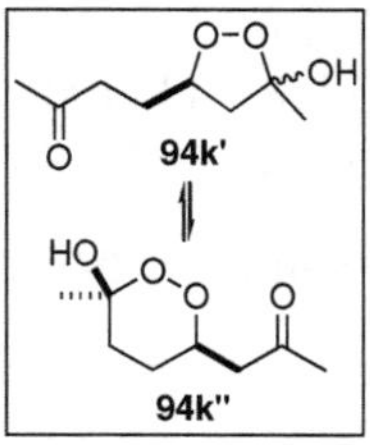

δ 5.33 (br s, 1H, O*H*), 5.30 (br s, 1H, O*H*), 4.37–4.28 (m, 1H, C*H*), 4.21–1.25 (m, 1H, C*H*), 2.64 (dd, J = 12.0, 7.0 Hz, 1H, C$H_{2, \text{cycl.}}$), 2.58–2.46 (m, 3H, C$H_{2, \text{cycl.}}$), 2.20–2.05 (m, 10H, CH_3C(=O) and CH_2(C=O)), 1.88–1.62 (m, 4H, CH_2CH_2CH), 1.40 (s, 3H, C$H_3$$Cq_{OH}$), 1.38 (s, 3H, C$H_3$$Cq_{OH}$); 3-*hydroxy*-1,2-*dioxane*: 5.34 (br s, 1H, O*H*), 4.37–4.28 (m, 1H, C*H*), 2.58–2.46 (m, 4H, C$H_2$$Cq_{OH}$ and CH_2C(=O)), 2.20–2.05 (m, 3H, CH_3C(=O)), 1.88–1.62 (m, 2H, CHC$H_{2, \text{cycl.}}$), 1.39 (s, 3H, C$H_3$$Cq_{OH}$).

^{13}C NMR (125 MHz, THF-d8) 3-*hydroxy*-1,2-*dioxolane* [*mixture of hemiketal epimers* (*dr* ~ 1:1)]: δ 206.8 (*C*=O), 206.5 (*C*=O), 105.9 (*C*q_{OH}), 105.0 (*C*q_{OH}), 81.8 (*C*H), 80.6 (*C*H), 53.5 (*C*$H_{2, \text{cycl.}}$), 53.4 (*C*$H_{2, \text{cycl.}}$), 40.3 (*C*H_2C(=O)), 40.2 (*C*H_2C(=O)), 29.8 (*C*H_3C(=O)), 29.8 (*C*H_3C(=O)), 29.7 (*C*H_2CH), 27.1 (*C*H_2CH), 24.3 (*C*$H_3$$Cq_{OH}$), 23.6 (*C*$H_3$$Cq_{OH}$); 3-*hydroxy*-1,2-*dioxane*: δ 204.4 (*C*=O), 99.2 (*C*q_{OH}), 77.5 (*C*H), 47.6 (*C*H_2C(=O)), 34.2 (*C*$H_2$$Cq_{OH}$), 31.0 (*C*$H_3$C(=O)), 26.9 (CH*C*$H_{2, \text{cycl.}}$), 26.7 (*C*$H_3$$Cq_{OH}$).

(5*R*)-5-(5-Hydroxypentyl)-3-methyl-1,2-dioxolan-3-ol ((5*R*)-**94l**):

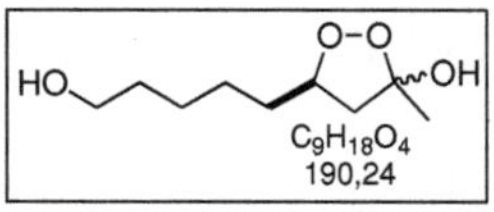

Peroxyhemiketal **94l** was prepared according to the general procedure. Catalyst [**13** · 2 TFA] (10 mol%) was used together with 50 wt% aqueous hydrogen peroxide (1.5 equiv) as oxidant. The reaction mixture was stirred for 36 h at 50 °C. Purification of the crude reaction mixture by flash column chromatography (silica gel, 10–40% Et_2O in CH_2Cl_2) provided the title compound **94l** (12 mg, 63.1 µmol, 25%) as a colorless oil along with a mixture of the corresponding epoxide **93l** (11 mg, 64.0 µmol, 26%; 98:2 *er*) and unreacted starting material **92 l** (10 mg, 65.0 µmol, 26%). *Characterized as a mixture of hemiketal epimers* (1:1 *dr*).

^{1}H NMR (400 MHz, THF-d8) δ 5.31 (s, 1H, CqO*H*), 5.27 (s, 1H, CqO*H*), 4.34–4.28 (m, 1H, C*H*), 4.17 (quint, J = 7.3 Hz, 1H, C*H*), 3.46 (t, J = 5.5 Hz,

4H, CH_2OH), 3.35 (br s, 2H, CH_2OH), 2.64–2.55 (m, 2H, $CH_{2,\ cycl.}$), 2.20–2.07 (m, 2H, $CH_{2,\ cycl.}$), 1.67–1.31 (m, 16H, –$(CH_2)_4$–), 1.40 (s overlapped, 3H, CH_3), 1.38 (s overlapped, 3H, CH_3).

^{13}C NMR (100 MHz, THF-d8) 105.8 (Cq_{OH}), 104.9 (Cq_{OH}), 82.7 (CH), 81.4 (CH), 62.7 (CH_2OH), 62.7 (CH_2OH), 54.0 ($CH_{2,\ cycl.}$), 53.6 ($CH_{2,\ cycl.}$), 35.9 (CH_2), 34.3 (CH_2), 34.2 (CH_2), 33.4 (CH_2), 27.6 (CH_2), 27.3 (CH_2), 27.3 (CH_2), 27.1 (CH_2), 24.4 (CH_3), 23.8 (CH_3).

MS (EI-DE) *m/z* (%) 157 (5), 139 (5), 125 (1), 121 (3), 111 (2), 103 (1), 97 (13), 81 (15), 79 (4), 71 (21), 55 (18), 43 (100), 41 (26), 31 (13).

HRMS (ESI+) calcd for $C_9H_{18}AgO_4$ [(M + Ag)$^+$] 297.0249, found 297.0249.

(5*R*)-3,5-Dipentyl-1,2-dioxolan-3-ol ((5*R*)-**94m**):

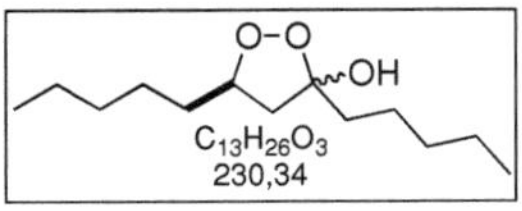

Peroxyhemiketal **94m** was prepared according to the general procedure. The reaction mixture was stirred for 48 h at 32 °C. Purification of the crude product by flash column chromatography (silica gel, 10–20% Et_2O in pentane) provided the title compound **94m** as a colorless oil (46 mg, 201 μmol, 40%; 97:3 *er*). The enantiomeric ratio was determined after converting dioxolane **94m** into the corresponding epoxide **93m**. The enantiomers were analyzed by GC using a chiral G-TA column 30 m (80 °C, 1 °C/min until 130 °C, 18 °C/min until 180 °C, 10 min at 320 °C, 0.9 bar H_2); major enantiomer: τ_R = 43.80 min, minor enantiomer: τ_R = 43.02 min. *Characterized as a mixture of hemiketal epimers* (1:1 *dr*).

^{1}H NMR (500 MHz, THF-d8) δ 5.20 (s, 1H, O*H*), 5.16 (s, 1H, O*H*), 4.33–4.28 (m, 1H, C*H*), 4.15–4.09 (m, 1H, C*H*), 2.61–2.52 (m, 2H, CH_2Cq_{OH}), 2.10–2.04 (m, 2H, CH_2Cq_{OH}), 1.68–1.26 (m, 32H, CH_2), 0.92–0.88 (m, 12H, CH_3).

^{13}C NMR (125 MHz, THF-d8) δ 107.7 (Cq_{OH}), 106.9 (Cq_{OH}), 82.6 (CH), 81.0 (CH), 52.4 ($CH_{2,\ cycl.}$), 52.1 ($CH_{2,\ cycl.}$), 38.3 (CH_2), 37.9 (CH_2), 35.8 (CH_2), 33.4 (CH_2), 33.4 (CH_2), 33.3 (CH_2), 33.1 (CH_2), 33.0 (CH_2), 27.4 (CH_2), 27.2 (CH_2), 25.9 (CH_2), 25.7 (CH_2), 23.8 (3C, CH_2), 23.7 (CH_2), 14.7 (2C, CH_3), 14.7 (CH_3), 14.6 (CH_3).

MS (EI) *m/z* (%) 230 [M$^+$] (1), 197 (100), 159 (1), 141 (3), 127 (30), 117 (5), 99 (50), 81 (6), 71 (30), 55 (30), 43 (90), 29 (24).

HRMS (EI-DE) calcd for $C_{13}H_{26}O_3$ [M$^+$] 230.1880, found 230.1882.

(5*R*)-3-Isopropyl-5-methyl-1,2-dioxolan-3-ol ((5*R*)-**94o**):

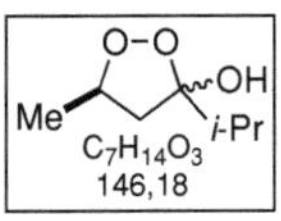

Peroxyhemiketal **94o** was obtained by the general procedure. Catalyst salt [**13** · 2 TCA] (20 mol%) were used and the reaction mixture was stirred for 48 h at 32 °C. Purification of the crude product by flash column chromatography (silica gel, 30–40% Et_2O in pentane) provided the title compound **94o** (57 mg, 391 μmol, 39%; 96:4 *er*) as a colorless oil. The enantiomeric ratio was determined after converting peroxyhemiketal **94o** into the corresponding epoxide **94o**. The enantiomers were analyzed by GC using a chiral Lipodex G column 25 m (50 °C, 1.2 °C/min until 80 °C, 18 °C/min until 220 °C, 10 min at 230 °C, 0.5 bar H_2); major enantiomer: $\tau_R = 8.36$ min, minor enantiomer: $\tau_R = 8.80$ min. *Characterized as a mixture of hemiketal epimers* (1:1 *dr*).

^{1}H NMR (400 MHz, THF-d8) δ 6.98 (s, 1H, O***H***), 6.97 (s, 1H, O***H***), 6.34 (app. sext, $J = 6.3$ Hz, 1H, C***H***CH_2), 6.10–6.03 (m, 1H, C***H***CH_2), 4.51 (dd, $J = 12.6$, 7.3 Hz, 1H, C***H***H), 4.33 (dd, $J = 12.1$, 6.8 Hz, 1H, C***H***H), 3.93 (dd, $J = 12.0$, 6.3 Hz, 1H, CH***H***), 3.81 (dd, $J = 12.6$, 8.5 Hz, 1H, CH***H***), 3.72 (hept, $J = 7.0$ Hz, 2H, C***H***Me_2), 3.09 (d, $J = 6.0$ Hz, 3H, C$\boldsymbol{H}_3$CH), 3.04 (d, $J = 6.3$ Hz, 3H, C$\boldsymbol{H}_3$CH), 2.81 (t, $J = 6.8$ Hz, 12H, CH(C$\boldsymbol{H}_3$)$_2$).

^{13}C NMR (125 MHz, THF-d8) δ 110.2 (**C**q_{OH}), 109.2 (**C**q_{OH}), 78.6 (**C**HCH_2), 76.9 (**C**HCH_2), 52.0 (**C**H_2), 51.8 (**C**H_2), 36.1 (**C**HMe_2), 35.9 (**C**HMe_2), 20.3 (**C**H_3), 19.1 (**C**H_3), 18.8 (**C**H_3), 18.3 (**C**H_3), 18.1 (**C**H_3), 17.4 (**C**H_3).

GC-MS (GC-EI) *m/z* (%) 146 [M^+] (trace), 128 (1), 113 (12), 103 (6), 85 (4), 71 (22), 61 (17), 43 (100), 27 (8).

HRMS (EI-FE) calcd for $C_7H_{14}O_3$ [M^+] 146.0941, found 146.0943.

3-Ethyl-5-nonyl-1,2-dioxolan-3-ol (94p):

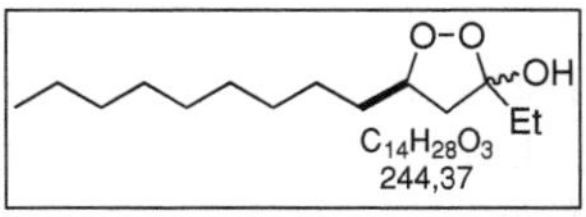

Peroxyhemiketal **94p** was prepared according to the general procedure. The reaction mixture was stirred for 48 h at 32 °C. Purification of the crude product by flash column chromatography (silica gel, 10–20% Et_2O in pentane) provided the title compound **94p** as a colorless oil (59 mg, 240 μmol, 48%; 98:2 *er*). The enantiomeric ratio was determined after converting dioxolane **94p** into the corresponding epoxide **93p**. The enantiomers were analyzed by GC using a chiral Lipodex E column 25 m (100 °C, 1.2 °C/min until 180 °C, 18 °C/min until 220 °C, 10 min at 320 °C, 0.5 bar H_2); major enantiomer: $\tau_R = 37.10$ min, minor enantiomer: $\tau_R = 35.28$ min. *Characterized as a mixture of hemiketal epimers* (1:1 *dr*).

^{1}H NMR (500 MHz, THF-d8) δ 5.18 (s, 1H, O***H***), 5.15 (s, 1H, O***H***), 4.33–4.28 (m, 1H, C***H***), 4.12 (quint, $J = 7.1$ Hz, 1H, C***H***), 2.61–2.50 (m, 2H, C$\boldsymbol{H}_2$Cq_{OH}), 2.10–2.02 (m, 2H, C$\boldsymbol{H}_2$Cq_{OH}), 1.71–1.52 (m, 7H, C$\boldsymbol{H}_2$), 1.42–1.36 (m, 3H, C$\boldsymbol{H}_2$), 1.35–1.26 (m, 26H, C$\boldsymbol{H}_2$), 0.97 (t, $J = 7.6$ Hz, 3H, Cq_{OH}CH$_2$C$\boldsymbol{H}_3$), 0.96 (t, $J = 7.6$ Hz, 3H, Cq_{OH}CH$_2$C$\boldsymbol{H}_3$), 0.89 (t, $J = 6.8$ Hz, 6H, C$\boldsymbol{H}_3$).

^{13}C NMR (125 MHz, THF-d8) δ 108.1 (**C**q_{OH}), 107.2 (**C**q_{OH}), 82.7 (**C**H), 81.1 (**C**H), 51.9 (**C**$H_{2,\ cycl.}$), 51.5 (**C**$H_{2,\ cycl.}$), 35.7 (**C**H_2), 33.3 (**C**H_2), 33.2 (2C, **C**H_2), 31.2 (**C**H_2), 30.9 (**C**H_2), 30.9 (**C**H_2), 30.9 (**C**H_2), 30.8 (**C**H_2), 30.8 (**C**H_2), 30.8

(CH_2), 30.8 (CH_2), 30.6 (2C, CH_2), 27.7 (CH_2), 27.5 (CH_2), 23.9 (2C, CH_2), 14.7 (2C, $(CH_2)_8CH_3$), 9.9 ($Cq_{OH}CH_2CH_3$), 9.7 ($Cq_{OH}CH_2CH_3$).

MS (EI) *m*/*z* (%) 244 [M^+] (1), 211 (77), 197 (1), 181 (1), 155 (9), 137 (2), 95 (8), 85 (50), 71 (17), 57 (100), 43 (46), 29 (34).

HRMS (EI-DE) calcd for $C_{14}H_{28}O_3$ [M^+] 244.2038, found 244.2038.

7.3.2 Synthesis of (R)-4-(Tert-butylperoxy)decan-2-one ((R)-95a)

The general procedure [8] described in Sect. 7.3.1.1 for the catalytic asymmetric hydroperoxidation of α, β-unsaturated ketones **92** was followed. *tert*-Butylhydroperoxide (70 wt%; 103 μL, 0.75 mmol, 1.5 equiv) was used under otherwise identical reaction conditions.

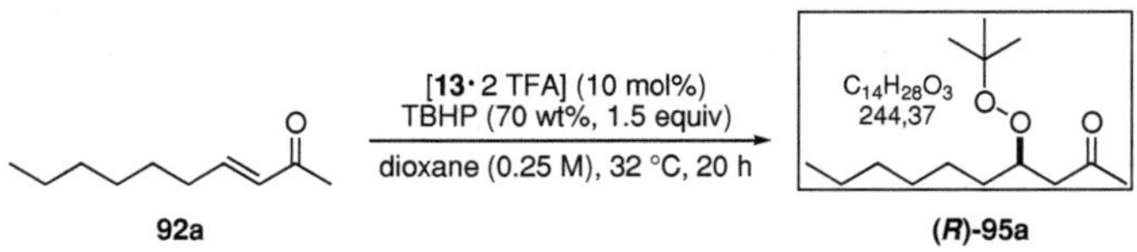

Purification by flash column chromatography (silica gel, 1–20% Et_2O in pentane) gave peroxide **95a** (67 mg, 274 μmol, 55%) along with the corresponding epoxide **93a** (~17 mg, 100 μmol, 20%; 99:1 *er*) (*vide infra*), both as colorless oils.

^{1}H NMR (500 MHz, CD_2Cl_2) δ 4.37–4.32 (m, 1H, C***H***), 2.79 (dd, $J = 15.2$, 6.6 Hz, 1H, C***H***HC(=O)), 2.43 (dd, $J = 15.7$, 5.2 Hz, 1H, CH***H***C(=O)), 2.15 (s, 3H, C(=O)CH_3), 1.56–1.25 (m, 10H, $(CH_2)_5$), 1.19 (s, 9H, $Cq(CH_3)_3$), 0.88 (t, $J = 6.9$ Hz, 3H, $(CH_2)_5CH_3$),

^{13}C NMR (125 MHz, CD_2Cl_2) δ 207.5 (***C***=O), 80.5 (***C***q) 80.4 (***C***H), 48.0 (CH_2C(=O)), 33.4 (CH_2), 31.2 (CH_2), 30.9 (CH_3C(=O)), 29.7 (CH_2), 26.5 (3C, $Cq(CH_3)_3$), 25.9 (CH_2), 23.0 (CH_2), 14.2 ($(CH_2)_5CH_3$).

MS (EI-DE) *m*/*z* (%) 244 [M^+] (trace), 188 (4), 171 (5), 155 (80), 137 (4), 127 (1), 113 (23), 95 (5), 85 (7), 71 (15), 57 (100), 43 (70), 29 (10).

HRMS (ESI+) calcd for $C_{14}H_{28}O_3Na_1$ [$(M + Na)^+$] 267.1928, found 267.1931.

7.3.3 Catalytic Asymmetric Epoxidation of Acyclic Enones

7.3.3.1 General Procedure

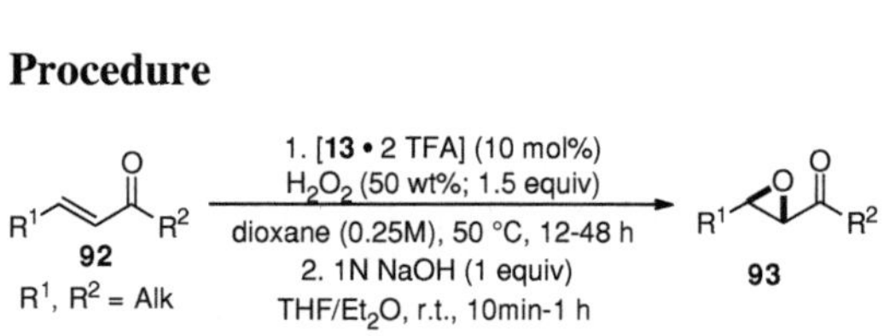

Reference [8] Catalyst salt [**13** · 2 TFA] was prepared *in situ* by the addition of amine **13** (32.3 mg, 0.1 mmol, 10 mol%) to a solution of trifluoroacetic acid (15.3 μL, 0.2 mmol, 20 mol%) in dioxane (4 mL). Then enone **92** (1.0 mmol, 1.0 equiv) was added, and 20 min later, aqueous hydrogen peroxide (50 wt%, 92 μL, 1.5 mmol, 1.5 equiv). After 12–48 h at 50 °C, the reaction mixture was extracted with Et_2O (3 × 25 mL). The combined organic phases were washed with brine and concentrated under reduced pressure to a volume of 5 mL. Aqueous 1 N NaOH solution (1.0 mL, 1.0 mmol, 1.0 equiv) was added and the reaction mixture stirred until TLC analysis indicated complete conversion to the epoxide (10 min to 1 h). The reaction mixture was extracted with Et_2O (3 × 25 mL), and the combined organic phases were washed with brine, dried (Na_2SO_4), and filtered. Evaporation of the solvent furnished the crude product, which was purified by flash column chromatography (silica gel, eluent: Et_2O-pentane) to afford pure α,β-epoxy ketones **93**.

7.3.3.2 Scope of Optically Active α,β-Epoxyketones

(*E*)-(3*S*,4*R*)-3,4-Epoxy-2-decanone (3*S*,4*R*)-**93a**):

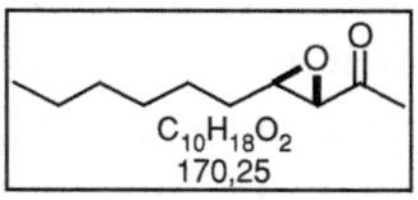

Epoxide (3*S*,4*R*)-**93a** was prepared according to the general procedure. The reaction mixture was stirred for 12 h at 50 °C. Purification by flash column chromatography (silica gel, 1–20% Et_2O in pentane) provided the title compound **93a** as a colorless oil (127 mg, 747 μmol, 75%; 98.5:1.5 *er*). The enantiomeric ratio was determined by GC using a chiral BGB-176/BGB-15 column 30 m (80 °C, 1.2 °C/min until 120 °C, 20 °C/min until 220 °C, 10 min at 220 °C, 0.6 bar H_2); major enantiomer: $\tau_R = 19.73$ min, minor enantiomer: $\tau_R = 21.60$ min.

(*E*)-(3*R*,4*S*)-3,4-Epoxy-2-decanone ((3*R*,4*S*)-**93a**):

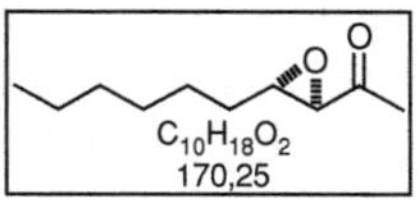

9-Amino(9-deoxy)*epi*quinidine (**67**) was used (instead of **13**) under otherwise identical conditions. The reaction mixture was stirred for 12 h at 50 °C. Purification by flash column chromatography (silica gel, 1–20% Et_2O in pentane) provided the title compound **93a** as a colorless oil (137 mg, 807 μmol, 81%; 96:4 *er*). The enantiomeric ratio was determined by GC using a chiral BGB-176/SE-52 column 30 m (80 °C, 1.2 °C/min until 115 °C, 20 °C/min until 220 °C, 10 min at 220 °C, 0.6 bar H_2); major enantiomer: $\tau_R = 25.06$ min, minor enantiomer: $\tau_R = 24.13$ min.

^{1}H NMR (500 MHz, CD_2Cl_2) δ 3.14 (d, *J* = 2.0 Hz, 1H, C$\boldsymbol{H}_{epo}$C(=O)), 3.05 (td, *J* = 5.5, 1.9 Hz, 1H, CH_2C$\boldsymbol{H}_{epo}$), 2.02 (s, 3H, C$\boldsymbol{H}_3$C(=O)), 1.64–1.55 (m, 2H,

$\boldsymbol{CH_2}CH_{epo}$), 1.49–1.41 (m, 2H, $\boldsymbol{CH_2}CH_2CH_{epo}$), 1.37–1.25 (m, 6H, $(\boldsymbol{CH_2})_3CH_3$), 0.89 (t, $J = 7.0$ Hz, 3H, $(CH_2)_5\boldsymbol{CH_3}$).

^{13}C NMR (125 MHz, CD_2Cl_2) δ 206.2 (***C*** =O), 60.4 ($\boldsymbol{C}H_{epo}C(=O)$), 58.6 ($CH_2\boldsymbol{C}H_{epo}$), 32.4 ($\boldsymbol{C}H_2CH_{epo}$), 32.2 ($\boldsymbol{C}H_2$), 29.5 ($\boldsymbol{C}H_2$), 26.3 ($\boldsymbol{C}H_2CH_2CH_{epo}$), 24.8 ($\boldsymbol{C}H_3C(=O)$), 23.1 ($\boldsymbol{C}H_2$), 14.4 ($(CH_2)_5\boldsymbol{C}H_3$).

GC-MS (GC-EI) *m*/*z* (%) 170 [M$^+$] (trace), 139 (7), 125 (1), 109 (4), 97 (12), 85 (54), 81 (9), 69 (21), 55 (41), 43 (100), 39 (9), 29 (17).

HRMS (CI-FE, *i*-butane) calcd for $C_{10}H_{19}O_2$ [(M + H)$^+$] 171.1384, found 171.1385.

(*E*)-(3*S*,4*R*)-3,4-Epoxy-6-phenyl-2-hexanone ((3*S*,4*R*)-**93b**):

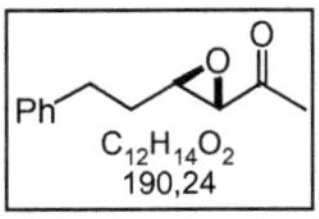

Epoxide (3*S*,4*R*)-**93b** was prepared according to the general procedure. The reaction mixture was stirred for 12 h at 50 °C. Purification by flash column chromatography (silica gel, 1–20% Et_2O in pentane) provided the title compound **93b** as a colorless oil (162 mg, 854 μmol, 85%; 98.5:1.5 *er*). The enantiomeric ratio was determined by GC using a chiral BGB-176/BGB-15 column 30 m (80 °C, 1.2 °C/min until 140 °C, 20 °C/min until 220 °C, 10 min at 220 °C, 0.6 bar H_2); major enantiomer: $\tau_R = 45.23$ min, minor enantiomer: $\tau_R = 46.65$ min.

(*E*)-(3*R*,4*S*)-3,4-Epoxy-6-phenyl-2-hexanone ((3*R*,4*S*)-**93b**):

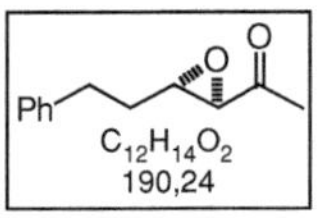

Epoxide (3*R*,4*S*)-**93b** was prepared according to the general procedure. 9-Amino(9-deoxy)*epi*quinidine (**67**) was used (instead of **13**) under otherwise identical conditions. The reaction mixture was stirred for 12 h at 50 °C. Purification by flash column chromatography (silica gel, 1–20% Et_2O in pentane) provided the title compound **93b** as a colorless oil (171 mg, 901 μmol, 90%; 95:5 *er*). The enantiomeric ratio was determined by GC using a chiral BGB-176/BGB-15 column 30 m (80 °C, 1.2 °C/min until 160 °C, 20 °C/min until 220 °C, 10 min at 220 °C, 0.6 bar H_2); major enantiomer: $\tau_R = 47.30$ min, minor enantiomer: $\tau_R = 46.86$ min.

^{1}H NMR (500 MHz, CD_2Cl_2) δ 7.32–7.28 (m, 2H, $C\boldsymbol{H}_{Ph}$), 7.23–7.20 (m, 3H, $C\boldsymbol{H}_{Ph}$), 3.15 (d, $J = 1.8$ Hz, 1H, $C\boldsymbol{H}_{epo}C(=O)$), 3.09 (td, $J = 5.6$, 1.8 Hz, 1H, $CH_2C\boldsymbol{H}_{epo}$), 2.85–2.72 (m, 2H, $PhC\boldsymbol{H_2}$), 1.99 (s, 3H, $C\boldsymbol{H_3}$), 1.96–1.91 (m, 2H, $C\boldsymbol{H_2}CH_{epo}$).

^{13}C NMR (125 MHz, CD_2Cl_2) δ 205.7 ($\boldsymbol{C}$=O), 141.3 ($\boldsymbol{C}q_{Ph}$), 129.1 (2C, $\boldsymbol{C}H_{Ph}$), 128.9 (2C, $\boldsymbol{C}H_{Ph}$), 126.7 ($\boldsymbol{C}H_{Ph,\ p}$), 60.3 ($\boldsymbol{C}H_{epo}C(=O)$), 58.1 ($CH_2\boldsymbol{C}H_{epo}$), 34.1 ($\boldsymbol{C}H_2$), 32.5 ($\boldsymbol{C}H_2$), 25.0 ($\boldsymbol{C}H_3$).

MS (EI-DE) *m/z* (%) 190 [M$^+$] (4), 172 (4), 157 (2), 147 (14), 134 (5), 129 (37), 117 (38), 104 (34), 91 (100), 85 (22), 77 (8), 65 (17), 57 (8), 51 (6), 43 (64), 27 (5).

HRMS (EI-FE) calcd for $C_{12}H_{14}O_2$ [M$^+$] 190.0992, found 190.0994.

(*E*)-(3*S*,4*R*)-3,4-Epoxy-5-phenyl-2-hexanone ((3*S*,4*R*)-**93q**):

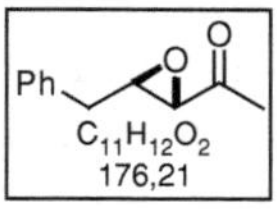

Epoxide (3*S*,4*R*)-**93q** was prepared according to the general procedure. Catalyst [**13** · 3 TFA] was used under otherwise identical conditions. After 12 h at 50 °C, the conversion was determined to be 94% by ^{1}H NMR of the crude mixture (37% epoxide **93q**, 33% 5-benzyl-3-methyl-1,2-dioxolan-3-ol (**94q**), 24% (*E*)-5-phenyl-pent-4-en-2-one (*iso*-**92q**, *vide infra*)). The enantiomeric ratio of epoxide **93q** was determined to be 99:1 *er* by GC using a chiral Lipodex E column 25 m (80 °C, 1.2 °C/min until 145 °C, 18 °C/min until 220 °C, 10 min at 220 °C, 0.5 bar H_2); major enantiomer: $\tau_R = 42.87$ min, minor enantiomer: $\tau_R = 41.72$ min.

^{1}H NMR (500 MHz, $CDCl_3$) δ 7.32–7.21 (m, 5H, $C_6\boldsymbol{H}_5$), 3.28 (td, $J = 5.2$, 1.9 Hz, 1H, $CH_2C\boldsymbol{H}_{epo}$), 3.22 (d, $J = 1.6$ Hz 1H, $C\boldsymbol{H}_{epo}$C(=O)), 2.93 (app. qd, $J = 14.4$, 5.2 Hz, 2H, $PhC\boldsymbol{H}_2$), 2.02 (s, 3H, $C\boldsymbol{H}_3$).

(*E*)-5-Phenylpent-4-en-2-one (*iso*-**92q**):

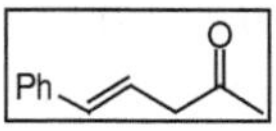

^{1}H NMR (500 MHz, $CDCl_3$) δ 7.35–7.21 (m, 5H, $C_6\boldsymbol{H}_5$), 6.45 (d, $J = 16.0$ Hz, 1H, PhC***H***=), 6.28 (dt, $J = 15.8$, 7.5 Hz, 1H, =C***H***CH_2), 3.31 (d, $J = 7.2$ Hz, 2H, =CHC$\boldsymbol{H}_2$), 2.19 (3H, $C\boldsymbol{H}_3$).

(*E*)-(3*S*,4*R*)-3,4-Epoxy-6-methyl-2-heptanone ((3*S*,4*R*)-**93d**):

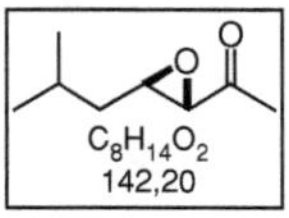

Epoxide (3*S*,4*R*)-**93d** was prepared according to the general procedure. The reaction mixture was stirred for 18 h at 50 °C. Purification by flash column chromatography (silica gel, 1–20% Et_2O in pentane) provided the title compound **93d** as a colorless oil (109 mg, 768 µmol, 77%; 98.5:1.5 *er*). The enantiomeric ratio was determined by GC using a chiral BGB-176/BGB-15 column 30 m (60 °C, 0.8 °C/min until 80 °C, 18 °C/min until 220 °C, 10 min at 220 °C, 0.6 bar H_2); major enantiomer: $\tau_R = 15.45$ min, minor enantiomer: $\tau_R = 17.76$ min.

^{1}H NMR (500 MHz, CD_2Cl_2) δ 3.12 (d, $J = 1.9$ Hz, 1H, $C\boldsymbol{H}_{epo}$C(=O)), 3.06 (td, $J = 5.8$, 1.9 Hz, 1H, $CH_2C\boldsymbol{H}_{epo}$), 2.03 (s, 3H, $C\boldsymbol{H}_3$C(=O)), 1.83 (hept,

$J = 6.7$ Hz, 1H, C***H***Me_2), 1.48 (dd, $J = 6.5$ Hz, 2H, C***H***$_2$), 0.98 (d, $J = 4.6$ Hz, 3H, CH(C***H***$_3$)$_2$), 0.97 (d, $J = 4.6$ Hz, 3H, CH(C***H***$_3$)$_2$).

^{13}C NMR (125 MHz, CD_2Cl_2) δ 206.2 (***C***=O), 60.5 (***C***H_{epo}C(=O)), 57.6 (CH_2***C***H_{epo}), 41.1 (***C***H_2), 27.0 (***C***HMe_2), 24.9 (***C***H_3C(=O)), 23.1 (CH(***C***H_3)$_2$), 22.6 (CH(***C***H_3)$_2$).

MS (EI-DE) *m/z* (%) 142 [M^+] (2), 127 (6), 100 (10), 85 (100), 81 (6), 74 (2), 69 (3), 57 (26), 55 (12), 43 (79), 41 (14), 27 (11).

HRMS (CI-FE, *i*-butane) calcd for $C_8H_{15}O_2$ [$(M + H)^+$] 143.1071, found 143.1072.

(*E*)-(3*S*,4*R*)-3,4-Epoxy-6,6-dimethyl-2-heptanone ((3*S*,4*R*)-**93r**):

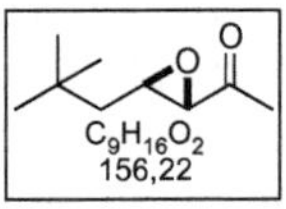

Epoxide (3*S*,4*R*)-**93r** was prepared according to the general procedure. The reaction mixture was stirred for 18 h at 50 °C. Purification by flash column chromatography (silica gel, 1–20% Et_2O in pentane) provided the title compound **93r** as a colorless oil (127 mg, 812 μmol, 81%; >99.5:0.5 *er*). The enantiomeric ratio was determined by GC using a chiral BGB-176/BGB-15 column 30 m (80 °C, 1.0 °C/min until 100 °C, 20 °C/min until 220 °C, 10 min at 220 °C, 0.5 bar H_2); major enantiomer: $\tau_R = 15.39$ min, minor enantiomer: $\tau_R = 16.03$ min.

^{1}H NMR (500 MHz, CD_2Cl_2) δ 3.10–3.07 (m, 2H, C***H***$_{epo}$), 2.04 (s, 3H, C***H***$_3$C(=O)), 1.53–1.44 (m, 2H, C***H***$_2$), 1.00 (s, 9 H, Cq(C***H***$_3$)$_3$).

^{13}C NMR (125 MHz, CD_2Cl_2) δ 206.2 (***C***=O), 60.2 (***C***H_{epo}C(=O)), 56.2 (CH_2***C***H_{epo}), 46.3 (***C***H_2), 31.1 (***C***qMe_3), 29.9 (3C, Cq(***C***H_3)$_3$), 25.0 (***C***H_3C(=O)).

GC-MS (GC-EI) *m/z* (%) 156 [M^+] (trace), 141 (1), 125 (1), 107 (1), 100 (11), 95 (4), 85 (53), 69 (16), 57 (97), 43 (100), 41 (41), 29 (24), 27 (7).

HRMS (CI-FE, *i*-butane) calcd for $C_9H_{17}O_2$ [$(M + H)^+$] 157.1227, found 157.1229.

(*E*)-(3*S*,4*R*)-3,4-Epoxy-5-methyl-2-hexanone ((3*S*,4*R*)-**93s**):

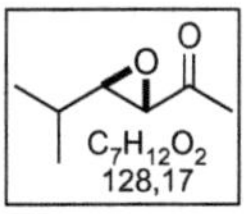

Epoxide (3*S*,4*R*)-**93s** was prepared according to the general procedure. 20 mol% of catalyst [**13** · 2 TFA] was used under otherwise identical conditions. The reaction mixture was stirred for 48 h at 50 °C. Purification by flash column chromatography (silica gel, 1–20% Et_2O in pentane) provided the title compound **93s** as a colorless oil [64 mg, 497 μmol, 50% (*reduced yield due to the high volatility of* **93 s**); 98:2 *er*]. The enantiomeric ratio was determined by GC using a chiral BGB-176/BGB-15 column 30 m (15 min at 60 °C, 20 °C/min until 220 °C,

10 min at 220 °C, 0.6 bar H_2); major enantiomer: $\tau_R = 8.34$ min, minor enantiomer: $\tau_R = 10.30$ min.

^{1}H NMR (500 MHz, CD_2Cl_2) δ 3.19 (d, $J = 1.9$ Hz, 1H, C***H***$_{epo}$C(=O)), 2.87 (dd, $J = 6.6$, 1.9 Hz, 1H, CHC***H***$_{epo}$), 2.03 (s, 3H, C***H***$_3$C(=O)), 1.62 (hept, $J = 6.7$ Hz, 1H, C***H***Me$_2$), 1.03 (d, $J = 6.6$ Hz, 3H, CH(C***H***$_3$)$_2$), 0.97 (d, $J = 7.0$, 3H, CH(C***H***$_3$)$_2$).

^{13}C NMR (125 MHz, CD_2Cl_2) δ 206.1 (***C***=O), 63.4 (CH***C***H$_{epo}$), 59.2 (***C***H$_{epo}$C(=O)), 30.8 (***C***HMe$_2$), 24.6 (***C***H$_3$C(=O)), 18.9 (CH(***C***H$_3$)$_2$), 18.2 (CH(***C***H$_3$)$_2$).

GC-MS (GC-EI) *m/z* (%) 128 [M$^+$] (trace), 113 (14), 97 (2), 95 (13), 85 (94), 83 (1), 69 (5), 67 (33), 65 (3), 59 (13), 55 (9), 45 (4), 43 (100), 41 (21), 39 (7), 29 (10).

(*E*)-(3*S*,4*R*)-3,4-Epoxy-4-cyclohexyl-2-butanone ((3*S*,4*R*)-**93e**):

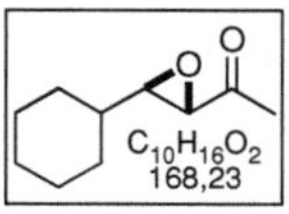

Epoxide (3*S*,4*R*)-**93e** was prepared according to the general procedure. 20 mol% of catalyst [**13** · 2 TFA] was used under otherwise identical conditions. The reaction mixture was stirred for 48 h at 50 °C. Purification by flash column chromatography (silica gel, 1–20% Et_2O in pentane) provided the title compound **93e** as a colorless oil (140 mg, 842 μmol, 84%; 98.5:1.5 *er*). The enantiomeric ratio was determined by GC using a chiral BGB-176/SE-52 column 29.5 m (80 °C, 1.2 °C/min until 135 °C, 18 °C/min until 220 °C, 10 min at 220 °C, 0.5 bar H_2); major enantiomer: $\tau_R = 33.22$ min, minor enantiomer: $\tau_R = 34.35$ min.

^{1}H NMR (500 MHz, CD_2Cl_2) δ 3.21 (d, $J = 2.3$ Hz, 1H, C***H***$_{epo}$C(=O)), 2.86 (dd, $J = 6.5$, 2.3 Hz, 1H, CHC***H***$_{epo}$), 2.02 (s, 3H, C***H***$_3$), 1.85–1.80 (m, 1H, C***H***$_2$), 1.78–1.72 (m, 2H, C***H***$_2$), 1.72–1.64 (m, 2H, C***H***$_2$), 1.35–1.07 (m, 6H, C***H***$_{cycl.}$ and C***H***$_2$).

^{13}C NMR (125 MHz, CD_2Cl_2) δ 206.3 (***C***=O), 62.7 (CH***C***H$_{epo}$), 59.3 (***C***H$_{epo}$ C(=O)), 40.4 (***C***H$_{cycl.}$), 30.0 (***C***H$_2$), 29.3 (***C***H$_2$), 26.7 (***C***H$_2$), 26.1 (***C***H$_2$), 26.0 (***C***H$_2$), 24.8 (***C***H$_3$).

MS (EI-DE) *m/z* (%) 168 [M$^+$] (2), 153 (1), 125 (1), 113 (3), 108 (7), 100 (2), 95 (11), 85 (100), 81 (22), 70 (5), 67 (17), 55 (24), 43 (45), 29 (10).

HRMS (ESI+) calcd for $C_{10}H_{16}NaO_2$ [(M + Na)$^+$] 191.1042, found 191.1042.

(*E*)-(3*S*,4*R*)-3,4-Epoxy-7-octen-2-one ((3*S*,4*R*)-**93f**):

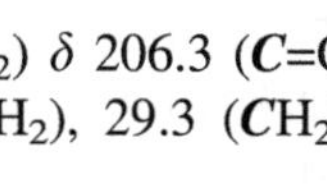

Epoxide (3*S*,4*R*)-**93f** was prepared according to the general procedure. The reaction mixture was stirred for 12 h at 50 °C. Purification by flash column chromatography (silica gel, 1–20% Et_2O in pentane) provided the title compound **93f** as a colorless oil

(107 mg, 764 μmol, 76%; 98.5:1.5 *er*). The enantiomeric ratio was determined by GC using a chiral BGB-176/BGB-15 column 30 m (80 °C, 0.8 °C/min until 100 °C, 20 °C/min until 220 °C, 10 min at 220 °C, 0.5 bar H_2); major enantiomer: τ_R = 11.06 min, minor enantiomer: τ_R = 12.59 min.

^{1}H NMR (500 MHz, CD_2Cl_2) δ 5.84 (ddt, J = 17.0, 10.3, 6.6 Hz, 1H, C***H***=CH_2), 5.07 (ddd, J = 17.1, 3.4, 1.7 Hz, 1H, CH=C$\boldsymbol{H}_{trans}$H), 5.01 (ddd, J = 10.2, 2.9, 1.3 Hz, 1H, CH=CH$\boldsymbol{H}_{cis}$), 3.18 (d, J = 1.9 Hz, 1H, C$\boldsymbol{H}_{epo}$C(=O)), 3.08 (td, J = 5.5, 1.9 Hz, 1H, CH_2C$\boldsymbol{H}_{epo}$), 2.28–2.17 (m, 2H, CH_2=CHC$\boldsymbol{H}_2$), 2.03 (s, 3H, C$\boldsymbol{H}_3$), 1.75–1.66 (m, 2H, C$\boldsymbol{H}_2$$CH_{epo}$).

^{13}C NMR (125 MHz, CD_2Cl_2) δ 205.9 (***C***=O), 137.7 (***C***H=CH_2), 115.9 (CH=$\boldsymbol{C}H_2$), 60.4 ($\boldsymbol{C}H_{epo}$C(=O)), 58.1 ($CH_2$$\boldsymbol{C}H_{epo}$), 31.6 ($\boldsymbol{C}H_2$), 30.5 ($\boldsymbol{C}H_2$), 25.0 ($\boldsymbol{C}H_3$).

MS (EI-DE) *m/z* (%) 140 [M^+] (trace), 107 (3), 97 (34), 85 (30), 79 (8), 73 (3), 67 (11), 57 (22), 55 (11), 43 (100), 41 (34), 39 (18), 27 (13).

HRMS (CI-FE, *i*-butane) calcd for $C_8H_{13}O_2$ [$(M + H)^+$] 141.0916, found 141.0916.

***(E)*-(3*S*,4*R*)-7-Bromo-3,4-epoxy-2-heptanone** ((3*S*,4*R*)-**93g**):

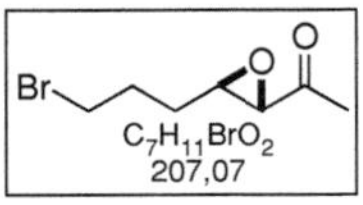

Epoxide (3*S*,4*R*)-**93g** was prepared according to the general procedure. The reaction mixture was stirred for 18 h at 50 °C. Purification by flash column chromatography (silica gel, 10–25% Et_2O in pentane) provided the title compound **93g** as a colorless oil (77 mg, 374 μmol, 75%; 98.5:1.5 *er*). The enantiomeric ratio was determined by GC using a chiral BGB-176/BGB-15 column 30 m (100 °C, 1.2 °C/min until 135 °C, 18 °C/min until 220 °C, 5 min at 220 °C, 0.5 bar H_2); major enantiomer: τ_R = 19.40 min, minor enantiomer: τ_R = 20.75 min. *Contains 11% (determined by GC) of (E)-7-chloro-3,4-epoxyheptan-2-one.*

^{1}H NMR (400 MHz, CD_2Cl_2) δ 3.47 (td, J = 6.6, 1.5 Hz, 2H, BrC$\boldsymbol{H}_2$), 3.20 (d, J = 2.1 Hz, 1H, C_{epo}***H***C(=O)), 3.11–3.08 (m, 1H, C_{epo}***H***CH_2), 2.05–1.98 (m, 2H, BrCH_2C$\boldsymbol{H}_2$), 2.03 (s, 3H, C$\boldsymbol{H}_3$), 1.92–1.83 (m, 1H, C***H***HC_{epo}H), 1.73–1.65 (m, 1H, CH***H***C_{epo}H).

^{13}C NMR (100 MHz, CD_2Cl_2): δ 205.2 (***C***=O), 59.5 ($\boldsymbol{C}_{epo}$HC(=O)), 57.1 ($\boldsymbol{C}_{epo}$HCH_2), 33.0 (Br$\boldsymbol{C}H_2$), 30.3 ($\boldsymbol{C}H_2$$C_{epo}$H), 29.1 (Br$CH_2$$\boldsymbol{C}H_2$), 24.4 ($\boldsymbol{C}H_3$).

MS (EI) *m/z* (%) 165 (35), 163 (36), 135 (2), 121 (1), 109 (2), 95 (1), 85 (67), 69 (1), 55 (32), 43 (100), 27 (15).

HRMS (CI-FE) calcd for $C_7H_{12}BrO_2$ [$(M + H)^+$] 207.0020, found 207.0021.

(*E*)-(3*S*,4*R*)-3,4-Epoxy-9-(tetrahydro-2*H*-pyran-2-yloxy)nonen-2-one ((3*S*,4*R*)-**93h**):

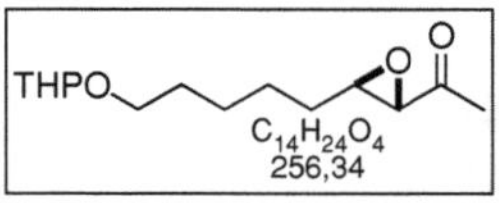

Epoxide (3*S*,4*R*)-**93h** was prepared according to the general procedure. The reaction mixture was stirred for 18 h at 50 °C. Purification by flash column chromatography (silica gel, 20–30% Et_2O in pentane) provided the title compound **93h** as a colorless oil (113 mg, 440 μmol, 88%). The enantiomeric ratio was determined after deprotection of the Thp ether (cf. epoxide **93l**; *vide infra*). *Characterized as mixture of Thp ether diastereomers.*

^{1}H NMR (500 MHz, CD_2Cl_2) δ 4.53 (t, J = 3.7 Hz, 1H, O–C***H***–O), 3.84–3.79 (m, 1H, OC***H***H_{THP}), 3.69 (dt, J = 9.5, 6.6 Hz, 1H, THPOC***H***H), 3.48–3.43 (m, 1H, OCH$\boldsymbol{H}_{THP}$), 3.35 (dt, J = 9.5, 6.6 Hz, 1H, THPOCH***H***), 3.15 (d, J = 1.9 Hz, 1H, C_{epo}***H***C(=O)), 3.06 (td, J = 5.2, 1.9 Hz, 1H, C_{epo}***H***CH_2), 2.03 (s, 3H, C$\boldsymbol{H}_3$), 1.83–1.75 (m, 1H, C$\boldsymbol{H}_2$), 1.69–1.39 (m, 13H, C$\boldsymbol{H}_2$).

^{13}C NMR (125 MHz, CD_2Cl_2) 205.7 (***C***=O), 98.9 (O–***C***H–O), 67.2 (THPO***C***H_2), 62.2 (O***C***$H_{2,\ THP}$), 59.9 ($\boldsymbol{C}_{epo}$HC(=O)), 58.0 ($\boldsymbol{C}_{epo}$HCH_2), 31.8 (***C***H_2), 30.9 (OCH***C***H_2), 29.7 (***C***H_2), 26.0 (***C***H_2), 25.7 (***C***H_2), 25.6 (***C***H_2), 24.3 (***C***H_3), 19.8 (***C***H_2).

MS (EI-DE) *m/z* (%) 256 [M^+] (trace), 239 (1), 225 (2), 185 (1), 156 (11), 140 (10), 126 (6), 111 (2), 97 (17), 85 (100), 81 (10), 67 (14), 55 (14), 43 (50), 29 (8).

HRMS (ESI+) calcd for $C_{14}H_{24}NaO_4$ [$(M + Na)^+$] 279.1565, found 279.1567.

(*E*)-(3*S*,4*R*)-3,4-Epoxy-9-hydroxy-2-nonanone (**93l**):

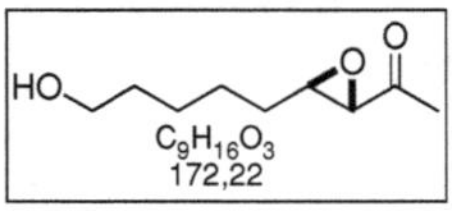

To a vial containing Thp ether **93h** (40.0 mg, 156 μmol) dissolved in acetonitrile (1.6 mL) and borate buffer (pH = 8) (1.6 mL), was added with stirring CAN (2.7 mg, 4.68 μmol, 3 mol%). After 3 h at 60 °C, the mixture was cooled to room temperature, diluted with water (4 mL) and repeatedly extracted with Et_2O (3 × 10 mL). The combined organic layers were washed with brine, dried over Na_2SO_4, filtered and concentrated under reduced pressure. After purification of the crude product by flash column chromatography (silica gel, 5–30% Et_2O in CH_2Cl_2) the alcohol **93l** (24 mg, 138 μmol, 88%; 98:2 *er*) was obtained as a colorless oil. The enantiomeric ratio was determined by GC using a chiral Ivadex 1 column 25 m (80 °C, 1.0 °C/min until 155 °C, 20 °C/min until 220 °C, 10 min at 320 °C, 0.6 bar H_2); major enantiomer: τ_R = 60.50 min, minor enantiomer: τ_R = 61.21 min.

^{1}H NMR (500 MHz, CD_2Cl_2) δ 3.60 (t, J = 6.5 Hz, 2H, C$\boldsymbol{H}_2$OH), 3.15 (d, J = 1.6 Hz, 1H, C_{epo}***H***C=O), 3.06 (td, J = 5.6, 1.6 Hz, 1H, CH_2C_{epo}***H***), 2.03 (s, 3H, C$\boldsymbol{H}_3$), 1.69–1.37 (m, 9H, C$\boldsymbol{H}_2$ and O***H***).

^{13}C NMR (125 MHz, CD_2Cl_2) 205.7 (*C*=O), 62.6 (*C*H_2OH), 59.9 (*C*$_{epo}$HC(=O)), 58.0 (CH_2*C*$_{epo}$H), 32.7 (*C*H_2), 31.8 (*C*H_2), 25.7 (*C*H_2), 25.5 (*C*H_2), 24.3 (*C*H_3).

MS (EI) *m/z* (%) 129 (1), 111 (2), 99 (3), 93 (4), 85 (36), 81 (13), 67 (8), 55 (31), 43 (100), 39 (9), 31 (18).

HRMS (CI-DE) calcd for $C_9H_{17}O_3$ [(M + H)$^+$] 173.1176, found 173.1178.

(*E*)-(3*S*,4*R*)-3,4-Epoxy-6-*tert*-butyldimethylsilyloxyhexan-2-one ((3*S*,4*R*)-**93i**):

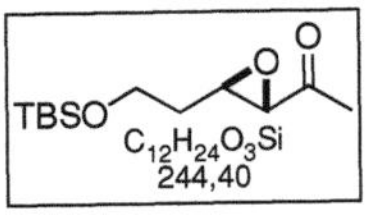

Epoxide **93i** was prepared according to the general procedure. The reaction mixture was stirred for 18 h at 50 °C. Purification by flash column chromatography (silica gel, 1–10% Et_2O in pentane) provided the title compound **93i** as a colorless oil (88 mg, 359 μmol, 72%; 98.5:1.5 *er*). The enantiomeric ratio was determined by GC using a chiral BGB-176/BGB-15 column 30 m (80 °C, 1.2 °C/min until 125 °C, 20 °C/min until 220 °C, 10 min at 220 °C, 0.5 bar H_2); major enantiomer: τ_R = 34.31 min, minor enantiomer: τ_R = 35.48 min.

^{1}H NMR (500 MHz, CD_2Cl_2) δ 3.76 (t, J = 6.3 Hz, 2H, TBSOC***H***$_2$), 3.22–3.19 (m, 2H, C$_{epo}$***H***), 2.03 (s, 3H, C***H***$_3$C(=O)), 1.87–1.81 (m, 1H, C$_{epo}$HC***H***H), 1.79–1.73 (m, 1H, C$_{epo}$HCH***H***), 0.90 (s, 9H, Cq(C***H***$_3$)$_3$), 0.07 (s, 3H, Si(C***H***$_3$)$_2$), 0.06 (s, 3H, Si(C***H***$_3$)$_2$).

^{13}C NMR (125 MHz, CD_2Cl_2) δ 205.4 (*C*=O), 59.9 (*C*$_{epo}$HC(=O)), 59.6 (TBSO*C*H_2), 55.9 (CH_2*C*$_{epo}$H), 35.1 (*C*H_2C$_{epo}$H), 25.7 (3C, C(*C*H_3)$_3$), 24.3 (*C*qMe$_3$), 18.2 (*C*H_3C(=O)), −5.7 (2C, Si(*C*H_3)$_2$).

MS (EI) *m/z* (%) 220 (1), 205 (2), 199 (4), 187 (11), 169 (1), 157 (100), 143 (8), 131 (2), 115 (7), 99 (4), 85 (3), 75 (16), 59 (4), 43 (9), 29 (1).

HRMS (ESI+) calcd for $C_{12}H_{24}O_3SiNa$ [(M + Na)$^+$] 267.1386, found 267.1387.

(*E*)-(3*S*,4*R*)-Ethyl 4,5-epoxy-6-oxoheptanoate ((3*S*,4*R*)-**93j**):

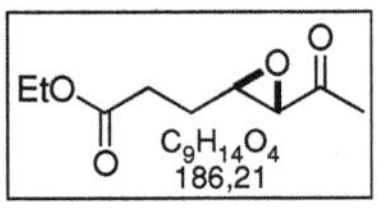

Epoxide (3*S*,4*R*)-**93j** was prepared according to the general procedure. The reaction mixture was stirred for 24 h at 50 °C. Base treatment of the crude product with sodium ethoxide in THF followed by purification by flash column chromatography (silica gel, 15–30% Et_2O in pentane) provided the title compound as a colorless oil (75 mg, 405 μmol, 81%; 98:2 *er*). The enantiomeric ratio was determined by GC using a chiral BGB-176/SE-52 column 30 m (80 °C, 1.2 °C/min until 130 °C, 20 °C/min until 220 °C, 10 min at 220 °C, 0.5 bar H_2); major enantiomer: τ_R = 35.76 min, minor enantiomer: τ_R = 37.27 min.

^{1}H NMR (500 MHz, CD_2Cl_2) δ 4.12 (q, J = 7.2 Hz, 2H, C***H***$_2$CH$_3$), 3.20 (d, J = 2.1 Hz, 1H, C$_{epo}$***H***C(=O)), 3.16–3.14 (m, 1H, C$_{epo}$***H***CH$_2$), 2.45 (t, J = 7.3 Hz, 2H, C***H***$_2$CO$_2$Et), 2.03 (s, 3H, C***H***$_3$C(=O)), 2.03–1.97 (m, 1H, C***H***HC$_{epo}$H), 1.90–1.82 (m, 1H, CH***H***C$_{epo}$H), 1.24 (t, J = 7.3 Hz, 3H, CH$_2$C***H***$_3$).

^{13}C NMR (125 MHz, CD_2Cl_2): δ 205.1 (***C***=O), 172.3 (***C***O$_2$Et), 60.7 (***C***H$_2$CH$_3$), 59.8 (***C***$_{epo}$HC(=O)), 56.9 (***C***$_{epo}$HCH$_2$), 30.1 (***C***H$_2$CO$_2$Et), 27.0 (***C***H$_2$C$_{epo}$H), 24.5 (***C***H$_3$C(=O)), 14.0 (CH$_2$***C***H$_3$).

MS (EI) *m/z* (%) 186 [M$^+$] (trace), 143 (13), 125 (1), 115 (29), 99 (15), 85 (30), 69 (7), 55 (13), 43 (100), 29 (29).

HRMS (CI-FE) calcd for $C_9H_{15}O_4$ [(M + H)$^+$] 187.0972, found 187.0970.

(*E*)-(3*S*,4*R*)-7-Oxo-3,4-epoxy-2-octanone ((3*S*,4*R*)-**93k**):

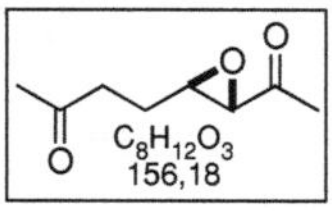

Epoxide (3*S*,4*R*)-**93k** was prepared according to the general procedure. The reaction mixture was stirred for 24 h at 50 °C. Base treatment of the crude product was omitted. Purification by flash column chromatography (silica gel, 10–30% Et_2O in pentane) provided the title compound as a colorless oil (31 mg, 202 μmol, 40%; 98.5:1.5 *er*). The enantiomeric ratio was determined by GC using a chiral BGB-176/SE-52 column 30 m (80 °C, 1.2 °C/min until 120 °C, 20 °C/min until 220 °C, 10 min at 220 °C, 0.5 bar H_2); major enantiomer: τ_R = 29.03 min, minor enantiomer: τ_R = 30.33 min.

^{1}H NMR (500 MHz, CD_2Cl_2) δ 3.17 (d, J = 1.8 Hz, 1H, C***H***$_{epo}$C(=O)), 3.12–3.10 (m, 1H, C$_{epo}$***H***CH$_2$), 2.56 (t, J = 7.0 Hz, 2H, C***H***$_2$C(=O)), 2.13 (s, 3H, CH$_2$C(=O)C***H***$_3$), 2.01 (s, 3H, CH$_{epo}$C(=O)C***H***$_3$), 2.00–1.93 (m, 1H, C***H***HC$_{epo}$H), 1.78–1.72 (m, 1H, CH***H***C$_{epo}$H).

^{13}C NMR (125 MHz, CD_2Cl_2) δ 207.1 (CH$_2$***C***=O), 205.6 (C$_{epo}$H***C***=O), 60.2 (***C***$_{epo}$HC(=O)), 57.3 (CH$_2$***C***$_{epo}$H), 39.2 (***C***H$_2$C(=O)), 30.0 (***C***H$_3$C(=O)CH$_2$), 25.9 (***C***H$_2$C$_{epo}$H), 24.7 (***C***H$_3$C(=O)CH$_{epo}$).

(*E*)-(4*S*,5*R*)-4,5-Epoxy-3-hexanone ((4*S*,5*R*)-**93c**):

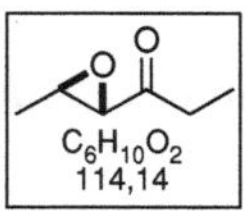

Epoxide (4*S*,5*R*)-**93c** was prepared according to the general procedure. The reaction mixture was stirred for 12 h at 50 °C. Purification by flash column chromatography (silica gel, 1–20% Et_2O in pentane) provided the title compound **93c** as a colorless oil [63 mg, 552 μmol, 55% (*reduced yield due to the high volatility of* **93c**); 98.5:1.5 *er*]. The enantiomeric ratio was determined by GC using a chiral BGB-178/OV-1701 30 m column (60 °C, 1.0 °C/min until 70 °C, 20 °C/min until 220 °C, 10 min at 220 °C, 0.4 bar H_2); major enantiomer: τ_R = 12.45 min, minor enantiomer: τ_R = 12.74 min.

^{1}H NMR (500 MHz, CD_2Cl_2) δ 3.15 (d, $J = 2.0$ Hz, 1H, C$\boldsymbol{H}_{epo}$C(=O)), 3.10 (qd, $J = 5.1, 2.0$ Hz, 1H, $CH_3C\boldsymbol{H}_{epo}$), 2.44 (dq, $J = 18.3, 7.4$ Hz, 1H, C$\boldsymbol{H}$H), 2.31 (dq, $J = 18.5, 7.2$ Hz, 1H, CH$\boldsymbol{H}$), 1.37 (d, $J = 5.2$ Hz, 3H, $C\boldsymbol{H}_3CH_{epo}$), 1.00 (t, $J = 7.3$ Hz, 3H, $CH_2C\boldsymbol{H}_3$).

^{13}C NMR (125 MHz, CD_2Cl_2) δ 208.3 ($\boldsymbol{C}$=O), 60.9 ($\boldsymbol{C}H_{epo}$C(=O)), 55.0 ($CH_3\boldsymbol{C}H_{epo}$), 31.2 ($\boldsymbol{C}H_2$), 17.9 ($\boldsymbol{C}H_3CH_{epo}$), 7.3 ($CH_2\boldsymbol{C}H_3$).

GC-MS (GC-EI) *m/z* (%) 114 [M^+] (1), 99 (17), 85 (14), 83 (2), 69 (70), 57 (75), 53 (6), 43 (14), 41 (42), 39 (18), 29 (100), 27 (26).

HRMS (EI-FE) calcd for $C_6H_{10}O_2$ [M^+] 114.0680, found 114.0681.

(*E*)-(4*S*,5*R*)-4,5-Epoxy-3-tetradecanone ((4*S*,5*R*)-**93p**):

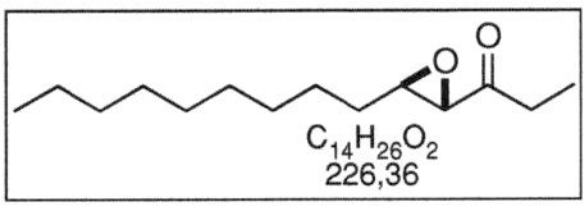

Epoxide (4*S*,5*R*)-**93p** was prepared according to the general procedure. The reaction mixture was stirred for 24 h at 50 °C. Purification by flash column chromatography (silica gel, 1–20% Et_2O in pentane) provided the title compound **93p** as a colorless oil (185 mg, 817 μmol, 82%; 99:1 *er*). The enantiomeric ratio was determined by GC using a chiral Lipodex E column 25 m (100 °C, 1.2 °C/min until 180 °C, 18 °C/min until 220 °C, 5 min at 220 °C, 0.5 bar H_2); major enantiomer: $\tau_R = 30.53$ min, minor enantiomer: $\tau_R = 29.18$ min.

^{1}H NMR (500 MHz, CD_2Cl_2) δ 3.19 (d, $J = 2.0$ Hz, 1H, C$\boldsymbol{H}_{epo}$C(=O)), 3.01 (td, $J = 5.5, 1.9$ Hz, 1H, $CH_2C\boldsymbol{H}_{epo}$), 2.44 (dq, $J = 18.3, 7.3$ Hz, 1H, C$\boldsymbol{H}$HC(=O)), 2.32 (dq, $J = 18.4, 7.2$ Hz, 1H, CH$\boldsymbol{H}$C(=O)), 1.62–1.55 (m, 2H, $C\boldsymbol{H}_2CH_{epo}$), 1.48–1.41 (m, 2H, $C\boldsymbol{H}_2CH_2CH_{epo}$), 1.35–1.26 (m, 12H, $(C\boldsymbol{H}_2)_6CH_3$), 1.01 (t, $J = 7.2$ Hz, 3H, $C(=O)CH_2C\boldsymbol{H}_3$), 0.88 (t, $J = 7.0$ Hz, 3H, $(CH_2)_8C\boldsymbol{H}_3$).

^{13}C NMR (125 MHz, CD_2Cl_2) δ 208.4 ($\boldsymbol{C}$=O), 60.0 ($\boldsymbol{C}H_{epo}$C(=O)), 59.0 ($CH_2\boldsymbol{C}H_{epo}$), 32.4 (2C, $\boldsymbol{C}H_2$), 31.2 ($\boldsymbol{C}H_2$), 30.0 (2C, $\boldsymbol{C}H_2$), 29.8 (2C, $\boldsymbol{C}H_2$), 26.3 ($\boldsymbol{C}H_2$), 23.2 ($\boldsymbol{C}H_2$), 14.4 ($(CH_2)_8\boldsymbol{C}H_3$), 7.4 ($C(=O)CH_2\boldsymbol{C}H_3$).

MS (EI-DE) *m/z* (%) 226 [M^+] (2), 197 (1), 179 (1), 169 (2), 151 (3), 127 (1), 109 (4), 99 (43), 95 (11), 81 (7), 69 (6), 57 (100), 41 (15), 29 (26).

HRMS (ESI+) calcd for $C_{14}H_{26}NaO_2$ [$(M + Na)^+$] 249.1826, found 249.1825.

(*E*)-(7*S*,8*R*)-7,8-Epoxy-6-tridecanone ((7*S*,8*R*)-**93m**):

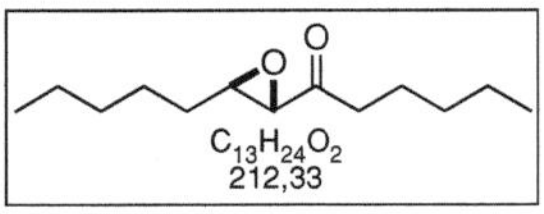

Epoxide (7*S*,8*R*)-**93m** was prepared according to the general procedure. The reaction mixture was stirred for 24 h at 50 °C. Purification by flash column chromatography (silica gel, 1–20% Et_2O in pentane) provided the title compound **93m** as a colorless oil (161 mg, 758 μmol, 76%; 99:1 *er*). The enantiomeric ratio was determined by GC using a chiral G-TA column 30 m (80 °C, 1.0 °C/min until

125 °C, 20 °C/min until 180 °C, 10 min at 180 °C, 0.9 bar H_2); major enantiomer: $\tau_R = 44.18$ min, minor enantiomer: $\tau_R = 43.42$ min.

^{1}H NMR (500 MHz, CD_2Cl_2) δ 3.18 (d, J = 2.0 Hz, 1H, C$\boldsymbol{H}_{epo}$C(=O)), 3.01 (td, J = 5.5, 2.0 Hz, 1H, $CH_2C\boldsymbol{H}_{epo}$), 2.41 (ddd, J = 17.3, 8.3, 6.5 Hz, 1H, C***H***HC(=O)), 2.28 (ddd, J = 17.3, 8.2, 6.8 Hz, 1H, CH***H***C(=O)), 1.62–1.52 (m, 4H, $C\boldsymbol{H}_2$), 1.48–1.42 (m, 2H, $C\boldsymbol{H}_2$), 1.34–1.23 (m, 8H, $C\boldsymbol{H}_2$), 0.91–0.87 (m, 6H, $C\boldsymbol{H}_3$).

^{13}C NMR (125 MHz, CD_2Cl_2) δ 208.1 (***C*** =O), 60.1 (***C***H_{epo}C(=O)), 58.9 ($CH_2\boldsymbol{C}H_{epo}$), 37.8 (***C***$H_2$C(=O)), 32.4 (***C***$H_2$), 32.0 (***C***$H_2$), 31.9 (***C***$H_2$), 26.0 (***C***$H_2$), 23.4 (***C***$H_2$), 23.1 (***C***$H_2$), 23.0 (***C***$H_2$), 14.3 (***C***$H_3$), 14.2 (***C***$H_3$).

MS (EI-DE) *m*/*z* (%) 212 [M^+] (6), 169 (1), 156 (3), 141 (58), 125 (4), 112 (7), 99 (98), 95 (12), 82 (8), 71 (76), 55 (18), 43 (100), 29 (30).

HRMS (ESI+) calcd for $C_{13}H_{24}NaO_2$ [$(M + Na)^+$] 235.1670, found 235.1668.

(*E*)-(5*S*,6*R*)-5,6-Epoxy-2-methyl-4-undecanone ((5*S*,6*R*)-**93n**):

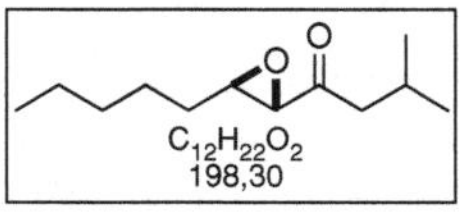

Epoxide (5*S*,6*R*)-**93e** was prepared according to the general procedure. 20 mol% of catalyst [**13** · 2 TFA] was used under otherwise identical conditions. The reaction mixture was stirred for 48 h at 50 °C. Base treatment of the crude product was omitted. Purification by flash column chromatography (silica gel, 1–20% Et_2O in pentane) provided the title compound **93n** as a colorless oil (161 mg, 812 µmol, 81%; >99.5:0.5 *er*). The enantiomeric ratio was determined by GC using a chiral G-TA column 30 m (80 °C, 1.0 °C/min until 115 °C, 20 °C/min until 180 °C, 10 min at 180 °C, 0.9 bar H_2); major enantiomer: $\tau_R = 30.89$ min, minor enantiomer: $\tau_R = 30.09$ min.

^{1}H NMR (500 MHz, CD_2Cl_2) δ 3.16 (d, J = 2.0 Hz, 1H, C$\boldsymbol{H}_{epo}$C(=O)), 2.99 (td, J = 5.5, 1.9 Hz, 1H, $CH_2C\boldsymbol{H}_{epo}$), 2.31 (dd, J = 16.0, 6.0 Hz, 1H, C***H***H*i*-Pr), 2.15 (dd, J = 16.2, 7.1 Hz, 1H, CH***H****i*-Pr), 2.11 (hept, J = 6.5 Hz, 1H, C***H***Me_2), 1.64–1.55 (m, 2H, $C\boldsymbol{H}_2CH_{epo}$), 1.48–1.42 (m, 2H, $C\boldsymbol{H}_2CH_2CH_{epo}$), 1.34–1.30 (m, 4H, $(C\boldsymbol{H}_2)_2CH_3$), 0.92–0.88 (m, 9H, $C\boldsymbol{H}_3$).

^{13}C NMR (125 MHz, CD_2Cl_2) δ 207.7 (***C***=O), 60.2 (***C***H_{epo}C(=O)), 58.7 ($CH_2\boldsymbol{C}H_{epo}$), 46.7 (***C***$H_2$*i*-Pr), 32.4 (***C***$H_2$), 32.0 (***C***$H_2$), 26.0 (***C***$H_2$), 24.6 (***C***$HMe_2$), 23.1 (***C***$H_2$), 22.9 ($CH(\boldsymbol{C}H_3)_2$), 22.8 ($CH(\boldsymbol{C}H_3)_2$), 14.3 ($(CH_2)_4\boldsymbol{C}H_3$).

MS (EI-DE) *m*/*z* (%) 198 [M^+] (1), 183 (1), 155 (1), 141 (2), 127 (33), 113 (5), 95 (6), 85 (75), 69 (5), 57 (100), 41 (30), 29 (18).

HRMS (ESI+) calcd for $C_{12}H_{22}NaO_2$ [$(M + Na)^+$] 221.1513, found 221.1512.

(*E*)-(4*S*,5*R*)-4,5-Epoxy-2-Methyl-3-hexanone ((4*S*,5*R*)-**93o**):

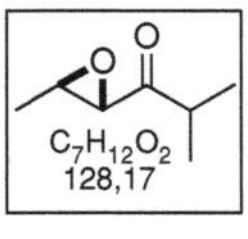

Epoxide (4*S*,5*R*)-**93o** was prepared according to the general procedure. 20 mol% of catalyst [**13** · 2 TFA] was used under otherwise identical conditions. The reaction mixture was stirred for 48 h at 50 °C. Purification by flash column chromatography (silica gel, 2–7% Et_2O in pentane) provided the title compound as a colorless liquid [77 mg, 604 μmol, 60% (*reduced yield due to the high volatility of* **93o**); 96.5:3.5 *er*]. THF (5 mL) was used as solvent for the base treatment with aqueous 1 N NaOH. The enantiomeric ratio was determined by GC using a chiral Lipodex G column 25 m (50 °C, 1.2 °C/min until 110 °C, 18 °C/min until 230 °C, 5 min at 230 °C, 0.5 bar H_2); major enantiomer: $\tau_R = 8.24$ min, minor enantiomer: $\tau_R = 8.76$ min.

^{1}H NMR (500 MHz, CD_2Cl_2) δ 3.25 (d, $J = 1.9$ Hz, 1H, $C_{epo}\boldsymbol{H}C(=O)$), 3.03 (dq, $J = 5.1$, 2.0 Hz, 1H, $MeC_{epo}\boldsymbol{H}$), 2.70 (hept, $J = 6.9$ Hz, 1H, $C\boldsymbol{H}Me_2$), 1.38 (d, $J = 5.0$ Hz, 3H, $C\boldsymbol{H}_3C_{epo}H$), 1.09 (d, $J = 7.3$ Hz, 3H, $CH(C\boldsymbol{H}_3)_2$), 1.06 (d, $J = 6.9$ Hz, 3H, $CH(C\boldsymbol{H}_3)_2$).

^{13}C NMR (125 MHz, CD_2Cl_2) δ 210.5 ($\boldsymbol{C}$=O), 59.5 ($\boldsymbol{C}_{epo}HC(=O)$), 55.0 ($Me\boldsymbol{C}_{epo}H$), 36.7 ($\boldsymbol{C}HMe_2$), 18.3 ($\boldsymbol{C}H_3C_{epo}H$), 17.7 ($CH(\boldsymbol{C}H_3)_2$), 17.4 ($CH(\boldsymbol{C}H_3)_2$).

GC-MS (GC-EI) *m/z* (%) 128 [M^+] (1), 113 (10), 85 (16), 83 (2), 71 (41), 69 (4), 58 (21), 55 (6), 45 (5), 43 (100), 41 (31), 39 (10), 29 (26).

HRMS (EI-FE) calcd for $C_7H_{12}O_2$ [M^+] 128.0836, found 128.0837.

7.3.4 Hydroperoxidation of 4,8-Dimethylnona-3,7-dien-2-one (100)

The general procedure [8] described in Sect. 7.3.1.1 for the catalytic asymmetric hydroperoxidation of α,β-unsaturated ketones **92** was adapted to β,β-disubstituted enone **100** by using 20 mol% catalyst [**13** · 2 TFA] and 3 equiv. of hydrogen peroxide (50 wt%) and the reaction temperature was increased to 70 °C.

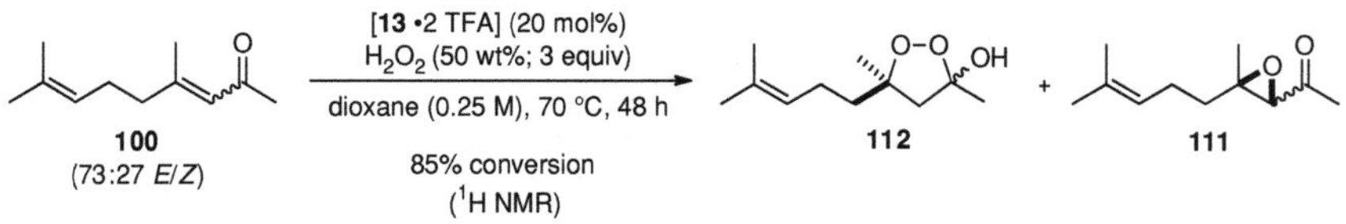

3,5-Dimethyl-5-(4-methylpent-3-enyl)-1,2-dioxolan-3-ol (112)

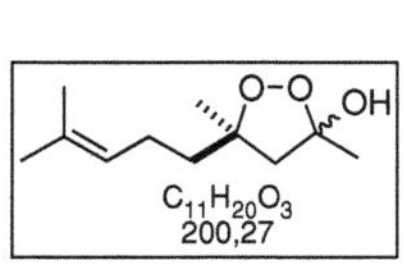

The title compound was isolated after 48 h. Purification by flash column chromatography (silica gel; 5–10% Et_2O in pentane) as a colorless oil (56 mg, 280 μmol, 56%; 78:22 *er*). The enantiomeric ratio was determined after converting peroxyhemiketal **112** into the corresponding epoxide **111** (1:1 *E*/*Z*). The enantiomers were analyzed by GC using a chiral BGB-176/SE-52 column 30 m (60 °C, 0.5 °C/min until 105 °C, 20 °C/min until 220 °C, 10 min at 320 °C, 0.4 bar H_2); (*E*)-**111**: major enantiomer: τ_R = 81.04 min, minor enantiomer: τ_R = 80.47 min; (*Z*)-**111**: major enantiomer: τ_R = 71.83 min, minor enantiomer: τ_R = 73.16 min. *Characterized as a mixture of hemiketal epimers.*

^{1}H NMR (500 MHz, $CDCl_3$) δ 5.10–5.06 (m, 2H, =C***H***CH_2), 2.93 (br s overlapped, 1H, O***H***), 2.90 (br s overlapped, 1H, O***H***), 2.53 (d, 1H, *J* = 12.8 Hz, C***H***$H_{cycl.}$), 2.44 (s, 2H, C$\boldsymbol{H}_{2,\ cycl.}$), 2.31 (d, 1H, *J* = 12.7 Hz, CH$\boldsymbol{H}_{cycl.}$), 2.09–1.94 (m, 4H, =CHC$\boldsymbol{H}_2$), 1.73 (ddd, 1H, *J* = 13.8, 11.4, 5.3 Hz, =CHCH_2C***H***H), 1.68–1.64 (m overlapped, 2H, =CHCH_2C$\boldsymbol{H}_2$), 1.66 (s overlapped, 6H, Cq_{ol}C$\boldsymbol{H}_3$), 1.59 (s, 6H, Cq_{ol}C$\boldsymbol{H}_3$), 1.54–1.48 (m overlapped, 1H, =CHCH_2CH***H***), 1.53 (s overlapped, 3H, Cq_{OH}C$\boldsymbol{H}_3$), 1.51 (s overlapped, 3H, Cq_{OH}C$\boldsymbol{H}_3$), 1.34 (s, 3H, Cq_{al}C$\boldsymbol{H}_3$), 1.33 (s, 3H, Cq_{al}C$\boldsymbol{H}_3$).

^{13}C NMR (125 MHz, $CDCl_3$) δ 132.3 (**C**q_{ol}), 131.9 (**C**q_{ol}), 123.8 (**C**H_{ol}), 123.5 (**C**H_{ol}), 105.9 (**C**q_{OH}), 105.7 (**C**q_{OH}), 86.3 (**C**q_{al}), 85.9 (**C**q_{al}), 57.7 (**C**$H_{2,\ cycl.}$), 57.1 (**C**$H_{2,\ cycl.}$), 39.8 ($CH_{ol}CH_2$**C**H_2), 37.9 ($CH_{ol}CH_2$**C**H_2), 25.7 (2C, Cq_{ol}(**C**$H_3)_2$), 25.2 (**C**H_3Cq_{al}), 23.5 (CH_{ol}**C**H_2), 23.2 (**C**H_3Cq_{OH}), 23.1 (**C**H_3Cq_{OH}), 23.0 (CH_{ol}**C**H_2), 22.3 (**C**H_3Cq_{al}), 17.7 (Cq_{ol}(**C**$H_3)_2$), 17.6 (Cq_{ol}(**C**$H_3)_2$).

MS (EI-DE) *m*/*z* (%) 185 (1), 167 (1), 139 (1), 123 (2), 109 (5), 97 (3), 82 (16), 69 (32), 55 (22), 43 (100), 29 (9).

HRMS (ESI+) calcd for $C_{11}H_{20}NaO_3$ [(M + Na)$^+$] 223.1307, found 223.1305.

3,4-Epoxy-4,8-dimethyl-7-nonen-2-one (**111**):

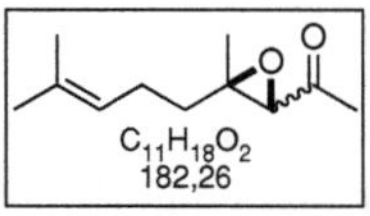

Epoxy ketone **111** was obtained as a colorless oil (9.1 mg, 50 μmol, 10%) [88:12 *dr*(*trans*/*cis*); 95.5:4.5 *er*(*trans*), 59.5:40.5 *er*(*cis*)]. Purification by flash column chromatography (silica gel; 1–10% Et_2O in pentane) provided pure samples of *trans*- and *cis*-isomers of epoxy ketone **111**. The enantiomeric ratio was determined by GC using a chiral BGB-176/SE-52 column 30 m (60 °C, 0.5 °C/min until 105 °C, 20 °C/min until 220 °C, 10 min at 320 °C, 0.4 bar H_2); (*E*)-**111**: major enantiomer: τ_R = 81.25 min, minor enantiomer: τ_R = 80.71 min; (*Z*)-**111**: major enantiomer: τ_R = 73.29 min, minor enantiomer: τ_R = 72.18 min. The analytical data were identical in all respects to those previously reported [9–11].

***trans*-3,4-Epoxy-4,8-dimethyl-7-nonen-2-one** (*trans*-**111**)

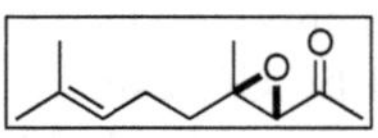

^{1}H NMR (500 MHz, CD_2Cl_2) δ 5.12 (t, *J* = 7.1 Hz, 1H, =C***H***CH_2), 3.40 (s, 1H, C_{epo}***H***), 2.17 (s, 3H, C$\boldsymbol{H}_3$C(=O)), 2.12 (app. q, 2H, =CHC$\boldsymbol{H}_2$), 1.77 (dd, *J* = 14.2,

7.1 Hz, 1H, C*H*HCq$_{epo}$), 1.70 (s, 3H, C*H*$_3$Cq$_{ol}$), 1.62 (s, 3H, C*H*$_3$Cq$_{ol}$), 1.56–1.49 (m, 1H, CH*H*Cq$_{epo}$), 1.20 (s, 3H, C*H*$_3$Cq$_{epo}$).

^{13}C NMR (125 MHz, CD_2Cl_2) δ 204.1 (*C*=O), 132.8 (*C*q$_{ol}$), 123.4 (*C*H$_{ol}$), 65.2 (*C*$_{epo}$H), 63.4 (*C*q$_{epo}$), 38.5 (CH$_{ol}$*C*H$_2$), 28.2 (*C*H$_3$C(=O)), 25.8 (*C*H$_3$), 24.1 (*C*H$_2$Cq$_{epo}$), 17.7 (*C*H$_3$), 16.1 (*C*H$_3$).

MS (EI-DE) *m/z* (%) 182 [M$^+$] (1), 164 (2), 149 (4), 139 (3), 121 (7), 109 (43), 99 (23), 82 (45), 69 (56), 67 (24), 55 (18), 43 (100), 27 (10).

HRMS (EI-FE) calcd for $C_{11}H_{18}O_2$ [M$^+$] 182.1307, found 182.1307.

***cis*-3,4-Epoxy-4,8-dimethyl-7-nonen-2-one** (*cis*-**111**)

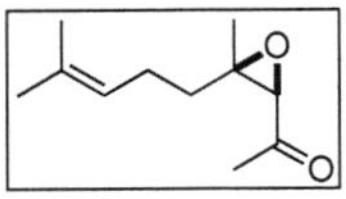

^{1}H NMR (500 MHz, CD_2Cl_2) δ 5.02 (t, J = 7.2 Hz, 1H, =C*H*CH$_2$), 3.37 (s, 1H, C$_{epo}$*H*), 2.19 (s, 3H, C*H*$_3$), 2.14–1.98 (m, 2H, =CHC*H*$_2$), 1.66 (s, 3H, C*H*$_3$Cq$_{ol}$), 1.59 (s, 3H, C*H*$_3$Cq$_{ol}$), 1.57–1.42 (m, 2H, C*H*$_2$Cq$_{epo}$), 1.40 (s, 3H, C*H*$_3$Cq$_{epo}$).

^{13}C NMR (125 MHz, CD_2Cl_2) δ 204.0 (*C*=O), 132.8 (*C*q$_{ol}$), 123.3 (*C*H$_{ol}$), 66.0 (*C*$_{epo}$H), 64.0 (*C*q$_{epo}$), 32.5 (CH$_{ol}$*C*H$_2$), 28.5 (*C*H$_3$C(=O)), 25.8 (*C*H$_3$), 24.4 (*C*H$_2$Cq$_{epo}$), 22.2 (*C*H$_3$), 17.6 (*C*H$_3$).

7.3.5 Hydroperoxidation of Cycloheptenone (46s) and Cyclooctenone (46v)

The reactions of 2-cycloheptenone (**46s**) and 2-cyclooctenone (**46v**) were performed according to the general procedure [8] described in Sect. 7.3.1.1 for the catalytic asymmetric hydroperoxidation of acyclic α,β-unsaturated ketones **92**.

7,8-Dioxabicyclo[4.2.1]nonan-1-ol/3-hydroperoxycycloheptanone (**116**):

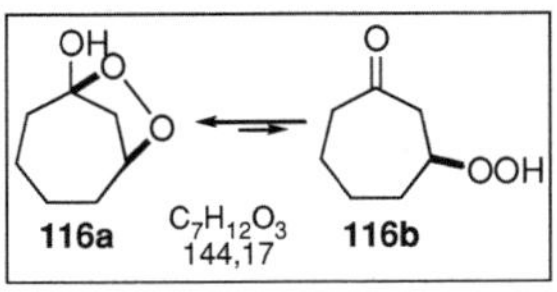

The reaction mixture was stirred for 24 h at 32 °C. Purification of the crude product by flash column chromatography (silica gel, 20–60% Et_2O in pentane) provided peroxide **116** as a clear oil (79 mg, 551 µmol, 55%; 95:5 *er*). The enantiomeric ratio was determined after converting peroxide **116** into 2,3-epoxycycloheptanone (**48s**). The enantiomers were analyzed by GC using a chiral Hydrodex-β-TBDAc column 25 m (80 °C, 1.5 °C/min until 135 °C, 20 °C/min until 220 °C, 10 min at 220 °C, 0.6 bar H_2); major enantiomer: τ_R = 24.31 min, minor enantiomer:

τ_R = 28.24 min. 7,8-*Dioxabicyclo*[4.2.1]*nonan*-1-*ol* (**116a**) *exists in equilibrium with* 3-*hydroperoxycycloheptanone* (**116b**): 80:20 (*in THF*).

^{1}H NMR (500 MHz, THF-d8) 7,8-*dioxabicyclo*[4.2.1]*nonan*-1-*ol* (**116a**): δ 5.43 (br s, 1H, O***H***), 4.42–4.40 (m, 1H, C***H***), 2.61 (d, J = 12.6 Hz, 1H, CHC***H***HCq), 2.50 (dd, J = 12.0, 6.7 Hz, 1H, CHCH***H***Cq), 1.93–1.83 (m, 2H, C$\boldsymbol{H}_2$), 1.77–1.74 (m, 3H, C$\boldsymbol{H}_2$), 1.61–1.53 (m, 2H, C$\boldsymbol{H}_2$), 1.43–1.37 (m, 1H, CH***H***CH); 3-*hydroperoxycycloheptanone* (**116b**): δ 9.39 (br s, 1H, OO***H***), 4.09–4.04 (m, 1H, C***H***), 2.77 (dd, J = 14.7, 3.0 Hz, 1H, CHC***H***HC(=O)), 2.68 (dd, J = 14.8, 8.8 Hz, 1H, CHCH***H***C(=O)), 2.42–2.31 (m, 2H, $CH_2C\boldsymbol{H}_2C$(=O)), 1.93–1.83 (m, 2H, C$\boldsymbol{H}_2$), 1.77–1.74 (m, 4H, C$\boldsymbol{H}_2$).

^{13}C NMR (125 MHz, THF-d8) 7,8-*dioxabicyclo*[4.2.1]*nonan*-1-*ol* (**116a**): δ 108.1 (***C***q), 81.3 (***C***H), 48.9 (CH***C***H_2Cq), 37.3 (***C***H_2), 35.9 (***C***H_2), 24.7 (***C***H_2), 24.3 (***C***H_2); 3-*hydroperoxycycloheptanone* (**116b**): δ 209.3 (***C***=O), 81.0 (***C***H), 48.4 (CH***C***H_2C(=O)), 45.1 (CH_2***C***H_2C(=O)), 35.0 (***C***H_2), 26.2 (***C***H_2), 25.4 (***C***H_2).

MS (EI-DE) *m*/*z* (%) 144 [M^+] (10), 126 (1), 111 (100), 97 (16), 83 (18), 69 (12), 55 (52), 41 (37), 39 (15), 29 (23).

HRMS (EI-FE) calcd for $C_7H_{12}O_3$ [M^+] 144.0787, found 144.0786.

3-Hydroperoxycyclooctanone/8,9-dioxabicyclo[5.2.1]decan-1-ol (117):

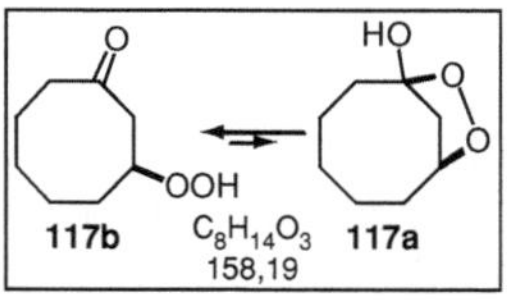

The reaction mixture was stirred for 24 h at 32 °C. Purification of the crude product by flash column chromatography (silica gel, 5–35% Et_2O in pentane) provided peroxide **117** as a clear oil (93 mg, 571 μmol, 59%; 97:3 *er*). The enantiomeric ratio was determined after converting peroxide **117** into 2,3-epoxycyclooctanone (**48v**). The enantiomers were analyzed by GC using a chiral BGB-176/SE-52 30 m (80 °C, 2 °C/min until 140 °C, 20 °C/min until 220 °C, 10 min at 220 °C, 0.6 bar H_2); major enantiomer: τ_R = 27.52 min, minor enantiomer: τ_R = 25.41 min. 3-*Hydroperoxycyclooctanone* (**117b**) *exists in equilibrium with* 8,9-*dioxabicyclo*[5.2.1]*decan*-1-*ol* (**117a**): 70:30 (*in THF*).

^{1}H NMR (500 MHz, CD_2Cl_2) 3-*Hydroperoxycyclooctanone* (**117b**): δ 8.98 (br s, 1H, OO***H***), 4.32-4.27 (m, 1H, C***H***), 2.91 (dd, J = 12.1, 3.6 Hz, 1H, CHC***H***HC(=O)), 2.87 (d, J = 11.7 Hz, 1H, CHCH***H***C(=O)), 2.43–2.32 (m, 2H, $CH_2C\boldsymbol{H}_2C$(=O)), 1.97–1.91 (m, 2H, C$\boldsymbol{H}_2$), 1.88–1.23 (m, 6H, –(C$\boldsymbol{H}_2$)$_n$–); 8,9-*Dioxabicyclo*[5.2.1]*decan*-1-*ol* (**117a**): δ 4.58–4.56 (m, 1H, C***H***), 3.02 (br s, 1H, O***H***), 2.88–2.84 (m, 2H, CHC$\boldsymbol{H}_2$Cq), 2.19–2.13 (m, 1H, C$\boldsymbol{H}_2$), 1.98–1.28 (m, 8H, –(C$\boldsymbol{H}_2$)$_n$–), 0.88 (t, J = 7.6 Hz, 1H, C$\boldsymbol{H}_2$).

^{13}C NMR (125 MHz, CD_2Cl_2) 3-*Hydroperoxycyclooctanone* (**117b**): δ 213.9 (***C***=O), 82.8 (***C***H), 44.0 (CH***C***H_2C(=O)), 42.4 (CH_2***C***H_2C(=O)), 29.6 (***C***H_2), 28.1 (***C***H_2), 22.9 (***C***H_2), 19.7 (***C***H_2); 8,9-*Dioxabicyclo*[5.2.1]*decan*-1-*ol* (**117a**): δ 105.5

(*C*q), 80.1 (*C*H), 52.6 (CH*C*H$_2$Cq), 37.3 (*C*H$_2$), 35.5 (*C*H$_2$), 28.0 (*C*H$_2$), 26.2 (*C*H$_2$), 25.3 (*C*H$_2$).

MS (EI-DE) *m/z* (%) 158 [M$^+$] (18), 141 (2), 125 (64), 107 (18), 97 (17), 83 (21), 69 (12), 55 (100), 43 (56), 41 (44), 29 (21).

HRMS (CI-FE, *i*-butane) calcd for $C_8H_{15}O_3$ [(M + H)$^+$] 159.1022, found 159.1021.

7.4 Synthetic Transformations of Optically Active Products

7.4.1 One-Pot Synthesis of Aldol Products

7.4.1.1 General Procedure

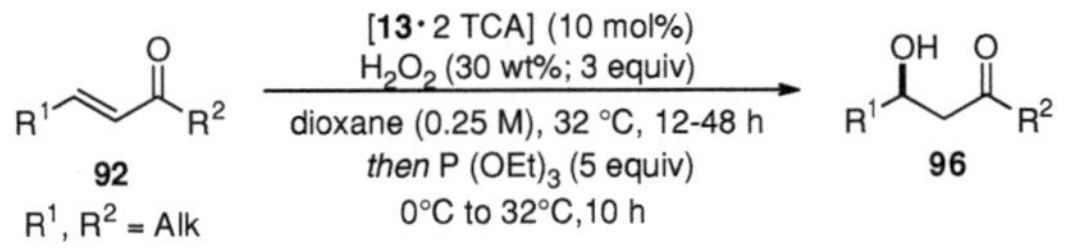

Reference [8] Catalyst salt [**13** · 2 TCA] was prepared *in situ* by the addition of 9-amino(9-deoxy)*epi*quinine (**13**; 32.3 mg, 0.1 mmol, 10 mol%) to a solution of trichloroacetic acid (32.6 mg, 0.2 mmol, 20 mol%) in dioxane (4 mL). Then, enone **92** (1.0 mmol, 1.0 equiv) was added, and 20 min later, aqueous hydrogen peroxide (30 wt%, 304 μL, 3.0 mmol, 3 equiv). After 36–48 h of stirring at 32 °C, triethylphosphite (519 μL, 3.0 mmol, 3 equiv) was added dropwise at 0 °C and the reaction mixture was stirred for 10 h at room temperature. Additional triethylphosphite (346 μL, 2 mmol, 2 equiv) was added and the reaction was further stirred until TLC analysis indicated complete reduction of the peroxyhemiketals (2 h). The reaction mixture was repeatedly extracted with Et_2O (3 × 25 mL) and the combined organic phases were washed with brine, dried (Na_2SO_4), and filtered. Removal of the volatiles under reduced pressure furnished the crude product, which was subjected to flash column chromatography (silica gel, eluent: Et_2O-pentane) to afford the corresponding pure aldol-type product **96**.

7.4.1.2 Scope of Optically Active Aldol Products

(*R*)-4-Hydroxy-2-decanone ((*R*)-**96a**):

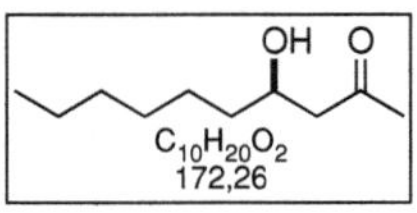

Aldol product (*R*)-**96a** was prepared according to the general procedure. The reaction mixture was stirred for 36 h at 32 °C followed by *in situ* reduction with triethylphosphite. Purification of the crude product by flash column chromatography (silica gel, 10–50% Et_2O in pentane) provided the title compound as a colorless oil (102 mg, 592 μmol, 59%; 97:3 *er*). The enantiomeric ratio was determined by GC using a chiral G-TA column 30 m (80 °C, 1.0 °C/min until 115 °C, 20 °C/min, until 180 °C, 10 min at 320 °C, 0.9 bar H_2); major enantiomer: $\tau_R = 31.94$ min, minor enantiomer: $\tau_R = 31.38$ min.

^{1}H NMR (500 MHz, CD_2Cl_2) δ 4.00–3.95 (m, 1H, C***H***(OH)), 2.81 (br s, 1H, O***H***), 2.61 (dd, *J* = 17.5, 2.7 Hz, 1H, C(=O)C***H***H), 2.48 (dd, *J* = 17.5, 9.3 Hz, 1H, C(=O)CH***H***), 2.14 (s, 3H, C(=O)C***H***$_3$), 1.46–1.35 (m, 2H, CH_2C***H***$_2$CH(OH)), 1.34–1.26 (m, 8H, CH_3(C***H***$_2$)$_4$), 0.88 (t, *J* = 6.9 Hz, 3H, C***H***$_3$$CH_2$).

^{13}C NMR (125 MHz, CD_2Cl_2) δ 210.3 (***C***(=O)), 67.9 (***C***H(OH)), 50.3 (C(=O)***C***H_2), 36.9 (CH_2***C***H_2CH(OH)), 32.2 (***C***H_2), 30.9 (C(=O)***C***H_3), 29.6 (***C***H_2), 25.8 (***C***H_2), 23.0 (***C***H_2), 14.2 ((CH_2)$_5$***C***H_3).

MS (EI-DE) *m*/*z* (%) 172 [M^+] (trace), 154 (1), 139 (1), 125 (1), 111 (1), 96 (6), 87 (44), 84 (2), 69 (5), 55 (17), 43 (100), 29 (11).

HRMS (CI-FE, NH_3) calcd for $C_{10}H_{24}NO_2$ [$(M + NH_4)^+$] 190.1805, found 190.1807.

(*R*)-4-Hydroxy-6-phenyl-2-hexanone ((*R*)-**96b**):

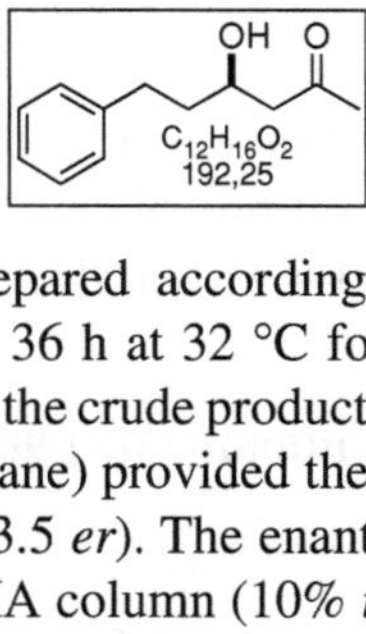

Aldol product (*R*)-**96b** was prepared according to the general procedure. The reaction mixture was stirred for 36 h at 32 °C followed by *in situ* reduction with triethylphosphite. Purification of the crude product by flash column chromatography (silica gel, 10–60% Et_2O in pentane) provided the title compound as a colorless oil (102 mg, 531 μmol, 53%; 96.5:3.5 *er*). The enantiomeric ratio was determined by HPLC using a chiral Chiralpak IA column (10% *i*-PrOH in heptane, 0.5 mL/min); major enantiomer: $\tau_R = 16.75$ min, minor enantiomer: $\tau_R = 15.41$ min.

Optical rotation $[\alpha]_D^{25} = 14.8$ (c = 1.1, $CHCl_3$, 96.5:3.5 *er* (*R*)) [Lit.: [3] (*R*)-**96b**, $[\alpha]_D^{rt} = 12.0$ (c = 1.1, $CHCl_3$, 94:6 *er*].

^{1}H NMR (500 MHz, CD_2Cl_2) δ 7.29–7.26 (m, 2H, C***H***$_{ar,\,m}$), 7.21–7.16 (m, 3H, C***H***$_{ar,\,o,\,p}$), 4.03–3.97 (m, 1H, C***H***(OH)), 2.95 (d, *J* = 3.5 Hz, 1H, O***H***), 2.81–2.75 (m, 1H, PhC***H***H), 2.69–2.64 (m, 1H, PhCH***H***), 2.63 (dd, *J* = 17.7, 2.9 Hz, 1H, C(=O)C***H***H), 2.53 (dd, *J* = 17.7, 9.2 Hz, 1H, C(=O)CH***H***), 2.14 (s, 3H, C***H***$_3$), 1.80–1.73 (m, 1H, PhCH_2C***H***H), 1.70–1.64 (m, 1H, PhCH_2CH***H***).

^{13}C NMR (125 MHz, CD_2Cl_2) δ 210.2 (***C***(=O)), 142.5 (***C***q$_{ar}$), 128.8 (2C, ***C***H$_{ar,\,o}$), 128.7 (2C, ***C***H$_{ar,\,m}$), 126.1 (***C***H$_{ar,\,p}$), 67.2 (***C***H(OH)), 50.3 (C(=O)***C***H_2), 38.5 (PhCH_2***C***H_2), 32.0 (Ph***C***H_2), 30.9 (***C***H_3).

MS (EI-DE) *m*/*z* (%) 192 (2), 174 (34), 159 (4), 131 (29), 117 (17), 104 (17), 91 (73), 87 (9), 77 (10), 65 (11), 58 (4), 51 (6), 43 (100), 39 (6), 27 (4).

HRMS (CI-FE, NH_3) calcd for $C_{12}H_{20}NO_2$ [(M + NH_4)$^+$] 210.1493, found 210.1494.

(*R*)-4-Hydroxy-6-methyl-2-heptanone ((*R*)-**96d**):

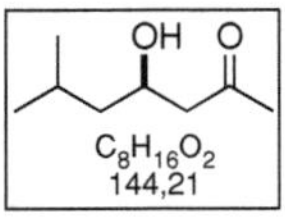

Aldol product (*R*)-**96d** was prepared according to the general procedure. The reaction mixture was stirred for 36 h at 32 °C followed by *in situ* reduction with triethylphosphite. Purification of the crude product by flash column chromatography (silica gel, 10–40% Et_2O in pentane) provided the title compound as a colorless oil (81 mg, 562 μmol, 56%; 96.5:3.5 *er*). The enantiomeric ratio was determined by HPLC using a chiral Chiralpak IA column (10% *i*-PrOH in heptane, 0.5 mL/min); major enantiomer: τ_R = 11.84 min, minor enantiomer: τ_R = 11.39 min.

^{1}H NMR (500 MHz, CD_2Cl_2) δ 4.10–4.04 (m, 1H, C***H***(OH)), 2.84 (br s, 1H, O***H***), 2.58 (dd, *J* = 17.7, 2.9 Hz, 1H, C***H***HC(=O)), 2.47 (dd, *J* = 17.7, 9.2 Hz, 1H, CH***H***C(=O)), 2.14 (s, 3H, C(=O)C***H***$_3$), 1.80–1.72 (m, 1H, C***H***Me$_2$), 1.41 (ddd, *J* = 13.6, 9.1, 5.6 Hz, 1H, *i*-PrC***H***H), 1.12 (ddd, *J* = 13.4, 8.6, 4.5 Hz, 1H, *i*-PrCH***H***), 0.90 (d, *J* = 6.6 Hz, 6H, CH(C***H***$_3$)$_2$).

^{13}C NMR (125 MHz, CD_2Cl_2) δ 210.3 (***C***(=O)), 66.0 (***C***H(OH)), 50.8 (C(=O)***C***H$_2$), 46.0 (*i*-Pr***C***H$_2$), 30.9 (C(=O)***C***H$_3$), 24.7 (***C***HMe$_2$), 23.4 (CH(***C***H$_3$)$_2$), 22.1 (CH(***C***H$_3$)$_2$).

MS (EI-DE) *m/z* (%) 144 [M$^+$] (trace), 126 (9), 111 (10), 108 (25), 93 (5), 87 (100), 83 (7), 69 (17), 58 (20), 43 (90), 29 (5).

HRMS (CI-FE, NH_3) calcd for $C_8H_{20}NO_2$ [(M + NH_4)$^+$] 162.1494, found 162.1494.

(*S*)-4-Cyclohexyl-4-hydroxy-2-butanone ((*S*)-**96e**):

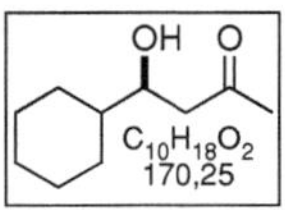

Aldol product (*S*)-**96e** was prepared according to the general procedure. The reaction mixture was stirred for 48 h at 32 °C followed by *in situ* reduction with triethylphosphite. Purification of the crude product by flash column chromatography (silica gel, 10–40% Et_2O in pentane) provided the title compound as a colorless oil (78 mg, 458 μmol, 46%; 96:4 *er*). The enantiomeric ratio was determined by HPLC using a chiral Chiralpak IA column (10% *i*-PrOH in heptane, 0.5 mL/min); major enantiomer: τ_R = 14.65 min, minor enantiomer: τ_R = 13.19 min.

^{1}H NMR (500 MHz, CD_2Cl_2) δ 3.78–3.73 (m, 1H, C***H***(OH)), 2.77 (br s, 1H, O***H***), 2.61 (dd, J = 17.3, 2.7 Hz, 1H, C***H***HC(=O)), 2.49 (dd, J = 17.4, 9.9 Hz, 1H, CH***H***C(=O)), 2.15 (s, 3H, C$\boldsymbol{H}_3$), 1.84–1.79 (m, 1H, C$\boldsymbol{H}_2$), 1.77–1.71 (m, 2H, C$\boldsymbol{H}_2$), 1.68–1.61 (m, 2H, C$\boldsymbol{H}_2$), 1.35–1.27 (m, 1H, C$\boldsymbol{H}_{cycl.}$), 1.26–1.10 (m, 3H, C$\boldsymbol{H}_2$), 1.07–0.94 (m, 2H, C$\boldsymbol{H}_2$).

^{13}C NMR (125 MHz, CD_2Cl_2) δ 210.6 (***C***(=O)), 72.0 (***C***H(OH)), 47.5 (C(=O)***C***H_2), 43.4 (***C***$H_{cycl.}$), 30.9 (***C***H_3), 29.2 (***C***H_2), 28.5 (***C***H_2), 26.9 (***C***H_2), 26.6 (***C***H_2), 26.5 (***C***H_2).

MS (EI-DE) *m/z* (%) 170 [M^+] (trace), 152 (3), 137 (2), 112 (5), 109 (3), 95 (12), 87 (80), 83 (6), 67 (14), 58 (9), 55 (26), 43 (100), 39 (9), 29 (10).

HRMS (CI-FE, NH_3) calcd for $C_{10}H_{22}NO_2$ [$(M + NH_4)^+$] 188.1650, found 188.1651.

4-Hydroxyoct-7-en-2-one (96f):

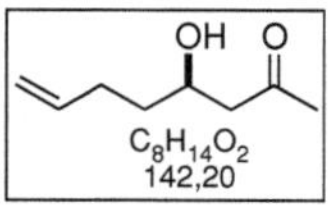

Aldol product **96f** was prepared according to the general procedure. The reaction mixture was stirred for 36 h at 32 °C followed by *in situ* reduction with triethylphosphite. Purification of the crude product by flash column chromatography (silica gel, 10–40% Et_2O in pentane) provided the title compound as a colorless oil (78 mg, 549 μmol, 55%; 96:4 *er*). The enantiomeric ratio was determined by HPLC using a chiral Chiralpak IA column (10% *i*-PrOH in heptane, 0.5 mL/min); major enantiomer: τ_R = 13.35 min, minor enantiomer: τ_R = 12.51 min.

^{1}H NMR (500 MHz, CD_2Cl_2) δ 5.84 (dddd, J = 17.0, 9.3, 7.7, 5.9 Hz, 1H, CH_2=C***H***), 5.04 (app. dq, 1H, J = 17.2, 1.8 Hz, C$\boldsymbol{H}_{trans}$H=CH), 4.96 (app. dq, 1H, J = 9.4, 1.5 Hz, 1H, CH$\boldsymbol{H}_{cis}$=CH), 4.03–3.98 (m, 1H, C***H***(OH)), 2.86 (d, J = 3.8 Hz, 1H, O***H***), 2.61 (dd, J = 17.3, 3.0 Hz, 1H, C***H***HC(=O)), 2.50 (dd, J = 17.7, 9.3 Hz, 1H, CH***H***C(=O)), 2.24–2.07 (m, 2H, C$\boldsymbol{H}_2$), 2.14 (s, 3H, C$\boldsymbol{H}_3$), 1.57–1.42 (m, 2H, C$\boldsymbol{H}_2$).

^{13}C NMR (125 MHz, CD_2Cl_2) δ 210.1 (***C***(=O)), 138.8 (CH_2=***C***H), 114.8 (***C***H_2=CH), 67.3 (***C***H(OH)), 50.3 (C(=O)***C***H_2), 35.9 (=CHCH_2***C***H_2), 30.9 (***C***H_3), 30.1 (=CH***C***H_2).

MS (EI-DE) *m/z* (%) 142 [M^+] (trace), 127 (1), 124 (2), 109 (4), 100 (7), 94 (3), 87 (11), 81 (6), 67 (8), 58 (10), 43 (100), 41 (16), 29 (10).

HRMS (CI-FE, NH_3) calcd for $C_8H_{18}NO_2$ [$(M + NH_4)^+$] 160.1337, found 160.1338.

7.4.2 Organoselenium-Mediated Reductive Epoxide Opening

7.4.2.1 General Procedure

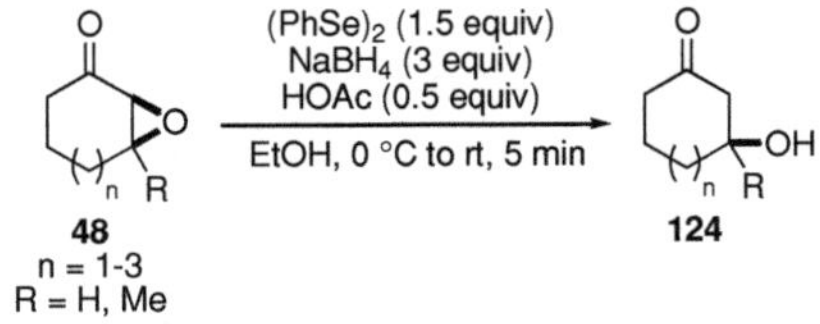

Reference [12] Under argon, acetic acid (15 μL, 0.25 mmol, 0.5 equiv) was added to an ethanolic solution of sodium phenylselenide, prepared by the reduction of diphenylselenide (234 mg, 0.75 mmol, 1.5 equiv) with sodium borohydride (57 mg, 1.5 mmol, 3.0 equiv) in ethanol (3 mL), and the mixture was stirred for 5 min at room temperature. The resulting solution was added at once to a solution of α,β-epoxy ketone **48** (0.5 mmol, 1.0 equiv) in ethanol (2 mL) at 0 °C. After 5 min at room temperature, the reaction mixture was diluted with Et_2O or EtOAc (15 mL) and washed with brine (5 mL). The aqueous layer was repeatedly extracted with Et_2O or EtOAc (2 × 10 mL). The combined organic layers were washed with brine, dried over Na_2SO_4, filtered and concentrated under reduced pressure. Purification of the crude product by flash column chromatography (silica gel, eluent: Et_2O–CH_2Cl_2) provided the corresponding β-hydroxy ketone **124**.

7.4.2.2 Scope of Optically Active Cyclic Aldol Products

(*S*)-3-Hydroxycyclohexanone ((*S*)-**124a**):

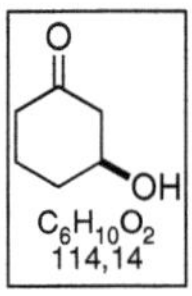

The title compound was prepared according to the general procedure and obtained as a colorless oil (54 mg, 470 μmol, 94%; 96:4 *er*) after purification by flash column chromatography (silica gel, 25–50% Et_2O in CH_2Cl_2). The enantiomeric ratio was determined by GC using a chiral Hydrodex-β-TBDAc column 25 m (35 min at 150 °C, 5 °C/min until 220 °C, 10 min at 220 °C, 0.5 bar H_2); major enantiomer: $\tau_R = 17.65$ min, minor enantiomer: $\tau_R = 21.68$ min.

^{1}H NMR (500 MHz, CD_2Cl_2) δ 4.18–4.13 (m, 1H, C***H***), 2.59 (dd, $J = 14.2$, 4.1 Hz, 1H, CHC***H***HC(=O)), 2.35 (dd, $J = 14.0$, 7.4 Hz, 1H, CHCH***H***C(=O)), 2.27 (t, $J = 6.6$ Hz, 2H, C***H***$_2$C(=O)), 2.12 (br s, 1H, O***H***), 2.07–1.95 (m, 2H, CH_2C***H***HCH and CH_2C***H***HCH_2), 1.77–1.65 (m, 2H, CH_2CH***H***CH and CH_2CH***H***CH_2).

^{13}C NMR (125 MHz, CD_2Cl_2): δ 209.5 (***C***=O), 69.8 (***C***H), 50.3 (CH***C***H_2C(=O)), 40.9 (***C***H_2C(=O)), 32.8 (CH_2***C***H_2CH), 20.7 (CH_2***C***H_2CH_2).

MS (EI-DE) *m/z* (%) 114 [M^+] (64), 96 (14), 86 (7), 81 (5), 73 (18), 68 (47), 60 (60), 58 (34), 55 (81), 44 (100), 42 (85), 31 (9), 27 (28).

HRMS (EI-FE) calcd for $C_6H_{10}O_2$ [M^+] 114.0680, found 114.0681.

(*S*)-3-Hydroxy-3-methylcyclohexanone ((*S*)-**124b**):

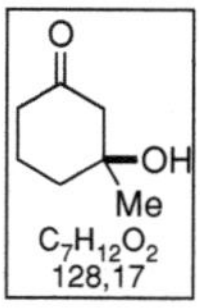

The title compound was prepared according to the general procedure and obtained as a white solid (46 mg, 357 μmol, 89%; 95.5:4.5 *er*) after purification by flash column chromatography (silica gel, 10–20% Et_2O in CH_2Cl_2). The enantiomeric ratio was determined by GC using a chiral Hydrodex-β-TBDAc column 25 m (60 min at 145 °C, 6 °C/min until 220 °C, 10 min at 220 °C, 0.6 bar H_2); major enantiomer: τ_R = 11.84 min, minor enantiomer: τ_R = 14.81 min.

^1H NMR (500 MHz, CD_2Cl_2) δ 2.40 (d, *J* = 14.0 Hz, 1H, Cq_{OH}C***H***HC(=O)), 2.36 (td, *J* = 13.9, 1.8 Hz, 1H, Cq_{OH}CH***H***C(=O)), 2.33–2.22 (m, 2H, C***H***$_2$C(=O)), 2.08–1.99 (m, 1H, CH_2C***H***HCH_2), 1.89–1.76 (m, 3H, CH_2CH***H***CH_2 and CH_2C***H***$_2Cq_{OH}$), 1.55 (br s, 1H, O***H***), 1.33 (s, 3H, C***H***$_3$).

^{13}C NMR (125 MHz, CD_2Cl_2): δ 209.8 (***C*** =O), 74.0 (***C***q_{OH}), 54.9 (Cq_{OH}***C***H_2C(=O)), 40.5 (***C***H_2C(=O)), 37.7 (CH_2***C***H_2Cq_{OH}), 30.1 (***C***H_3), 21.4 (CH_2***C***H_2CH_2).

MS (EI-DE) *m/z* (%) 128 [M^+] (29), 110 (16), 99 (7), 86 (34), 71 (57), 68 (74), 60 (17), 57 (86), 55 (100), 43 (97), 41 (71), 31 (13), 29 (43).

HRMS (EI-FE) calcd for $C_7H_{12}O_2$ [M^+] 128.0836, found 128.0837.

(*S*)-3-Hydroxycycloheptanone ((*S*)-**124c**):

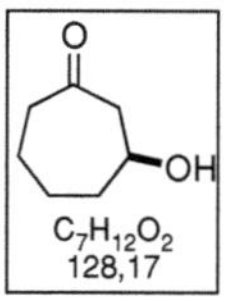

The title compound was prepared according to the general procedure and obtained as a colorless oil (31 mg, 242 μmol, 84%; 99:1 *er*) after purification by flash column chromatography (silica gel, 40–60% Et_2O in CH_2Cl_2). The enantiomeric ratio was determined by GC using a chiral Ivadex 7/PS086 column 25 m (40 min at 105 °C, 10 °C/min until 220 °C, 10 min at 320 °C, 0.5 bar H_2); major enantiomer: τ_R = 27.66 min, minor enantiomer: τ_R = 29.55 min.

^1H NMR (500 MHz, CD_2Cl_2) δ 4.06–4.02 (m, 1H, C***H***), 2.77–2.69 (m, 2H, CHC***H***$_2$C(=O)), 2.47–2.36 (m, 2H, C***H***$_2$C(=O)), 2.03 (br s, 1H, O***H***), 1.90–1.78

(m, 3H, $CH_2C\boldsymbol{H}HCH$ and $CH_2C\boldsymbol{H}_2CH_2$), 1.77–1.72 (m, 2H, $C\boldsymbol{H}_2CH_2C(=O)$), 1.61–1.55 (m, 1H, $CH_2CH\boldsymbol{H}CH$).

^{13}C NMR (125 MHz, CD_2Cl_2) δ 211.5 ($\boldsymbol{C}$=O), 67.6 ($\boldsymbol{C}$H), 51.7 ($CH\boldsymbol{C}H_2C(=O)$), 44.3 ($\boldsymbol{C}H_2C(=O)$), 38.9 ($CH_2\boldsymbol{C}H_2CH_2$), 24.4 ($CH_2\boldsymbol{C}H_2CH$), 23.8 ($\boldsymbol{C}H_2CH_2C(=O)$).

MS (EI) *m/z* (%) 128 [M^+] (26), 113 (6), 100 (16), 95 (8), 85 (15), 71 (44), 69 (17), 58 (60), 55 (25), 43 (100), 27 (13).

HRMS (EI-DE) calcd for $C_7H_{12}O_2$ [M^+] 128.0836, found 128.0837.

(*R*)-3-Hydroxycyclooctanone ((*R*)-**124d**):

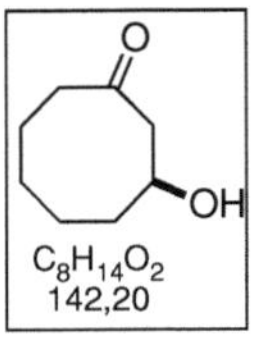

The title compound was prepared according to the general procedure. The reaction was stirred for 19 h at room temperature (90% conversion). Aldol product **124d** was obtained as a colorless oil (5.4 mg, 38 µmol, 53%; 86:14 *er*) after purification by flash column chromatography (silica gel, 40–60% Et_2O in CH_2Cl_2). The enantiomeric ratio was determined by GC using a chiral Ivadex 7 column 25 m (80 °C, 1 °C/min until 160 °C, 20 °C/min until 220 °C, 10 min at 220 °C, 0.5 bar H_2); major enantiomer: τ_R = 41.2 min, minor enantiomer: τ_R = 42.3 min.

^{1}H NMR (500 MHz, $CDCl_3$) δ 4.07–4.02 (m, 1H, $C\boldsymbol{H}$), 2.77 (dd, J = 11.7, 3.8 Hz, 1H, $CHC\boldsymbol{H}HC(=O)$), 2.66 (dd, J = 11.4, 8.6 Hz, 1H, $CHCH\boldsymbol{H}C(=O)$), 2.37–2.34 (m, 3H, $C\boldsymbol{H}_2$ and $O\boldsymbol{H}$), 1.99–1.86 (m, 2H, $C\boldsymbol{H}_2$), 1.83–1.77 (m, 1H, $C\boldsymbol{H}_2$), 1.66–1.44 (m, 4H, $C\boldsymbol{H}_2$), 1.28–1.20 (m, 1H, $C\boldsymbol{H}_2$).

^{13}C NMR (125 MHz, $CDCl_3$) δ 214.2 ($\boldsymbol{C}$=O), 70.5 ($\boldsymbol{C}$H), 47.4 ($CH\boldsymbol{C}H_2C(=O)$), 44.4 ($\boldsymbol{C}H_2C(=O)$), 34.6 ($CH_2\boldsymbol{C}H_2CH$), 28.1 ($\boldsymbol{C}H_2$), 22.8 ($\boldsymbol{C}H_2$), 19.7 ($\boldsymbol{C}H_2$).

MS (EI-DE) *m/z* (%) 142 [M^+] (12), 124 (13), 109 (2), 99 (21), 86 (30), 81 (24), 71 (30), 60 (8), 57 (64), 55 (100), 43 (88), 41 (55), 39 (33), 29 (36).

HRMS (EI-FE) calcd for $C_8H_{14}O_2$ [M^+] 142.0996, found 142.0996.

7.4.3 Synthesis of (3R,5S)-3-Hexyl-5-methyl-1,2-dioxolane (127)

The ketalization of peroxyhemiketal **94a** was conducted according to the protocol of Dussault and Liu [13].

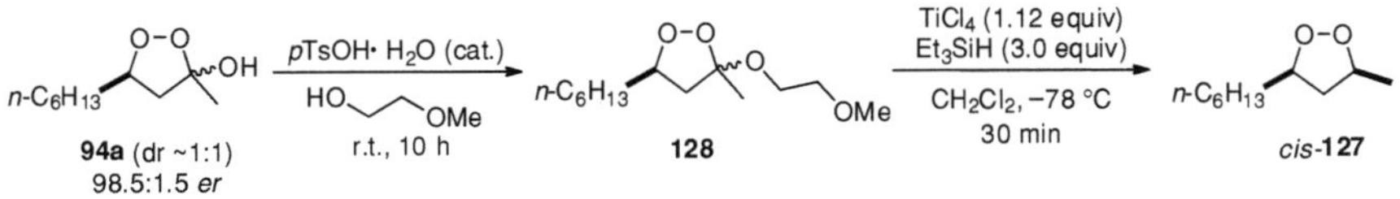

(5*R*)-5-Hexyl-3-(2-methoxyethoxy)-3-methyl-1,2-dioxolane (128) [13]:

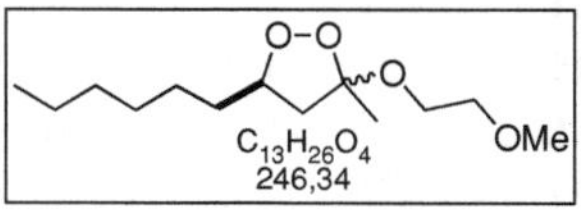

p-Toluenesulfonic acid monohydrate (26 mg, 0.14 mmol) was added to a solution of peroxyhemiketal **94a** (302 mg, 1.60 mmol; 98.5:1.5 *er*) in 2-methoxyethanol (15 mL). After stirring for 12 h at room temperature, the solvent was removed *in vacuo* and the residue subjected to flash column chromatography (silica gel, 5–15% Et_2O in pentane) to afford peroxyketal **128** (236 mg, 0.96 mmol, 60%) as a colorless oil. *Characterized as a mixture of diastereoisomers* (1:1 *dr*).

^{1}H NMR (500 MHz, CD_2Cl_2) *isomer* 1: δ 4.27 (quint, *J* = 7.1 Hz, 1H, C***H***), 3.69–3.64 (m, 1H, OC***H***$_2$), 3.61–3.57 (m, 1H, OC***H***$_2$), 3.52–3.46 (m, 2H, OC***H***$_2$), 3.34 (s, 3H, OC***H***$_3$), 2.57 (dd, *J* = 12.6, 7.6 Hz, 1H, C***H***HCq), 2.34 (dd, *J* = 12.6, 8.2 Hz, 1H, CH***H***Cq), 1.69–1.63 (m, 1H, CH_2C***H***HCH), 1.60–1.54 (m, 1H, CH_2CH***H***CH), 1.45 (s, 3H, CqC***H***$_3$), 1.34–1.24 (m, 8H, (C***H***$_2$)$_4$$CH_3$), 0.88 (t, *J* = 7.1 Hz, 3H, $(CH_2)_5$C***H***$_3$); *isomer* 2: δ 4.37 (quint, *J* = 6.5 Hz, 1H, C***H***), 3.69–3.64 (m, 1H, OC***H***$_2$), 3.61–3.55 (m, 1H, OC***H***$_2$), 3.52–3.45 (m, 2H, OC***H***$_2$), 3.33 (s, 3H, OC***H***$_3$), 2.81 (dd, *J* = 12.5, 7.1 Hz, 1H, C***H***HCq), 2.15 (dd, *J* = 12.8, 6.2 Hz, 1H, CH***H***Cq), 1.70–1.62 (m, 1H, CH_2C***H***HCH), 1.47 (s, 3H, CqC***H***$_3$), 1.39–1.24 (m, 9H, CH_2CH***H***CH and (C***H***$_2$)$_4$$CH_3$), 0.88 (t, *J* = 7.1 Hz, 3H, $(CH_2)_5$C***H***$_3$).

^{13}C NMR (125 MHz, THF-d8) δ 108.6 (**C**q), 107.6 (**C**q), 82.7 (**C**H), 81.6 (**C**H), 73.1 (O**C**H_2), 73.1 (O**C**H_2), 62.2 (O**C**H_2), 62.1 (O**C**H_2), 59.2 (O**C**H_3), 59.1 (O**C**H_3), 53.6 (**C**H_2Cq), 53.3 (**C**H_2Cq), 36.0 (**C**H_2), 33.1 (**C**H_2), 33.0 (**C**H_2), 32.9 (**C**H_2), 30.5 (**C**H_2), 30.4 (**C**H_2), 27.5 (**C**H_2), 27.3 (**C**H_2), 23.8 (**C**H_2), 23.8 (**C**H_2), 20.5 (Cq**C**H_3), 19.4 (Cq**C**H_3), 14.7 (2C, $(CH_2)_5$**C**H_3).

MS (EI-DE) *m*/*z* (%) 246 [M^+] (trace), 213 (37), 171 (2), 155 (4), 129 (4), 113 (7), 95 (9), 87 (9), 71 (8), 59 (88), 43 (100), 29 (20).

HRMS (ESI+) calcd for $C_{13}H_{26}NaO_4$ [$(M + Na)^+$] 269.1720, found 269.1723.

***cis*-(3*R*,5*S*)-3-Hexyl-5-methyl-1,2-dioxolane (*cis*-127)**:

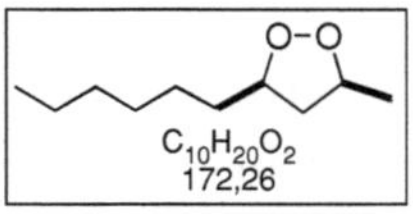

Peroxyketal **128** (20.0 mg, 81.2 μmol) and triethylsilane (38.8 μL, 244 μmol, 3.0 equiv) were dissolved in CH_2Cl_2 (1 mL). At −78 °C, $TiCl_4$ (90.9 μL, 90.9 μmol, 1.12 equiv; 1.0 M in CH_2Cl_2) was added dropwise. After stirring for 1 h, the reaction was quenched by the addition of saturated aqueous $NaHCO_3$ solution (1.5 mL) and repeatedly extracted with Et_2O (2 × 7 mL). The combined organic layers were dried (Na_2SO_4), filtered, and evaporated. Dioxolane **127** [11.4 mg, 66.6 μmol, 80%; 3:1 *dr* (*cis*/*trans*)] was obtained by flash column chromatography (silica gel, 0–5% Et_2O in pentane).

^{1}H NMR (400 MHz, CD_2Cl_2) *cis-isomer*: δ 4.37–4.28 (m, 1H, C***H***Me), 4.25–4.14 (m, 1H, *n*-HexC***H***), 2.79 (dd, J = 11.6, 7.2 Hz, 1H, CHC***H***HCH), 1.76 (dd, J = 11.9, 6.8 Hz, 1H, CHCH***H***CH), 1.70–1.59 (m, 1H, CH_2CH***H***CH), 1.53–1.43 (m, 1H, CH_2CH***H***CH), 1.37–1.22 (m, 8H, $(C\boldsymbol{H}_2)_4CH_3$), 1.25 (d overlapped, 3H, J = 6.1 Hz, 1H, $CHC\boldsymbol{H}_3$), 0.88 (t, J = 6.8 Hz, 3H, $(CH_2)_5C\boldsymbol{H}_3$); *trans-isomer*: δ 4.37–4.28 (m, 1H, C***H***Me), 4.25–4.14 (m, 1H, *n*-HexC***H***), 2.31–2.18 (m, 2H, $CHC\boldsymbol{H}_2CH$), 1.70–1.59 (m, 1H, CH_2CH***H***CH), 1.53–1.43 (m, 1H, CH_2CH***H***CH), 1.37–1.22 (m, 8H, $(C\boldsymbol{H}_2)_4CH_3$), 1.23 (d overlapped, 3H, J = 6.1 Hz, 1H, $CHC\boldsymbol{H}_3$), 0.88 (t, J = 6.8 Hz, 3H, $(CH_2)_5C\boldsymbol{H}_3$).

^{13}C NMR (100 MHz, CD_2Cl_2) *cis-isomer*: δ 81.7 (*n*-Hex**C**H), 77.3 (**C**HMe), 48.1 ($CH\boldsymbol{C}H_2CH$), 34.5 ($CH_2\boldsymbol{C}H_2CH$), 32.1 ($\boldsymbol{C}H_2$), 29.6 ($\boldsymbol{C}H_2$), 26.7 ($\boldsymbol{C}H_2$), 23.0 ($\boldsymbol{C}H_2$), 19.1 ($CH\boldsymbol{C}H_3$), 14.2 ($(CH_2)_5\boldsymbol{C}H_3$); *trans-isomer*: δ 81.4 (*n*-Hex**C**H), 77.1 (**C**HMe), 47.4 ($CH\boldsymbol{C}H_2CH$), 33.6 ($CH_2\boldsymbol{C}H_2CH$), 32.1 ($\boldsymbol{C}H_2$), 29.6 ($\boldsymbol{C}H_2$), 26.5 ($\boldsymbol{C}H_2$), 23.0 ($\boldsymbol{C}H_2$), 18.5 ($CH\boldsymbol{C}H_3$), 14.2 ($(CH_2)_5\boldsymbol{C}H_3$).

GC-MS (GC-EI) *m*/*z* (%) 172 [M^+] (1), 131 (1), 113 (3), 95 (15), 87 (10), 81 (8), 69 (28), 55 (36), 43 (100), 29 (15).

HRMS (EI-FE) calcd for $C_{10}H_{20}O_2$ [M^+] 172.1461, found 172.1463.

7.5 Preparation of Starting Materials

7.5.1 Preparation of Cyclic α,β-Unsaturated Ketones

7.5.1.1 General Procedure A: Protocol of Woods

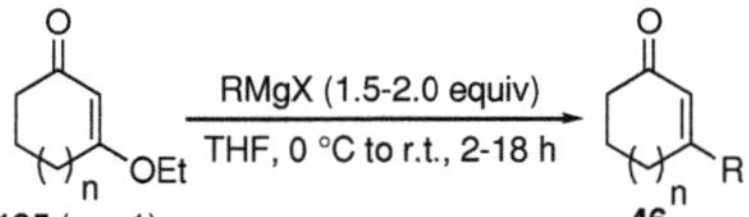

135 (n = 1)
136 (n = 2)
46

Reference [14, 15] 3-Ethoxy-2-cyclohexenone or –heptenone (**135** or **136**; 21.4 mmol, 1.0 equiv) in THF (15 mL) was added dropwise to a solution of a *Grignard* reagent (1 M in Et_2O or THF, 1.5–2.0 equiv) at 0 °C under argon. Once the addition was complete, the resulting solution was allowed to warm to room temperature and stirred until TLC indicated complete disappearance of the starting material (2–18 h). The reaction was slowly quenched with diluted aqueous acid (1 N HCl or 5% H_2SO_4) at 0 °C. The layers were separated, and the aqueous layer extracted with Et_2O (3 × 50 mL). The combined organic layers were washed successively with saturated aqueous $NaHCO_3$-solution, water, and brine, dried (Na_2SO_4), filtered, and evaporated under reduced pressure. The crude product was purified by flash column chromatography (silica gel, eluent: Et_2O-pentane) to yield 3-substituted 2-cyclohexenone and –heptenone derivatives **46**.

Scope of Cyclic Enones Prepared According to General Procedure A

3-Isopropyl-2-cyclohexenone (46i):

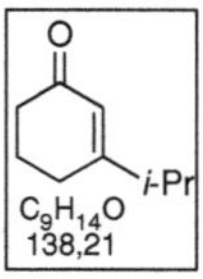

The title compound was prepared according to the general procedure **A** from 3-ethoxy-2-cyclohexenone (**135**) and isopropyl magnesium chloride (2 M in Et_2O). Purification by flash column chromatography (silica gel, 20% Et_2O in pentane) gave enone **46i** (2.10 g, 15.2 mmol, 71%) as a pale yellow liquid.

^{1}H NMR (300 MHz, CD_2Cl_2) δ 5.81 (q, J = 1.4 Hz, 1H, =C***H***), 2.41 (hept, J = 6.9 Hz, 1H, C***H***Me_2), 2.33–2.28 (m, 4H, C$\boldsymbol{H}_2$C(=O) and C$\boldsymbol{H}_2$$Cq_{ol}$), 1.96 (quint, J = 6.2 Hz, 2H, $CH_2C\boldsymbol{H}_2CH_2$), 1.09 (d, J = 7.2 Hz, 6H, $CH(C\boldsymbol{H}_3)_2$).

^{13}C NMR (75 MHz, $CDCl_3$) δ 200.3 (***C***=O), 171.8 (***C***q=CH), 123.6 (Cq=***C***H), 37.6 (***C***H_2C(=O)), 35.7 (***C***HMe_2), 27.7 (***C***H_2Cq_{ol}), 22.9 ($CH_2\boldsymbol{C}H_2CH_2$), 20.6 (2C, $CH(\boldsymbol{C}H_3)_2$).

GC-MS (GC-EI) m/z 138 [M^+].

HRMS (EI-FE) calcd for $C_9H_{14}O$ [M^+] 138.1043, found 138.1045.

3-Allyl-2-cyclohexenone (46k):

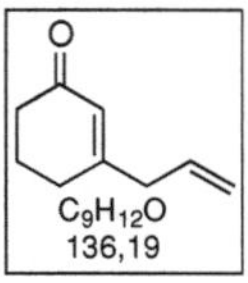

The title compound was prepared according to the general procedure **A** from 3-ethoxy-2-cyclohexenone (**135**) and allyl magnesium chloride (2 M in THF). Purification by flash column chromatography (silica gel, 20% Et_2O in pentane) gave enone **46k** (893 mg, 6.55 mmol, 47%) as a colorless liquid [16].

^{1}H NMR (500 MHz, $CDCl_3$) δ 5.87 (s, 1H, Cq=C***H***), 5.76 (ddt, J = 16.7, 10.3, 6.8 Hz, 1H, C***H***=CH_2), 5.14–5.09 (m, 2H, CH=C$\boldsymbol{H}_2$), 2.92 (d, J = 7.1 Hz, 2H, C$\boldsymbol{H}_2$CH=CH_2), 2.34 (t, J = 6.8 Hz, 2H, C$\boldsymbol{H}_2$C(=O)), 2.27 (br t, J = 6.1 Hz, 2H, $CH_2C\boldsymbol{H}_2Cq_{ol}$), 1.97 (quint, J = 6.4 Hz, 2H, $CH_2C\boldsymbol{H}_2CH_2$).

^{13}C NMR (125 MHz, $CDCl_3$) δ 199.8 (***C***=O), 164.1 (***C***q=CH), 133.2 (***C***H=CH_2), 126.3 (Cq=***C***H), 118.3 (CH=***C***H_2), 42.2 (***C***H_2CH=CH_2), 37.3 (C=O***C***H_2), 29.5 ($CH_2\boldsymbol{C}H_2Cq_{ol}$), 22.6 ($CH_2\boldsymbol{C}H_2CH_2$).

MS (EI-DE) m/z (%) 136 [M^+] (74), 121 (8), 108 (42), 93 (8), 79 (100), 77 (18), 74 (1), 67 (19), 53 (9), 41 (19), 39 (43), 29 (2), 27 (13).

HRMS (EI-FE) calcd for $C_9H_{12}O$ [M^+] 136.0888, found 136.0888.

3-Benzyl-2-cyclohexenone (46m):

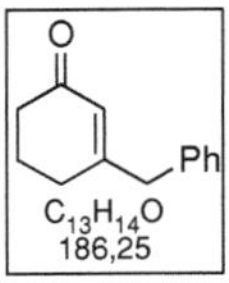

The title compound was prepared according to the general procedure **A** from 3-ethoxy-2-cyclohexenone (**135**) and benzyl magnesium chloride (1 M in Et_2O). Purification by flash column chromatography (silica gel, 20% Et_2O in pentane) gave enone **46m** (3.02 g, 16.2 mmol, 76%) as a colorless liquid.

^{1}H NMR (500 MHz, CD_2Cl_2) δ 7.33–7.30 (m, 2H, C$\boldsymbol{H}_{Ph,\ m}$), 7.28–7.24 (m, 1H, C$\boldsymbol{H}_{Ph,\ p}$), 7.21–7.18 (m, 2H, C$\boldsymbol{H}_{Ph,\ o}$), 5.80 (s, 1H, =C***H***), 3.52 (s, 2H, PhC$\boldsymbol{H}_2$), 2.31 (t, J = 6.8 Hz, 2H, C$\boldsymbol{H}_2$C(=O)), 2.25 (br t, J = 6.0 Hz, 2H, C$\boldsymbol{H}_2$Cq_{ol}), 1.94 (quint, J = 6.4 Hz, 2H, CH_2C$\boldsymbol{H}_2$$CH_2$).

^{13}C NMR (125 MHz, CD_2Cl_2) δ 199.6 (***C***=O), 164.9 (***C***q_{ol}), 137.8 (***C***q_{Ph}), 129.5 (2C, ***C***$H_{Ph,\ o}$), 129.0 (2C, ***C***$H_{Ph,\ m}$), 127.1 (***C***$H_{Ph,\ p}$), 127.0 (=***C***H), 44.7 (Ph***C***H_2), 37.7 (***C***H_2C(=O)), 29.6 (***C***H_2Cq_{ol}), 23.1 (CH_2***C***H_2CH_2).

MS (EI-DE) *m/z* (%) 186 [M^+] (100), 168 (8), 158 (91), 142 (9), 129 (58), 115 (34), 102 (3), 91 (30), 77 (9), 67 (26), 51 (15), 39 (46), 27 (11).

HRMS (EI-DE) calcd for $C_{13}H_{14}O$ [M^+] 186.1043, found 186.1045.

3-Phenyl-2-cyclohexenone (46n):

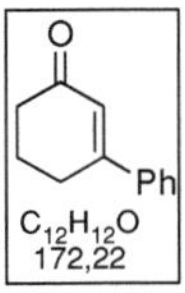

The title compound was prepared according to the general procedure **A** from 3-ethoxy-2-cyclohexenone (**135**) and phenyl magnesium chloride (2 M in THF). Purification by flash column chromatography (silica gel, 30% Et_2O in pentane) gave enone **46n** (3.60 g, 20.9 mmol, 98%) as a white solid [17].

^{1}H NMR (500 MHz, CD_2Cl_2) δ 7.57–7.55 (m, 2H, C$\boldsymbol{H}_{Ph}$), 7.43–7.40 (m, 3H, C$\boldsymbol{H}_{Ph}$), 6.37 (t, J = 1.5 Hz, 1H, =C***H***), 2.77 (td, J = 6.0, 1.4 Hz, 2H, C$\boldsymbol{H}_2$Cq_{ol}), 2.44 (app. t, J = 6.8 Hz, 2H, C$\boldsymbol{H}_2$C(=O)), 2.14 (quint, J = 6.3 Hz, 2H, CH_2C$\boldsymbol{H}_2$$CH_2$).

^{13}C NMR (125 MHz, CD_2Cl_2) δ 199.6 (***C***=O), 159.9 (***C***q=CH), 139.3 (***C***q_{Ph}), 130.1 (***C***$H_{Ph,\ p}$), 129.0 (2C, ***C***H_{Ph}), 126.4 (2C, ***C***H_{Ph}), 125.6 (Cq=***C***H), 37.6 (***C***H_2C(=O)), 28.4 (***C***H_2Cq_{ol}), 23.2 (CH_2***C***H_2CH_2).

3-Vinyl-2-cyclohexenone (46o):

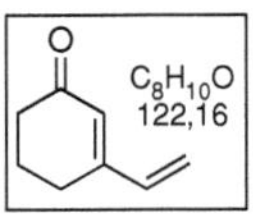

The title compound was prepared according to the general procedure **A** from 3-ethoxy-2-cyclohexenone (**135**) and vinyl magnesium bromide (1 M in THF). Purification by flash column chromatography (silica gel, 15–25% Et_2O in pentane) gave enone **46o** (1.97 g, 16.1 mmol, 75%) as a colorless oil.

^{1}H NMR (500 MHz, CD_2Cl_2) δ 6.50 (dd, J = 17.6, 10.7 Hz, 1H, C***H***=CH$_2$), 5.89 (s, 1H, Cq=C***H***), 5.70 (d, J = 17.7 Hz, 1H, CH=C$\boldsymbol{H}_{trans}$H), 5.45 (d, J = 10.7 Hz, 1H, CH=CH$\boldsymbol{H}_{cis}$), 2.47 (br t, J = 6.1 Hz, 2H, C$\boldsymbol{H}_2$Cq$_{ol}$), 2.37 (t, J = 6.8 Hz, 2H, C$\boldsymbol{H}_2$C(=O)), 2.02 (quint, J = 6.4 Hz, 2H, CH$_2$C$\boldsymbol{H}_2$CH$_2$).

^{13}C NMR (126 MHz, CD_2Cl_2) δ 200.1 (***C***=O), 157.1 (***C***q=CH), 138.3 (***C***H=CH$_2$), 128.4 (Cq=***C***H), 120.7 (CH=***C***H$_2$), 37.8 (***C***H$_2$C(=O)), 24.6 (***C***H$_2$), 22.6 (***C***H$_2$).

GC-MS (GC-EI) *m/z* 122 [M$^+$].

HRMS (EI-FE) calcd for $C_8H_{10}O$ [M$^+$] 122.0730, found 122.0732.

3-Ethynyl-2-cyclohexenone (**46p**):

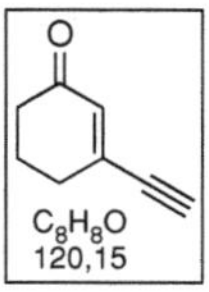

The title compound was prepared according to the general procedure **A** from 3-ethoxy-2-cyclohexenone (**135**) and ethynyl magnesium bromide (0.5 M in THF). Purification by flash column chromatography (silica gel, 10–25% Et_2O in pentane) gave enone **46p** (1.33 g, 11.1 mmol, 77%) as a colorless liquid.

^{1}H NMR (500 MHz, CD_2Cl_2) δ 6.21 (t, J = 1.6 Hz, 1H, =C***H***), 3.59 (s, 1H, $\equiv$C***H***), 2.45 (td, J = 6.1, 1.7 Hz, 2H, C$\boldsymbol{H}_2$Cq$_{ol}$), 2.34 (t, J = 6.7 Hz, 2H, C$\boldsymbol{H}_2$C(=O)), 2.02 (quint, J = 6.3 Hz, 2H, CH$_2$C$\boldsymbol{H}_2$CH$_2$).

^{13}C NMR (125 MHz, CD_2Cl_2) δ 198.4 (***C***=O), 142.3 (***C***q=CH), 134.2 (Cq=***C***H), 87.0 (***C***q$\equiv$CH), 82.8 (Cq$\equiv$***C***H), 37.6 (***C***H$_2$C(=O)), 30.5 (***C***H$_2$Cq$_{ol}$), 22.9 (CH$_2$***C***H$_2$CH$_2$).

GC-MS (GC-EI) *m/z* (%) 120 [M$^+$] (44), 92 (100), 89 (2), 77 (4), 64 (54), 50 (12), 42 (5), 39 (15), 27 (3).

HRMS (EI-FE) calcd for C_8H_8O [M$^+$] 120.0576, found 120.0575.

3-Ethyl-2-cycloheptenone (**46t**):

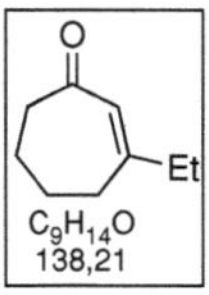

The title compound was prepared according to the general procedure **A** from 3-ethoxy-2-cycloheptenone (**136**) and ethyl magnesium bromide (3 M in THF).

Purification by flash column chromatography (silica gel, 20% Et_2O in pentane) gave enone **46t** (570 mg, 4.12 mmol, 61%) as a pale yellow liquid. 3-Ethoxy-2-cycloheptenone (**136**) was obtained by heating a mixture of 1,3-cycloheptanedione (0.85 g, 6.74 mmol), dry ethanol (1.7 mL), and a catalytic amount of *p*-toluene sulfonic acid monohydrate (17 mg) in benzene (40 mL) under reflux with a *Dean-Stark* trap. After 12 h, the mixture was allowed to cool to room temperature and the solvent was evaporated under reduced pressure. The resulting crude 3-ethoxy-2-cycloheptenone (**136**; ~90% purity) was used without further purification for the synthesis of 3-ethyl-2-cycloheptenone (**46t**).

^{1}H NMR (400 MHz, CD_2Cl_2) δ 5.83 (s, 1H, =C***H***), 2.52 (t, J = 6.1 Hz, 2H, C***H***$_2$C(=O)), 2.41 (t, J = 5.8 Hz, 2H, C***H***$_2$Cq$_{ol}$), 2.22 (qd, J = 7.4, 1.2 Hz, 2H, C***H***$_2$CH$_3$), 1.81–1.72 (m, 4H, CH$_2$(C***H***$_2$)$_2$CH$_2$), 1.08 (t, J = 7.4 Hz, 3H, CH$_2$C***H***$_3$).

^{13}C NMR (75 MHz, CD_2Cl_2) δ 203.9 (***C***=O), 163.8 (***C***q=CH), 128.2 (Cq=***C***H), 42.5 (***C***H$_2$C(=O)), 34.1 (Cq$_{ol}$***C***H$_{2,\ cycl.}$), 32.9 (***C***H$_2$CH$_3$), 25.5 (***C***H$_{2,\ cycl.}$), 21.7 (***C***H$_{2,\ cycl.}$), 12.3 (***C***H$_3$).

MS (EI-DE) *m/z* (%) 138 [M$^+$] (29), 109 (100), 96 (18), 81 (32), 79 (10), 67 (17), 53 (17), 39 (20), 27 (17).

HRMS (EI-FE) calcd for $C_9H_{14}O$ [M$^+$] 138.1045, found 138.1045.

3-Benzyl-2-cycloheptenone (**46u**):

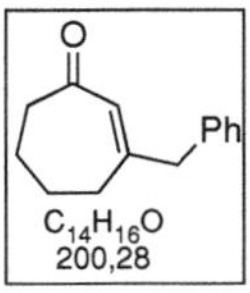

The title compound was prepared according to the general procedure **A** from 3-ethoxy-2-cycloheptenone (**136** *vide supra*) and benzyl magnesium chloride (1 M in Et_2O). Purification by flash column chromatography (silica gel, 20% Et_2O in pentane) gave enone **46u** (551 mg, 2.75 mmol, 41%) as a pale yellow liquid.

^{1}H NMR (400 MHz, CD_2Cl_2) δ 7.34–7.19 (m, 5H, C$_6$***H***$_5$), 5.89 (s, 1H, =C***H***), 3.49 (s, 2H, C***H***$_2$Ph), 2.53 (t, J = 6.4 Hz, 2H, C***H***$_2$C(=O)), 2.37 (t, J = 5.9 Hz, 2H, C***H***$_2$Cq$_{ol}$), 1.78–1.64 (m, 4H, CH$_2$(C***H***$_2$)$_2$CH$_2$).

^{13}C NMR (75 MHz, CD_2Cl_2) δ 203.8 (***C***=O), 160.2 (***C***q$_{ol}$), 138.4 (***C***q$_{Ph}$), 130.9 (=***C***H), 129.5 (2C, ***C***H$_{Ph}$), 128.9 (2C, ***C***H$_{Ph}$), 127.0 (***C***H$_{Ph,\ p}$), 47.3 (***C***H$_2$Ph), 42.6 (***C***H$_2$C(=O)), 32.7 (***C***H$_2$Cq$_{ol}$), 25.6 (***C***H$_2$CH$_2$Cq$_{ol}$), 21.7 (***C***H$_2$CH$_2$C(=O)).

MS (EI-DE) *m/z* (%) 200 [M$^+$] (14), 182 (1), 171 (4), 158 (4), 143 (3), 129 (13), 115 (12), 109 (100), 91 (15), 81 (24), 65 (12), 53 (11), 39 (14), 27 (6).

HRMS (EI-DE) calcd for $C_{14}H_{16}O$ [M$^+$] 200.1199, found 200.1201.

7.5.1.2 General Procedure B: Saegusa Oxidation

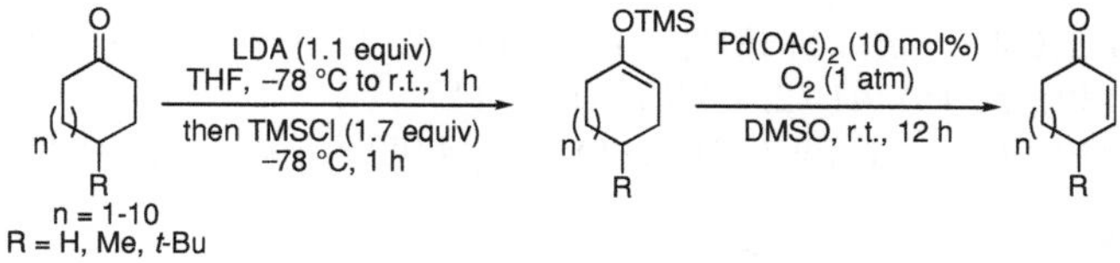

Reference [18] To a solution of LDA, freshly prepared from *n*-BuLi (13.2 ml, 33.0 mmol, 1.1 equiv, 2.5 M in hexanes) and diisopropylamine (5.06 mL, 36.0 mmol, 1.2 equiv) in THF (46 mL) at 0 °C, was added a cyclic ketone (30.0 mmol, 1.0 equiv) dissolved in THF (18 ml) at −78 °C. After stirring for 30 min, the solution was warmed to room temperature and stirred for additional 30 min at room temperature. After re-cooling to −78 °C, TMSCl (6.47 ml, 51.0 mmol, 1.7 equiv) was added, and stirring was continued for 1 h. The reaction was quenched with saturated aqueous NH_4Cl solution and extracted with Et_2O (3 × 150 mL). The combined organic layers were dried (Na_2SO_4), filtered, and concentrated *in vacuo*. The crude silyl enol ether was dissolved in DMSO (600 mL), and $Pd(OAc)_2$ (673 mg, 3.00 mmol, 10 mol%) was added. The flask was carefully evacuated, purged with O_2, and equipped with an O_2 ballon. After stirring for 12 h at room temperature, the reaction mixture was poured into ice water (300 mL), and repeatedly extracted with Et_2O (2 × 300 mL). The combined organic phases were washed with brine, dried (Na_2SO_4), filtered, and concentrated under reduced pressure. The crude product was purified by flash column chromatography (silica gel, eluent: Et_2O-pentane) to afford the corresponding pure cyclic α,β-unsaturated ketones **46**.

Scope of Cyclic Enones Prepared According to General Procedure B

2-Cyclooctenone (46v):

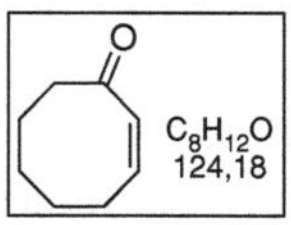

The title compound was prepared according to the general procedure **B** from cyclooctanone (3.79 g, 30.0 mmol). Purification by flash column chromatography (silica gel, 5–10% Et_2O in pentane) gave 2-cyclooctenone (**46v**; 2.38 g, 19.2 mmol, 64%) as a clear oil.

^{1}H NMR (500 MHz, CD_2Cl_2) δ 6.33 (dt, J = 12.5, 7.1 Hz, 1H, $CH_2C\boldsymbol{H}$=), 5.94 (d, J = 12.5 Hz, 1H, $CH_2CH=C\boldsymbol{H}$), 2.60 (t, J = 6.9 Hz, 2H, $C\boldsymbol{H}_2C(=O)$), 2.52–2.47 (m, 2H, $C\boldsymbol{H}_2CH$=), 1.82–1.77 (m, 2H, $C\boldsymbol{H}_2CH_2C(=O)$), 1.63–1.53 (m, 4H, $C\boldsymbol{H}_2CH_2CH_{ol}$ and $C\boldsymbol{H}_2(CH_2)_2C(=O)$).

^{13}C NMR (125 MHz, CD_2Cl_2) δ 205.4 (***C*** =O), 141.5 (CH_2***C***H=), 132.2 (CH_2CH=***C***H), 42.7 (***C***H_2C(=O)), 28.6 (***C***H_2CH=), 25.3 (***C***$H_2(CH_2)_2$C(=O)), 23.2 (***C***H_2CH_2CH=), 22.7 (***C***H_2CH_2C(=O)).

GC-MS (GC-EI) *m/z* 124 [M^+] (8), 109 (1), 95 (11), 91 (2), 81 (100), 68 (40), 65 (4), 53 (39), 51 (5), 41 (23), 39 (34), 27 (17).

HRMS (EI-FE) calcd for $C_8H_{12}O$ [M^+] 124.0887, found 124.0888.

4-Methyl-2-cyclohexenone (80a):

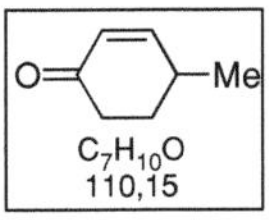

The title compound was prepared according to the general procedure **B** from 4-methylcyclo-hexanone (1.84 mL, 15.0 mmol). Purification by flash column chromatography (silica gel, 10–20% Et_2O in pentane) gave cyclohexenone **80a** (1.36 g, 12.3 mmol, 82%) as a clear oil.

^{1}H NMR (500 MHz, CD_2Cl_2) δ 6.80 (ddd, J = 10.2, 2.6, 1.3 Hz, 1H, =C***H***CH), 5.88 (dd, J = 10.1, 2.5 Hz, 1H, C***H***=CHCH), 2.58–2.50 (m, 1H, C***H***Me), 2.42 (dt, J = 16.7, 4.7 Hz, 1H, C***H***HC(=O)), 2.33 (ddd, J = 16.8, 12.2, 4.7 Hz, 1H, CH***H***C(=O)), 2.09 (dqd, J = 13.6, 5.0, 1.3 Hz, 1H, C***H***HCH), 1.68–1.61 (m, 1H, CH***H***CH), 1.14 (d, J = 7.3 Hz, 3H, C***H***$_3$).

^{13}C NMR (125 MHz, CD_2Cl_2) δ 199.6 (***C***=O), 156.5 (=***C***HCH), 128.8 (***C***H=CHCH), 37.2 (***C***H_2C(=O)), 31.5 (***C***HMe), 31.3 (***C***H_2CH), 20.3 (***C***H_3).

GC-MS (GC-EI) *m/z* 110 [M^+] (51), 95 (4), 82 (100), 79 (4), 68 (67), 65 (6), 54 (56), 41 (18), 39 (40), 27 (20).

HRMS (EI-FE) calcd for $C_7H_{10}O$ [M^+] 110.0731, found 110.0732.

4-*tert*-Butyl-2-cyclohexenone (80b):

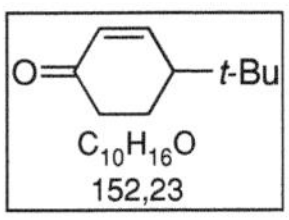

The title compound was prepared according to the general procedure **B** from 4-*tert*-butylcyclohexanone (2.31 g, 15.0 mmol). Purification by flash column chromatography (silica gel, 5% EtOAc in hexanes) gave cyclohexenone **80b** (1.52 g, 9.97 mmol, 66%) as a clear oil.

^{1}H NMR (500 MHz, $CDCl_3$) δ 6.99 (dt, J = 10.6, 2.1 Hz, 1H, =C***H***CH), 6.01 (dd, J = 10.4, 2.8 Hz, 1H, C***H***=CHCH), 2.50 (dt, J = 16.9, 3.9 Hz, 1H, C***H***HC(=O)), 2.32 (ddd, J = 16.4, 14.4, 4.9 Hz, 1H, CH***H***C(=O)), 2.20–2.16 (m, 1H, C***H***HCH), 2.11–2.05 (m, 1H, CH***H***CH), 1.76–1.67 (m, 1H, C***H****t*-Bu), 0.96 (s, 9H, C***H***$_3$).

^{13}C NMR (125 MHz, $CDCl_3$) δ 200.90 (***C***=O), 152.9 (=***C***HCH), 130.0 (***C***H=CHCH), 46.8 (***C***H*t*-Bu), 37.8 (***C***H_2C(=O)), 32.9 (***C***qMe_3), 27.3 (3C, ***C***H_3), 24.4 (***C***H_2CH).

MS (EI-DE) *m/z* 152 [M$^+$] (5), 137 (5), 119 (1), 109 (6), 96 (100), 91 (2), 81 (5), 67 (11), 57 (83), 53 (6), 41 (53), 39 (23), 29 (36), 27 (19).

HRMS (EI-FE) calcd for $C_{10}H_{16}O$ [M$^+$] 152.1202, found 152.1201.

(*E*)-2-Cyclododecenone (**46x**):

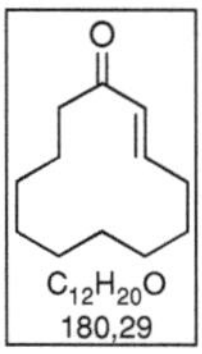

The title compound was prepared according to the general procedure **B** from cyclododecanone (2.73 g, 15.0 mmol). Stoichiometric $Pd(OAc)_2$ (3.37 g, 15.0 mmol) in acetonitrile (20 mL) was used in the oxidation step. Purification by flash column chromatography (silica gel, 2% Et_2O in pentane) gave pure (*E*)-enone **46x** (1.40 g, 7.80 mmol, 52%) as a clear oil.

^{1}H NMR (500 MHz, CD_2Cl_2) δ 6.74 (td, J = 15.3, 7.7 Hz, 1H, =C***H***CH$_2$), 6.29 (d, J = 16.1 Hz, 1H, C***H***=CHCH$_2$), 2.46–2.44 (m, 2H, C***H***$_2$C(=O)), 2.28–2.24 (m, 2H, =CHC***H***$_2$), 1.71–1.67 (m, 2H, C***H***$_2$), 1.62–1.57 (m, 2H, C***H***$_2$), 1.42–1.37 (m, 2H, C***H***$_2$), 1.33–1.29 (m, 4H, C***H***$_2$), 1.28–1.21 (m, 4H, C***H***$_2$).

^{13}C NMR (125 MHz, CD_2Cl_2) δ 203.1 (***C***=O), 146.7 (= ***C***HCH$_2$), 131.4 (***C***H=CHCH$_2$), 40.2 (***C***H$_2$C(=O)), 32.8 (=CH***C***H$_2$), 26.8 (***C***H$_2$), 25.6 (***C***H$_2$), 25.4 (***C***H$_2$), 25.4 (***C***H$_2$), 25.0 (***C***H$_2$), 24.8 (***C***H$_2$), 24.0 (***C***H$_2$).

MS (EI) *m/z* (%) 180 [M$^+$] (49), 162 (2), 151 (6), 137 (11), 123 (11), 109 (45), 98 (51), 95 (29), 84 (39), 81 (100), 68 (65), 55 (78), 41 (76), 27 (23).

HRMS (EI-FE) calcd for $C_{12}H_{20}O$ [M$^+$] 180.1515, found 180.1514.

(*E*)-2-Cyclopentadecenone (**46w**):

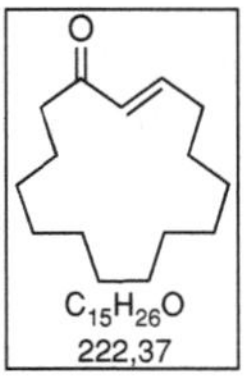

The title compound was prepared according to the general procedure **B** from cyclopentadecanone (3.37 g, 15.0 mmol). Stoichiometric $Pd(OAc)_2$ (3.37 g, 15.0 mmol) in acetonitrile (20 mL) was used in the oxidation step. Purification by flash column chromatography (silica gel, 2% Et_2O in pentane) gave pure (*E*)-enone **46w** (2.16 g, 9.71 mmol, 65%) as a clear oil.

^{1}H NMR (500 MHz, CD_2Cl_2) δ 6.78 (td, J = 15.3, 7.7 Hz, 1H, =C***H***CH$_2$), 6.16 (d, J = 16.1 Hz, 1H, C***H***=CHCH$_2$), 2.47–2.45 (m, 2H, C***H***$_2$C(=O)), 2.29–2.25 (m, 2H, =CHC***H***$_2$), 1.68–1.62 (m, 2H, C***H***$_2$), 1.56–1.51 (m, 2H, C***H***$_2$), 1.35–1.22 (m, 16H, C***H***$_2$).

^{13}C NMR (125 MHz, CD_2Cl_2) δ 201.5 (*C*=O), 148.0 (=*C*HCH$_2$), 130.8 (*C*H=CHCH$_2$), 40.1 (*C*H$_2$C(=O)), 31.7 (=CH*C*H$_2$), 27.1 (*C*H$_2$), 27.0 (*C*H$_2$), 26.8 (3C, –(*C*H$_2$)$_n$–), 26.6 (*C*H$_2$), 26.3 (*C*H$_2$), 26.1 (*C*H$_2$), 25.5 (*C*H$_2$), 25.3 (*C*H$_2$).

MS (EI) *m*/*z* (%) 222 [M$^+$] (92), 207 (2), 193 (2), 179 (4), 164 (20), 151 (7), 135 (12), 121 (14), 109 (53), 96 (63), 81 (52), 68 (57), 55 (100), 41 (78), 29 (24).

HRMS (EI-DE) calcd for $C_{15}H_{26}O$ [M$^+$] 222.1982, found 222.1984.

7.5.1.3 Synthesis of 3-*tert*-butylcyclohex-2-enone (46j)

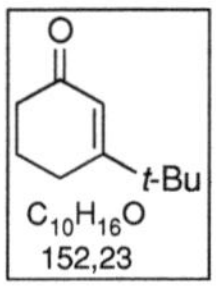

Reference [19] A flask was charged with 10% Pd/C (96 mg, 0.09 mmol, 2.5 mol%), CH_2Cl_2 (70 mL), TBHP (3.29 mL, 18.1 mmol, 5.5 M in decane, 5 equiv), K_2CO_3 (124 mg, 0.91 mmol, 25 mol%), and 1-*tert*-butylcyclohexene (500 mg, 3.62 mmol, 54 uL, 0.32 mmol, 1 equiv) under argon. The mixture was stirred at 0 °C for 12 h. After removal of the solvent at 0 °C under reduced, the crude product was purified by flash column chromatography (silica gel, 30–60% Et_2O in pentane) to provide the title compound (320 mg, 2.10 mmol, 58%) as a clear liquid.

^{1}H NMR (500 MHz, CD_2Cl_2) δ 5.88 (t, J = 1.3 Hz, 1H, =C***H***), 2.34 (td, J = 6.0, 1.3 Hz, 2H, C***H***$_2$Cq$_{ol}$), 2.30 (app. t, J = 6.6 Hz, 2H, C***H***$_2$C(=O)), 1.94 (quint, J = 6.3 Hz, 2H, CH$_2$C***H***$_2$CH$_2$), 1.12 (s, 9H, C***H***$_3$).

^{13}C NMR (125 MHz, CD_2Cl_2) δ 200.4 (*C*=O), 173.9 (*C*q=CH), 123.1 (Cq=*C*H), 37.8 (*C*H$_2$C(=O)), 36.9 (*C*qMe$_3$), 28.4 (3C, C(*C*H$_3$)$_3$), 26.3 (*C*H$_2$), 23.7 (*C*H$_2$).

GC-MS (GC-EI) *m*/*z* (%) 152 [M$^+$] (31), 137 (11), 124 (35), 109 (100), 96 (67), 81 (39), 79 (13), 67 (32), 65 (7), 57 (20), 41 (32), 29 (7).

HRMS (EI-FE) calcd for $C_{10}H_{16}O$ [M$^+$] 152.1202, found 152.1201.

7.5.1.4 Synthesis of 2-Methyl-2-cyclohexenone (79)

The iron-catalyzed cross coupling reaction of 2-iodo-2-cyclohexenone (**137**) was conducted according to the protocol of Cahiez and Avedissian [20].

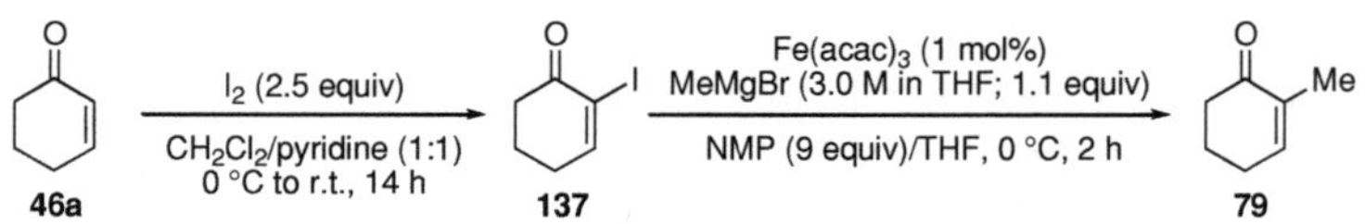

2-Iodo-2-cyclohexenone (137):

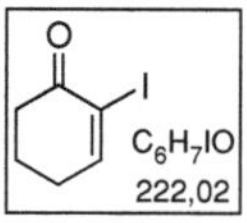

2-Cyclohexenone (0.97 mL, 10.0 mmol) was dissolved in CH_2Cl_2 (20 mL) under an atmosphere of argon, and cooled to 0 °C. Pyridine was added (20 mL) followed by iodine (6.35 g, 25.0 mmol, 2.5 equiv). The reaction mixture was allowed to warm to room temperature, and stirred for 14 h. EtOAc (100 mL) was added, and the organic layer was washed successively with saturated aqueous $Na_2S_2O_3$ solution (2 × 30 mL), water (30 mL), aqueous 10% $CuSO_4$ solution (6 × 30 mL), water (30 mL), and brine (30 mL). The organic layer was dried over Na_2SO_4, filtered, and concentrated. Purification by flash column chromatography (silica gel, 15–20% Et_2O in pentane) provided the title compound (2.01 g, 9.06 mmol; 91%) as a yellow solid.

^{1}H NMR (500 MHz, CD_2Cl_2) δ 7.77 (t, $J = 4.5$ Hz, 1H, C***H***$_{ol}$), 2.62 (t, $J = 6.7$ Hz, 2H, C***H***$_2$C(=O)), 2.42 (t, $J = 5.5$ Hz, 2H, C***H***$_2$CH$_{ol}$), 2.08–2.03 (m, 2H, CH$_2$C***H***$_2$CH$_2$).

^{13}C NMR (125 MHz, CD_2Cl_2) 192.1 (***C***=O), 159.9 (***C***H$_{ol}$), 103.6 (***C***q$_{ol}$), 37.3 (***C***H$_2$C(=O)), 30.0 (***C***H$_2$Cq$_{ol}$), 22.9 (CH$_2$***C***H$_2$CH$_2$).

MS (EI-DE) *m/z* (%) 222 [M$^+$] (100), 194 (52), 181 (5), 166 (10), 153 (1), 127 (2), 95 (8), 67 (9), 55 (9), 39 (29).

HRMS (EI-DE) calcd for C_6H_7OI [M$^+$] 221.9540, found 221.9542.

2-Methyl-2-cyclohexenone (79):

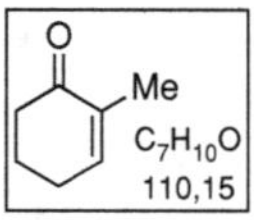

To a solution of 2-iodo-2-cyclohexenone (**137**) (0.5 g, 2.25 mmol) and Fe(acac)$_3$ (8.0 mg, 0.0225 mmol, 1 mol%) in a solvent mixture of THF (4.5 mL) and NMP (2.25 mL, 9 equiv) was added dropwise (within 15 min) at 0 °C, methylmagnesium bromide (0.85 mL, 2.55 mmol, 3.0 M in THF, 1.1 equiv), and stirring was continued for 2 h. Then the reaction mixture was hydrolyzed with aqueous 1 M HCl (7 mL), and the aqueous layer repeatedly extracted with Et_2O (3 × 20 mL). The combined organic phases were washed successively with saturated aqueous $NaHCO_3$ solution, water, and brine. The organic layer was dried over Na_2SO_4, filtered, and concentrated. After purification of the crude product by flash column chromatography (silica gel, 5–10% Et_2O in pentane) the title compound [84 mg, 760 μmol; 34% (*reduced yield due to the high volatility of* **79**)] was obtained as a colorless oil.

^{1}H NMR (400 MHz, CD_2Cl_2) δ 6.75–6.73 (m, 1H, C***H***$_{ol}$), 2.39–2.36 (m, 2H, C***H***$_2$C(=O)), 2.33–2.28 (m, 2H, C***H***$_2$CH$_{ol}$), 1.99–1.93 (m, 2H, CH$_2$C***H***$_2$CH$_2$), 1.73–1.72 (m, 3H, C***H***$_3$).

^{13}C NMR (100 MHz, CD_2Cl_2) 199.5 (*C*=O), 145.5 (*C*H$_{ol}$), 135.5 (*C*q$_{ol}$), 38.4 (*C*H$_2$C(=O)), 26.1 (*C*H$_2$CH$_{ol}$), 23.4 (CH$_2$*C*H$_2$ CH$_2$), 15.8 (*C*H$_3$).

GC-MS (GC-EI) *m/z* 110 [M$^+$].

HRMS (EI-FE) calcd for $C_7H_{10}O$ [M$^+$] 110.0731, found 110.0732.

7.5.1.5 Synthesis of (*S*)-3-Methyl-5-phenyl-2-cyclohexenone (*S*)-(83)

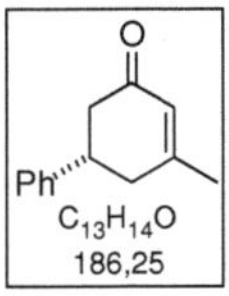

Reference [21] Acetic acid (36.0 mg, 0.6 mmol, 60 mol%) was added to a solution of 9-amino(9-deoxy)*epi*quinine (**13**; 64.7 mg, 0.2 mmol, 20 mol%) in toluene. After cooling to −15 °C, 4-phenyl-2,6-heptandione (204 mg, 1.0 mmol) was added. The resulting mixture was stirred at −15 °C for 48 h. The reaction mixture was then directly subjected to flash column chromatography (silica gel, 20% Et_2O in pentane) to afford cyclohexenone derivative **83** (159 mg, 842 μmol, 84%; 95.5:4.5 *er*) as a colorless solid. The enantiomeric ratio was determined by HPLC using a chiral Chiralcel OJ-H column (10% *i*-PrOH in heptane, 0.5 mL/min); major enantiomer: $\tau_R = 21.71$ min, minor enantiomer: $\tau_R = 3.86$ min.

^{1}H NMR (500 MHz, $CDCl_3$) δ 7.36–7.33 (m, 2H, C*H*$_{Ph}$), 7.27–7.23 (m, 3H, C*H*$_{Ph}$), 5.92 (s, 1H, =C*H*), (hept, $J = 5.5$ Hz, 1H, C*H*$_{al}$), 2.60–2.47 (m, 4H, C*H*$_2$), 2.00 (s, 3H, C*H*$_3$).

^{13}C NMR (125 MHz, CD_2Cl_2) δ 198.8 (*C*=O), 162.0 (*C*q=), 144.1 (*C*q$_{Ph}$), 129.0 (2C, *C*H$_{Ph}$), 127.2 (2C, *C*H$_{Ph}$), 127.1 (*C*H), 126.6 (*C*H), 44.3 (*C*H$_2$C(=O)), 41.2 (*C*H$_{al}$), 39.2 (*C*H$_2$Cq$_{ol}$), 24.4 (*C*H$_3$).

MS (EI-DE) *m/z* (%) 186 [M$^+$] (41), 171 (2), 158 (1), 142 (6), 128 (4), 115 (4), 104 (14), 91 (4), 82 (100), 77 (7), 65 (3), 54 (10), 39 (10), 27 (2).

HRMS (EI-DE) calcd for $C_{13}H_{14}O$ [M$^+$] 186.1043, found 186.1044.

7.5.2 Preparation of Acyclic α,β-Unsaturated Ketones

7.5.2.1 General Procedure A: Cross Metathesis (CM)

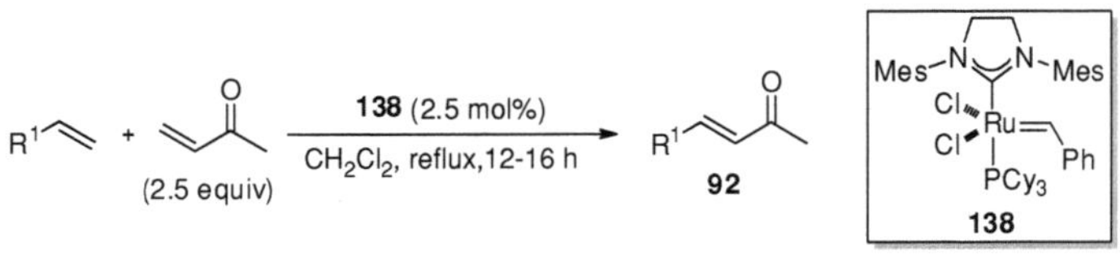

Methyl vinyl ketone (MVK; 12.5 mmol, 2.5 equiv) and *Grubbs*' second generation catalyst (**138**; 106 mg, 0.125 mmol, 2.5 mol%) were successively added to the solution of a terminal alkene (5 mmol, 1 equiv) in CH_2Cl_2 (50 mL). The reaction mixture was heated to reflux overnight (12–16 h) and then allowed to cool to room temperature. The volatiles were removed under reduced pressure and the crude product was purified by flash column chromatography (silica gel, eluent: Et_2O-pentane or Et_2O–CH_2Cl_2) to obtain the corresponding pure (*E*)-enone **92**.

Scope of *α,β*-Unsaturated Ketones Prepared According to General Procedure A

(*E*)-7-Bromo-3-hepten-2-one (92g):

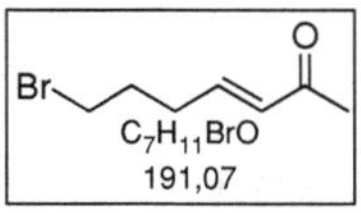

The title compound was prepared according to the general procedure **A** from MVK and 5-bromo-1-pentene. The crude product was purified by flash column chromatography (silica gel, 5–15% Et_2O in pentane) to give enone **92g** (748 mg, 3.94 mmol, 78%) as a pale yellow oil. *Contains* 4% *of* (*E*)-7-*chloro*-3-*hepten*-2-*one as determined by GC.*

^{1}H NMR (500 MHz, CD_2Cl_2) δ 6.74 (dt, J = 15.9, 6.9 Hz, 1H, =C***H***CH_2), 6.09 (dt, J = 16.1, 1.4 Hz, 1H, C***H***=CHCH_2), 3.44 (t, J = 6.6 Hz, 2H, BrC$\boldsymbol{H}_2$), 2.41–2.36 (m, 2H, =CHC$\boldsymbol{H}_2$), 2.21 (s, 3H, C$\boldsymbol{H}_3$), 2.06–2.00 (m, 2H, CH_2C$\boldsymbol{H}_2$$CH_2$).

^{13}C NMR (125 MHz, CD_2Cl_2): δ 198.0 (***C***=O), 145.7 (=***C***HCH_2), 132.1 (***C***H=CHCH_2), 32.9 (BrCH_2), 31.0 ($CH_2CH_2CH_2$), 30.8 (=CHCH_2), 26.8 (***C***H_3).

GC-MS (GC-EI) *m/z* 190 [M^+].

HRMS (EI-FE) calcd for $C_7H_{11}BrO$ [M^+] 189.9992, found 189.9993.

(*E*)-Ethyl 6-oxohept-4-enoate (92j):

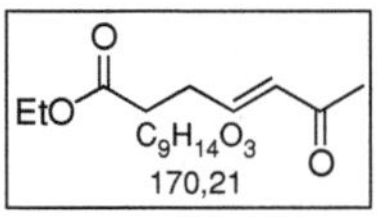

The title compound was prepared according to the general procedure **A** from MVK and ethyl 4-pentenoate. The crude product was purified by flash column chromatography (silica gel, 40–55% Et_2O in pentane) to give enone **92j** (679 mg, 3.99 mmol, 80%) as a pale yellow oil.

^{1}H NMR (300 MHz, CD_2Cl_2) δ 6.77 (dt, J = 15.9, 6.3 Hz, 1H, =C***H***CH_2), 6.08 (dt, J = 15.9, 1.5 Hz, 1H, C***H***=CHCH_2), 4.11 (q, J = 7.2 Hz, 2H, C$\boldsymbol{H}_2$$CH_3$), 2.57–2.43 (m, 4H, C$\boldsymbol{H}_2$), 2.20 (s, 3H, C$\boldsymbol{H}_3$C(=O)), 1.23 (t, J = 7.1 Hz, 3H, CH_2C$\boldsymbol{H}_3$).

^{13}C NMR (100 MHz, CD_2Cl_2) δ 198.0 (***C***=O), 172.2 (***C***O_2Et), 145.7 (=***C***HCH_2), 131.7 (***C***H=CHCH_2), 60.5 (***C***H_2CH_3), 32.5 (CH_2CO_2Et), 27.5 (=CHCH_2), 26.7 (***C***H_3C(=O)), 14.0 (CH_2***C***H_3).

MS (EI-DE) *m/z* 170 [M$^+$] (26), 155 (3), 141 (5), 124 (76), 109 (18), 97 (41), 83 (40), 68 (3), 55 (20), 43 (100), 29 (27).

HRMS (EI-FE) calcd for $C_9H_{14}O_3$ [M$^+$] 170.0941, found 170.0943.

(*E*)-3-Octene-2,7-dione (92k):

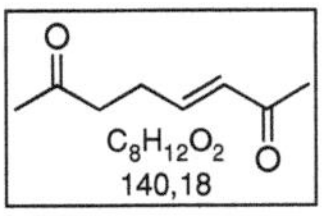

The title compound was prepared according to the general procedure **A** from MVK and 5-hexen-2-one. The crude product was purified by flash column chromatography (silica gel, 30–55% Et_2O in pentane) to give enone **92k** (698 mg, 4.98 mmol, 99%) as a pale yellow oil.

^{1}H NMR (300 MHz, CD_2Cl_2) δ 6.76 (dt, J = 15.9, 6.7 Hz, 1H, =C***H***CH_2), 6.03 (dt, J = 16.1, 1.6 Hz, 1H, C***H***=CHCH_2), 2.63–2.58 (m, 2H, C***H***$_2$), 2.49–2.41 (m, 2H, C***H***$_2$), 2.19 (s, 3H, CH_{ol}C(=O)C***H***$_3$), 2.13 (s, 3H, C***H***$_3$C(=O)CH_2).

^{13}C NMR (100 MHz, CD_2Cl_2) δ 206.6 (CH_2***C***(=O)), 198.1 (CH_{ol}***C***(=O)), 146.3 (=***C***HCH_2), 131.6 (***C***H=CHCH_2), 41.5 (***C***H_2C(=O)), 29.7 (***C***H_3C(=O)CH_2), 26.7 (CH_{ol}C(=O)***C***H_3), 26.2 (=CH***C***H_2).

MS (EI-DE) *m/z* (%) 140 [M$^+$] (15), 122 (2), 97 (82), 83 (7), 79 (6), 69 (3), 55 (5), 43 (100), 41 (6), 27 (4).

HRMS (EI-FE) calcd for $C_8H_{12}O_2$ [M$^+$] 140.0836, found 140.0837.

(*E*)-9-hydroxy-3-nonen-2-one (92l):

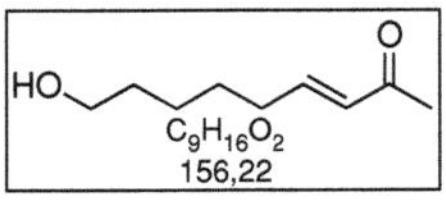

The title compound was prepared according to the general procedure **A** from MVK and 7-hydroxy-1-heptene. Flash column chromatography (silica gel, 10–30% Et_2O in CH_2Cl_2) afforded enone **92l** (503 mg, 3.22 mmol, 64%) as a greenish oil (*contains ruthenium trace impurities*).

^{1}H NMR (400 MHz, CD_2Cl_2) δ 6.79 (dt, J = 15.9, 7.0 Hz, 1H, CH_2C***H***=), 6.03 (dt, J = 16.2, 1.4 Hz, 1H, CH_2CH=C***H***), 3.59 (t, J = 6.6 Hz, 2H, C***H***$_2$OH), 2.21 (app. qd overlapped, J = 7.1, 1.4 Hz, 2H, C***H***$_2$), 2.20 (s, 3H, C***H***$_3$), 1.75 (br s, 1H, O***H***), 1.56–1.46 (m, 4H, C***H***$_2$), 1.41–1.34 (m, 2H, C***H***$_2$).

^{13}C NMR (100 MHz, CD_2Cl_2) δ 198.8 (***C***=O), 148.6 (CH_2***C***H=), 131.6 (CH_2CH=***C***H), 62.9 (***C***H_2OH), 32.9(***C***H_2), 32.7 (***C***H_2), 28.3 (***C***H_2), 26.9 (***C***H_3), 25.7 (***C***H_2).

MS (EI-DE) *m/z* (%) 156 [M$^+$] (1), 138 (11), 123 (9), 113 (7), 95 (55), 84 (8), 81 (32), 71 (42), 67 (32), 58 (9), 55 (49), 53 (15), 43 (100), 41 (27), 31 (19).

HRMS (CI-FE, *i*-butane) calcd for $C_9H_{17}O_2$ [(M + H)$^+$] 157.1230, found 157.1229.

7.5.2.2 General Procedure B: Wittig Reaction

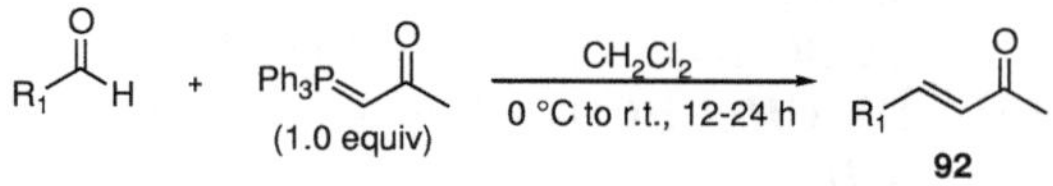

An aldehyde (20 mmol, 1 equiv) was dissolved in CH_2Cl_2 (30 mL) and 1-(triphenylphosphoranylidene)-2-propanone (6.37 g, 20 mmol, 1 equiv) was added at 0 °C. The reaction mixture was allowed to warm to room temperature and stirred until TLC or GC/MS analysis indicated complete consumption of the starting aldehyde (12–24 h). Then silica gel was added and the solvent was removed under reduced pressure. The residue was purified by flash column chromatography (silica gel, eluent: Et_2O-pentane) to afford the corresponding (*E*)-α,β-enone **92**.

Scope of α,β-Unsaturated Ketones Prepared According to General Procedure B

(*E*)-6-Phenyl-3-hexen-2-one (92b):

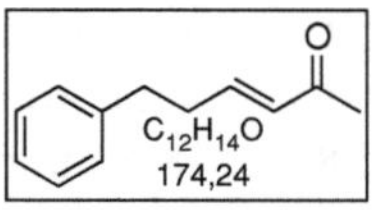

The title compound was prepared according to the general procedure **B** from hydrocinnam-aldehyde and 1-(triphenyl-phos-phor-anylidene)-2-propanone. The crude product was purified by flash column chromatography (silica gel, 10% Et_2O in pentane) to give enone **92b** (2.62 g, 15.0 mmol, 75%) as a pale yellow oil.

^{1}H NMR (500 MHz, $CDCl_3$) δ 7.30–7.27 (m, 2H, C$H_{Ph,\ m}$), 7.21–7.16 (m, 3H, C$H_{Ph,\ o,\ p}$), 6.80 (dt, *J* = 15.9, 6.9 Hz, 1H, =C*H*CH_2), 6.08 (dt, *J* = 16.1, 1.4 Hz, 1H, C*H*=CHCH_2), 2.78 (t, *J* = 7.7 Hz, 2H, PhCH_2), 2.54 (app. qd, *J* = 7.3, 1.4 Hz, 2H, =CHCH_2), 2.21 (s, 3H, CH_3).

^{13}C NMR (125 MHz, $CDCl_3$) δ 198.6 (*C*=O), 147.1 (=*C*HCH_2), 140.7 (Cq_{Ph}), 131.7 (*C*H=CHCH_2), 128.5 (2C, CH_{Ph}), 128.3 (2C, CH_{Ph}), 126.2 ($CH_{Ph,\ p}$), 34.4 (CH_2), 34.1 (CH_2), 26.9 (CH_3).

MS (EI-DE) *m/z* (%) 174 [M^+] (5), 159 (4), 131 (5), 116 (16), 104 (2), 91 (100), 77 (2), 65 (10), 51 (3), 43 (9), 27 (2).

HRMS (EI-FE) calcd for $C_{12}H_{14}O$ [M^+] 174.1045, found 174.1045.

(*E*)-6-Methylhept-3-en-2-one (92d):

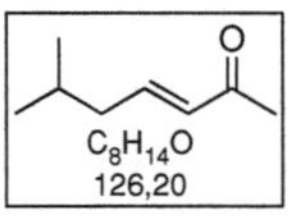

The title compound was prepared according to the general procedure **B** from isovaleraldehyde with 1-(triphenylphosphoranylidene)-2-propanone. The crude product was purified by flash column chromatography (silica gel, 5% Et_2O in pentane) to give enone **92d** [1.51 g, 12.0 mmol, 60% (*reduced yield due to the high volatility of **92d***)] as a pale yellow oil.

^{1}H NMR (500 MHz, $CDCl_3$) δ 6.74 (dt, J = 16.0, 7.4 Hz, 1H, =C***H***CH_2), 6.03 (dt, J = 16.0, 1.4 Hz, 1H, C***H***=$CHCH_2$), 2.21 (s, 3H, C***H***$_3$C(=O)), 2.08 (app. td, J = 7.1, 1.3 Hz, 2H, C***H***$_2$), 1.74 (hept, J = 6.7 Hz, 1H, C***H***Me_2), 0.90 (d, J = 6.7 Hz, 6H, CH(C***H***$_3$)$_2$).

^{13}C NMR (125 MHz, $CDCl_3$) δ 198.6 (***C***=O), 147.3 (=***C***HCH_2), 132.3 (***C***H=$CHCH_2$), 41.7 (***C***H_2), 27.8 (***C***HMe_2), 26.8 (***C***H_3C(=O)), 22.3 (2C, CH(***C***H_3)$_2$).

GC-MS (GC-EI) *m/z* 126 [M^+].

HRMS (EI-FE) calcd for $C_8H_{14}O$ [M^+] 126.1043, found 126.1045.

(*E*)-4-Cyclohexylbut-3-en-2-one (92e):

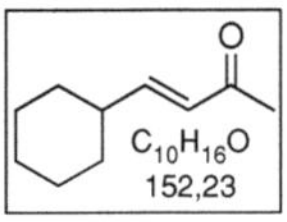

The title compound was prepared according to the general procedure **B** from cyclohexanecarbaldehyde with 1-(triphenylphosphoranylidene)-2-propanone. The crude product was purified by flash column chromatography (silica gel, 10% Et_2O in pentane) to give enone **92e** (2.80 g, 18.4 mmol, 92%) as a pale yellow oil.

^{1}H NMR (500 MHz, CD_2Cl_2) δ 6.71 (dd, J = 16.1, 6.8 Hz, 1H, =C***H***CH), 5.98 (dd, J = 16.3, 1.2 Hz, 1H, C***H***=CHCH), 2.20 (s, 3H, C***H***$_3$), 2.18–2.11 (m, 1H, =CHC***H***), 1.79–1.74 (m, 4H, C***H***$_2$), 1.70–1.66 (m, 1H, C***H***$_2$), 1.35–1.27 (m, 2H, C***H***$_2$), 1.24–1.11 (m, 3H, C***H***$_2$).

^{13}C NMR (125 MHz, CD_2Cl_2) δ 199.0 (***C***=O), 153.5 (=***C***HCH), 129.1 (***C***H=CHCH), 41.0 (=CH***C***H), 32.2 (2C, CH***C***H_2), 26.9 (***C***H_3), 26.3 (***C***H_2), 26.1 (2C, ***C***H_2).

GC-MS (GC-EI) *m/z* 152 [M^+].

HRMS (EI-FE) calcd for $C_{10}H_{16}O$ [M^+] 152.1200, found 152.1201.

(*E*)-Octa-3,7-dien-2-one (92f):

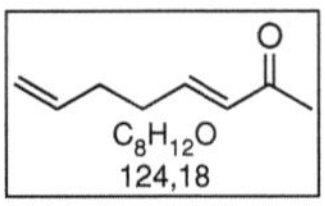

The title compound was prepared according to the general procedure **B** from 4-pentenal and 1-(triphenylphosphor-anylidene)-2-propa-none. The crude product was purified by flash column chromatography (silica gel, 5–10% Et_2O in pentane) to give enone **92f** (2.19 g, 17.6 mmol, 88%) as a pale yellow oil.

^{1}H NMR (500 MHz, CD_2Cl_2) δ 6.77 (dt, J = 16.1, 6.8 Hz, 1H, CH_2CH=C***H***), 6.05 (dt, J = 15.9, 1.5 Hz, 1H, CH_2C***H***=CH), 5.82 (ddt, J = 16.9, 11.7, 6.0 Hz, 1H, CH_2=C***H***), 5.06 (app. dq, J = 17.0, 1.7 Hz, 1H, C$\boldsymbol{H}_{trans}$H=), 5.01 (app. dq, J = 10.5, 1.5 Hz, 1H, CH$\boldsymbol{H}_{cis}$=), 2.35–2.30 (m, 2H, C***H***$_2$CH=CH), 2.25–2.21 (m, 2H, CH_2=CHC***H***$_2$), 2.20 (s, 3H, C***H***$_3$).

^{13}C NMR (125 MHz, CD_2Cl_2) δ 198.5 (***C***=O), 147.6 (CH_2***C***H=CH), 137.7 (CH_2=***C***H), 131.9 (CH_2CH=***C***H), 115.5 (***C***H_2=CH), 32.5 (CH_2=CH***C***H_2), 32.0 (***C***H_2CH=CH), 26.9 (***C***H_3).

GC-MS (GC-EI) *m/z* 124 [M$^+$] (trace), 122 (4), 109 (14), 95 (12), 91 (3), 81 (58), 79 (35), 11 (12), 66 (12), 55 (34), 53 (18), 51 (6), 43 (100), 41 (34), 27 (10).

HRMS (CI-FE, *i*-butane) calcd for $C_8H_{13}O$ [(M + H)$^+$] 125.0966, found 125.0966.

(*E*)-9-(Tetrahydro-2*H*-pyran-2-yloxy)-3-nonen-2-one (92h):

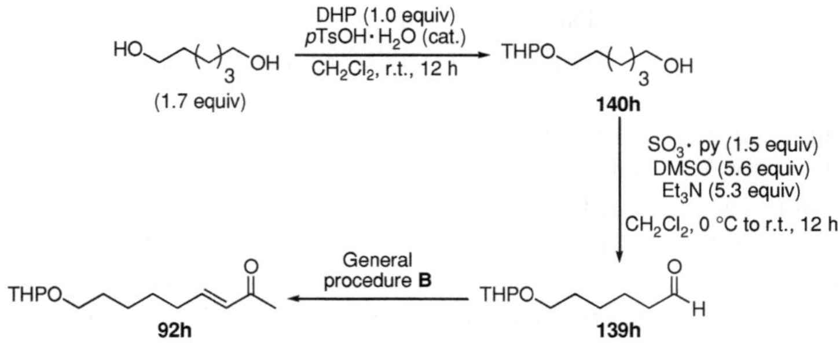

6-(Tetrahydro-2*H*-pyran-2-yloxy)-1-hexanol (140h):

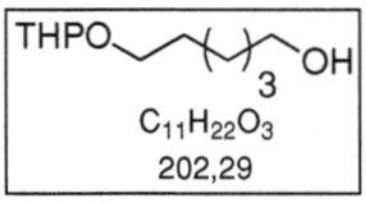

A solution of 1,6-hexanediol (2.36 g, 20.0 mmol, 1.7 equiv) and dihydropyrane (1.10 mL, 12.0 mmol) in THF (100 mL) in the presence of *p*-toluenesulfonic acid monohydrate (190 mg, 1.0 mmol) was stirred at 0 °C for 12 h. Then the solution was poured on saturated aqueous $NaHCO_3$ solution and extracted with Et_2O. The combined organic layers were dried (Na_2SO_4), filtered, and concentrated under reduced pressure. The crude product was purified by flash column chromatography (silica gel, 40% EtOAc in hexanes) to provide THP-mono-protected alcohol **140h** (1.44 g, 7.13 mmol, 59%) as a colorless oil.

^{1}H NMR (500 MHz, CD_2Cl_2) δ 4.55 (t, J = 3.4 Hz, 1H, OC***H***O), 3.88–3.81 (m, 1H, OC***H***H_{THP}), 3.72 (dt, J = 9.8, 6.8 Hz, 1H, THPOC***H***H), 3.62 (t, J = 6.6 Hz, 2H, C***H***$_2$OH), 3.51–3.44 (m, 1H, OCH***H***$_{THP}$), 3.37 (dt, J = 9.6, 6.5 Hz, 1H, THPOCH***H***), 1.86–1.35 (m, 15H, C***H***$_2$ and O***H***).

^{13}C NMR (125 MHz, CD_2Cl_2) δ 98.9 (O***C***HO), 67.5 (THPO***C***H_2), 62.9 (O***C***H_2), 62.4 (O***C***H_2), 32.7 (***C***H_2), 30.8 (***C***H_2), 29.7 (***C***H_2), 26.0 (***C***H_2), 25.6 (***C***H_2), 25.5 (***C***H_2), 19.7 (***C***H_2).

MS (EI-DE) *m/z* (%) 202 [M$^+$] (trace), 129 (2), 117 (4), 101 (27), 85 (100), 67 (10), 55 (43), 41 (22), 29 (9).

HRMS (CI-FE, *i*-butane) calcd for $C_{11}H_{23}O_3$ [$(M + H)^+$] 203.1644, found 203.1647.

6-(Tetrahydro-2*H*-pyran-2-yloxy)hexanal (139h):

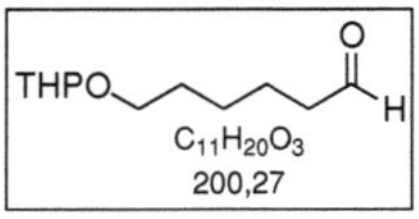

THP-mono-protected alcohol **140h** (1.20 g, 5.93 mmol) was dissolved in CH_2Cl_2 (15 mL) and anhydrous DMSO (2.36 mL, 33.2 mmol, 5.6 equiv) and Et_3N (4.36 mL, 31.4 mmol, 5.3 equiv) were added sequentially at room temperature. The resulting solution was cooled to 0 °C and $SO_3 \cdot$ py complex (1.42 g, 8.90 mmol, 1.5 equiv) was added in several portions. The reaction mixture was kept at 0 °C for 1 h, and was then stirred at room temperature overnight (12 h). The reaction was quenched by the addition of water (10 mL) and extracted with Et_2O (2 × 25 mL). The combined organic layers were washed with water and brine, dried (Na_2SO_4), filtered, and concentrated *in vacuo*. The crude product was purified by flash column chromatography (silica gel, 20–30% Et_2O in pentane) to afford aldehyde **139h** (754 mg, 3.77 mmol, 64%) as a colorless liquid.

^{1}H NMR (500 MHz, CD_2Cl_2) δ 9.73 (t, J = 1.7 Hz, 1H, C***H***O), 4.53 (t, J = 3.8 Hz, 1H, C***H***$_{THP}$), 3.83–3.79 (m, 1H, OC***H***H$_{THP}$), 3.69 (dt, J = 9.6, 6.7 Hz, 1H, THPOC***H***H), 3.47–3.43 (m, 1H, OCH***H***$_{THP}$), 3.35 (dt, J = 9.6, 6.4 Hz, 1H, THPOCH***H***), 2.42 (td, J = 7.3, 1.7 Hz, 2H, C***H***$_2$CHO), 1.83–1.74 (m, 1H, C***H***$_2$), 1.69–1.47 (m, 9H, C***H***$_2$), 1.42–1.36 (m, 2H, C***H***$_2$).

^{13}C NMR (125 MHz, CD_2Cl_2) δ 202.9 (***C***=O), 99.2 (O–***C***H–O), 67.5 (THPO***C***H$_2$), 62.5 (O***C***H$_{2,\ THP}$), 44.2 (***C***H$_2$CHO), 31.2 (***C***H$_2$), 29.9 (***C***H$_2$), 26.3 (***C***H$_2$), 26.0 (***C***H$_2$), 22.3 (***C***H$_2$), 20.1 (***C***H$_2$).

(*E*)-9-(Tetrahydro-2*H*-pyran-2-yloxy)-3-nonen-2-one (92h):

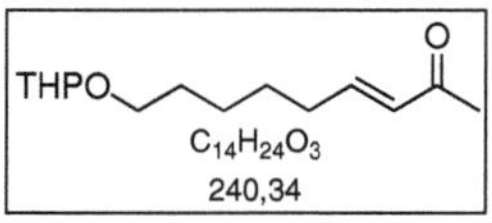

The title compound was prepared according to the general procedure **B** from 6-(tetrahydro-2*H*-pyran-2-yloxy)-hexanal (**139h**) with 1-(triphenylphosphoranylidene)–2-propanone. The crude product was purified by flash column chromatography (silica gel, 10–30% Et_2O in pentane) to afford enone **92h** (613 mg, 2.55 mmol, 85%) as a colorless oil.

^{1}H NMR (500 MHz, CD_2Cl_2) δ 6.79 (dt, J = 15.9, 7.0 Hz, 1H,=C***H***CH$_2$), 6.04 (dt, J = 15.9, 1.5 Hz, 1H, C***H***=CHCH$_2$), 4.54–4.52 (br t, 1H, OC***H***O), 3.84–3.79 (m, 1H, OC***H***H$_{THP}$), 3.71–3.67 (m, 1H, THPOC***H***H), 3.47–3.43 (m, 1H, OCH***H***$_{THP}$), 3.37–3.33 (m, 1H, THPOCH***H***), 2.26–2.21 (m, 2H,=CHC***H***$_2$), 2.20 (s, 3H, C***H***$_3$), 1.82–1.75 (m, 1H, C***H***$_2$), 1.69–1.64 (m, 1H, C***H***$_2$), 1.60–1.47 (m, 8H, –(C***H***$_2$)$_n$–), 1.43–1.36 (m, 2H, C***H***$_2$).

^{13}C NMR (125 MHz, CD_2Cl_2) δ 198.3 (*C*=O), 148.2 (= *C*HCH$_2$), 131.4 (*C*H=CHCH$_2$), 98.9 (O*C*HO), 67.3 (THPO*C*H$_2$), 62.2 (O*C*H$_{2,\ THP}$), 32.4 (=CH*C*H$_2$), 30.1 (*C*H$_2$), 29.6 (*C*H$_2$), 28.1 (*C*H$_2$), 26.6 (*C*H$_3$), 25.9 (*C*H$_2$), 25.7 (*C*H$_2$), 19.8 (*C*H$_2$).

MS (EI-DE) *m*/*z* (%) 225 (2), 185 (1), 156 (11), 140 (10), 126 (6), 111 (2), 97 (17), 85 (100), 81 (10), 67 (14), 55 (14), 43 (50), 29 (8).

HRMS (ESI+) calcd for $C_{14}H_{24}NaO_3$ [(M + Na)$^+$] 263.1615, found 263.1618.

(*E*)-6-(*tert*-Butyldimethylsilyloxy)-3-hexen-2-one (92i):

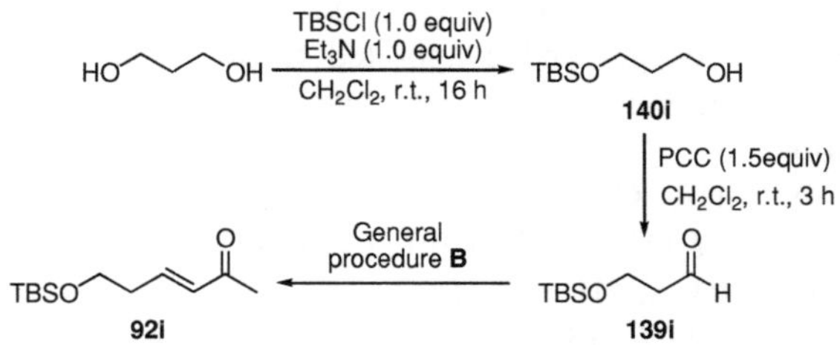

3-(*tert*-Butyldimethylsilyloxy)-1-propanol (140i):

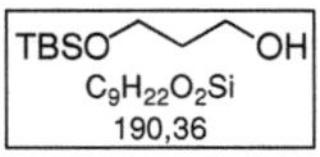

1,3-Propanediol (3.62 mL, 50.0 mmol, 1.0 equiv) was dissolved in CH_2Cl_2 (150 mL) and triethylamine (6.93 mL, 50.0 mmol, 1.0 equiv) and a solution of *tert*-butyldimethylsilyl chloride (7.54 g, 50.0 mmol, 1.0 equiv) in CH_2Cl_2 (20 mL) were added. After stirring for 16 h at room temperature, the reaction mixture was successively extracted with 10% aqueous $NaHCO_3$ (50 mL), water (50 mL), and brine (50 mL). The organic layer was dried (Na_2SO_4), filtered, and concentrated under reduced pressure. The crude product was purified by flash column chromatography (silica gel, 10–30% EtOAc in hexane) to afford the title compound (9.00 g, 47.3 mmol, 95%) as a colorless liquid.

^{1}H NMR (500 MHz, $CDCl_3$) δ 3.81 (t, J = 5.5 Hz, 2H, TBSOC***H***$_2$), 3.78 (app. br q, J = 5.1 Hz, 2H, C***H***$_2$OH), 2.59 (br s, 1H, OH), 1.75 (quint, J = 5.6 Hz, 2H, CH$_2$C***H***$_2$CH$_2$), 0.88 (s, 9H, Cq(C***H***$_3$)$_3$), 0.05 (s, 6H, Si(C***H***$_3$)$_2$).

^{13}C NMR (125 MHz, $CDCl_3$) δ 63.0 (O*C*H$_2$), 62.5 (O*C*H$_2$), 34.2 (CH$_2$*C*H$_2$CH$_2$), 25.9 (3C, Cq(*C*H$_3$)$_3$), 18.1 (*C*qMe$_3$), −5.5 (2C, Si(*C*H$_3$)$_2$).

GC-MS (GC-EI) *m*/*z* (%) 133 (14), 115 (3), 105 (41), 91 (3), 75 (100), 59 (5), 45 (6), 29 (2).

HRMS (ESI+) calcd for $C_9H_{22}NaO_2Si$ [(M + Na)$^+$] 213.1282, found 213.1281.

3-(*tert*-Butyldimethylsilyloxy)propanal (139i):

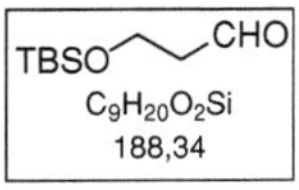

PCC (13.1 g, 60.6 mmol, 1.5 equiv) was added in one portion to a solution of 3-(*tert*-butyldimethylsilyloxy)-1-propanol (**140i**; 7.69 g, 40.4 mmol, 1.0 equiv) in CH_2Cl_2 (130 mL). The mixture was stirred for 3 h at room temperature. Then the solution was separated from the black insoluble material, which was repeatedly extracted with Et_2O (3 × 70 mL). The combined organic layers were filtered through a plug of silica gel and evaporated to afford aldehyde **139i** (5.20 g, 27.6 mmol) as a colorless liquid. The aldehyde was used in the next step without further purification.

^{1}H NMR (500 MHz, $CDCl_3$) δ 9.78 (t, J = 2.0 Hz, 1H, C***H***O), 3.97 (t, J = 6.0 Hz, 2H, C***H***$_2$OTBS), 2.58 (td, J = 6.0, 2.3 Hz, 2H, C***H***$_2$CHO), 0.86 (s, 9H, Cq(C***H***$_3$)$_3$), 0.04 (s, 6H, Si(C***H***$_3$)$_2$).

GC-MS (GC-EI) *m*/*z* (%) 131 (72), 117 (7), 101 (100), 89 (3), 75 (33), 73 (9), 59 (27), 45 (9), 29 (4).

(*E*)-6-(*tert*-Butyldimethylsilyloxy)-3-hexen-2-one (92i):

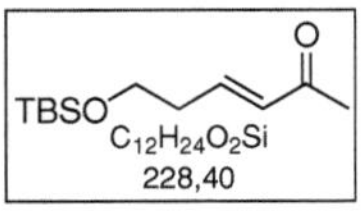

The title compound was prepared according to the general procedure **B** from 3-(*tert*-butyl-dimethyl-silyloxy)propanal (**139i**) with 1-(triphenylphosphoranylid-ene)-2-propanone. The crude product was purified by flash column chromatography (silica gel, 5% Et_2O in pentane) to afford enone **92i** (819 mg, 3.59 mmol, 72%) as a pale yellow oil.

^{1}H NMR (500 MHz, CD_2Cl_2) δ 6.78 (dt, J = 16.1, 7.1 Hz, 1H, =C***H***CH$_2$), 6.08 (dt, J = 16.1, 1.6 Hz, 1H, C***H***=CHCH$_2$), 3.74 (t, J = 6.3 Hz, 2H, TBSOC***H***$_2$), 2.42 (app. qd, J = 6.6, 1.3 Hz, 2H, =CHC***H***$_2$), 2.21 (s, 3H, C***H***$_3$C(=O)), 0.89 (s, 9H, Cq(C***H***$_3$)$_3$), 0.05 (s, 6H, Si(C***H***$_3$)$_2$).

^{13}C NMR (125 MHz, CD_2Cl_2) δ 198.4 (***C***=O), 145.4 (=***C***HCH$_2$), 133.1 (***C***H=CHCH$_2$), 61.9 (TBSO***C***H$_2$), 36.3 (=CH***C***H$_2$), 26.8 (***C***H$_3$C(=O)), 26.0 (3C, Cq(***C***H$_3$)$_3$), 18.5 (***C***qMe$_3$), −5.3 (2C, Si(***C***H$_3$)$_2$).

MS (EI-DE) *m*/*z* (%) 228 [M$^+$] (trace), 213 (3), 198 (5), 183 (2), 171 (93), 141 (100), 127 (33), 115 (4), 103 (6), 89 (11), 75 (27), 59 (4), 43 (6), 29 (2).

HRMS (ESI+) calcd for $C_{12}H_{24}NaO_2Si$ [(M + Na)$^+$] 251.1437, found 251.1438.

(*E*)-5-Phenyl-3-penten-2-one (92q):

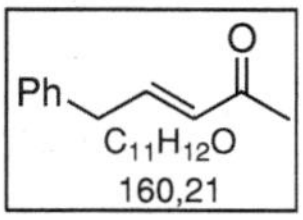

The title compound was prepared according to the general procedure **B** from phenylacetaldehyde with 1-(triphenylphosphoranylidene)-2-propanone. The reaction was constantly maintained at 0 °C. The crude product was purified by flash column chromatography (silica gel, 5–10% Et_2O in pentane) to afford enone **92q** (414 mg, 2.58 mmol, 26%) as a pale yellow oil.

^{1}H NMR (500 MHz, $CDCl_3$) δ 7.32–7.28 (m, 2H, C$\boldsymbol{H}_{Ph}$), 7.24–7.21 (m, 1H, C$\boldsymbol{H}_{Ph,\ p}$), 7.16–7.15 (m, 2H, C$\boldsymbol{H}_{Ph}$), 6.89 (dt, J = 15.9, 6.9 Hz, 1H, =C***H***CH_2), 6.05 (dt, J = 16.1, 1.6 Hz, 1H, C***H***=CHCH$_2$), 3.52 (dd, J = 6.6, 1.3 Hz, 2H, PhC$\boldsymbol{H}_2$), 2.02 (s, 3H, C$\boldsymbol{H}_3$).

^{13}C NMR (125 MHz, $CDCl_3$) δ 198.5 (***C***=O), 146.3 (=***C***HCH$_2$), 137.6 (***C***q$_{Ph}$), 132.0 (***C***H=CHCH$_2$), 128.8 (2C, ***C***H$_{Ph}$), 128.7 (2C, ***C***H$_{Ph}$), 126.8 (***C***H$_{Ph,\ p}$), 38.8 (***C***H$_2$), 26.9 (***C***H$_3$).

MS (EI-DE) *m*/*z* (%) 160 [M$^+$] (59), 145 (27), 127 (31), 117 (100), 102 (3), 91 (38), 89 (8), 77 (5), 65 (17), 58 (9), 51 (11), 43 (57), 39 (15), 27 (3).

HRMS (EI-FE) calcd for $C_{11}H_{12}O$ [M$^+$] 160.0886, found 160.0888.

(*E*)-6,6-Dimethylhept-3-en-2-one (92r):

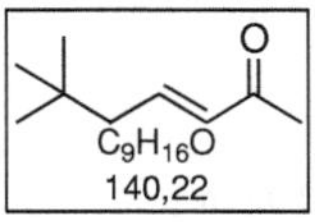

The title compound was prepared according to the general procedure **B** from 3,3-dimethylbutanal with 1-(triphenylphosphoranylidene)-2-propanone. The crude product was purified by flash column chromatography (silica gel, 5% Et_2O in pentane) to give enone **92r** [1.58 g, 11.3 mmol, 56% (*reduced yield due to the high volatility of* ***92r***)] as a pale yellow oil.

^{1}H NMR (500 MHz, $CDCl_3$) δ 6.79 (dt, J = 15.6, 7.9 Hz, 1H, =C***H***CH$_2$), 6.04 (dt, J = 15.9, 1.4 Hz, 1H, C***H***=CHCH$_2$), 2.22 (s, 3H, C$\boldsymbol{H}_3$C(=O)), 2.08 (dd, J = 7.8, 1.2 Hz, 2H, C$\boldsymbol{H}_2$), 0.92 (s, 9H, Cq(C$\boldsymbol{H}_3$)$_3$).

^{13}C NMR (125 MHz, $CDCl_3$) δ 198.4 (***C***=O), 145.8 (= ***C***HCH$_2$), 133.2 (***C***H=CHCH$_2$), 46.9 (***C***H$_2$), 31.4 (***C***q), 29.4 (3C, Cq(***C***H$_3$)$_3$), 26.9 (***C***H$_3$C(=O)).

GC-MS (GC-EI) *m*/*z* 140 [M$^+$].

HRMS (CI-FE, *i*-butane) calcd for $C_9H_{17}O$ [(M + H)$^+$] 141.1280, found 141.1279.

(*E*)-5-Benzyloxy-3-penten-2-one (105):

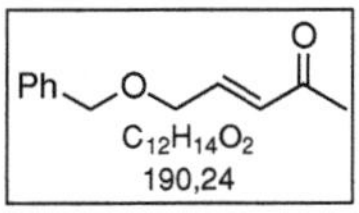

The title compound was prepared according to the general procedure **B** from benzyloxyacetaldehyde with 1-(triphenylphosphoranylidene)-2-propanone. The crude product was purified by flash column chromatography (silica gel, 5–15% Et_2O in pentane) to afford enone **105** (950 mg, 4.99 mmol, 88%) as a pale yellow oil.

^{1}H NMR (500 MHz, $CDCl_3$) δ 7.39–7.35 (m, 4H, CH_{Ph}), 7.33–7.29 (m, 1H, CH_{Ph}), 6.80 (dt, $J = 16.1$, 4.5 Hz, 1H, =C*H*CH_2), 6.31 (d, $J = 16.1$, 1H, C*H*=CHCH_2), 4.46 (s, 2H, PhCH_2), 4.21 (dd, $J = 4.5$, 1.7 Hz, 2H, CH_2CH=), 2.24 (s, 3H, CH_3).

^{13}C NMR (125 MHz, $CDCl_3$) δ 198.4 (*C*=O), 143.6 (=*C*HCH_2), 138.6 (Cq_{Ph}), 130.7 (*C*H=CHCH_2), 128.9 (2C, CH_{Ph}), 128.3 ($CH_{Ph,\ p}$), 128.2 (2C, CH_{Ph}), 73.4 (CH_2), 69.4 (CH_2), 27.6 (CH_3).

MS (EI-DE) *m/z* (%) 190 [M^+] (trace), 160 (6), 145 (5), 132 (9), 117 (1), 107 (2), 99 (2), 91 (100), 84 (10), 65 (9), 51 (3), 43 (13), 29 (2).

HRMS (ESI+) calcd for $C_{12}H_{14}O_2Na$ [$(M + Na)^+$] 213.0885, found 213.0886.

7.5.2.3 Synthesis of (*E*)-2-methyl-4-hexen-3-one (92o)

(*E*)-2-Methyl-4-hexen-3-one (**92o**) was prepared according to a procedure of Oare et al. [22].

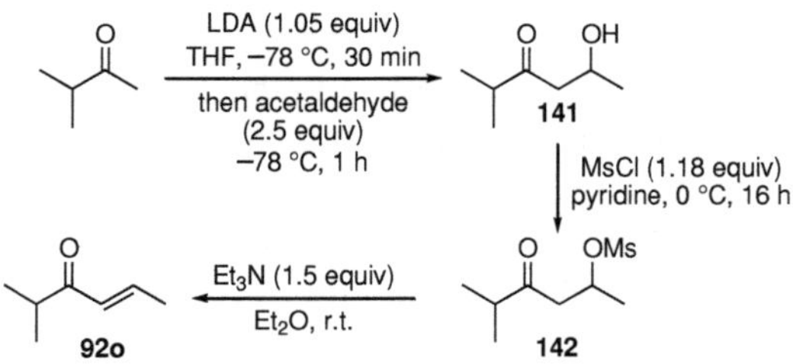

5-Hydroxy-2-methylhexan-3-one (141):

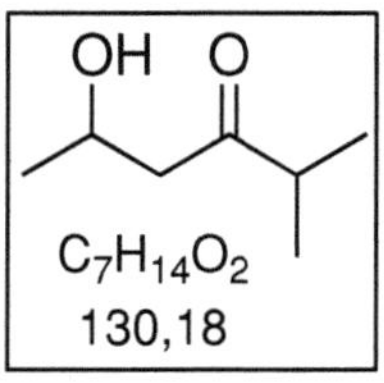

To a solution of LDA, freshly prepared by addition of *n*-BuLi (26.8 ml, 67.0 mmol, 1.05 equiv, 2.5 M in hexanes) to diisopropylamine (9.40 mL,

67.0 mmol, 1.05 equiv) in THF (100 mL) at 0 °C, was added isopropylmethylketone (6.87 mL, 63.8 mmol) at −78 °C. After 30 min, acetaldehyde (9.01 mL, 159.5 mmol, 2.5 equiv) was added, and after stirring for 1 h, the reaction was quenched by addition of saturated aqueous $NaHCO_3$-solution (40 mL) −78 °C. Then, the reaction mixture was allowed to warm to room temperature, and the layers were separated. The aqueous layer was extracted with Et_2O (2 × 75 mL) and the combined organic phases were successively washed with cold 1% aqueous HCl (80 mL), saturated aqueous $NaHCO_3$-solution (80 mL), and brine (80 mL), dried over Na_2SO_4, filtered, and concentrated under reduced pressure to give aldol product **141** (7.91 g, 60.8 mmol, 95%) as a colorless oil.

^{1}H NMR (500 MHz, $CDCl_3$) δ 4.21–4.14 (m, 1H, C***H***(OH)), 3.23 (d, J = 3.3 Hz, 1H, O***H***), 2.62 (dd, J = 18.1, 3.3 Hz, 1H, C***H***H), 2.58–2.49 (m, 2H, CH***H*** and C***H***$(CH_3)_2$), 1.16 (d, J = 6.1 Hz, 3H, CH(OH)C***H***$_3$), 1.08 (d, J = 7.0 Hz, 3H, CH(C***H***$_3$)$_2$), 1.07 (d, J = 6.9 Hz, 3H, CH(C***H***$_3$)$_2$).

^{13}C NMR (125 MHz, $CDCl_3$) δ 216.1 (***C***=O), 63.9 (***C***H(OH)), 47.9 (***C***H_2), 41.4 (***C***$H(CH_3)_2$), 22.3 (CH(OH)***C***H_3), 18.0 (CH(***C***H_3)$_2$), 17.9 (CH(***C***H_3)$_2$).

5-Methyl-4-oxohexan-2-yl methanesulfonate (142):

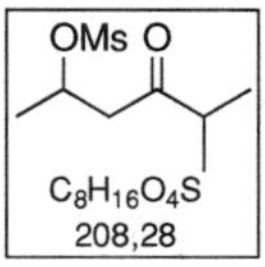

The crude aldol product **141** (7.91 g, 60.8 mmol) was dissolved in pyridine (60 mL) and methanesulfonyl chloride (5.59 mL, 71.7 mmol, 1.18 equiv) was added at 0 °C. The solution was kept at room temperature for 16 h. To work-up the reaction, water (120 mL) was added, and the mixture repeatedly extracted with Et_2O (3 × 100 mL). The combined organic layers were washed with saturated aqueous $CuSO_4$-solution (4 × 75 mL), and brine (100 mL), dried over Na_2SO_4, filtered, and concentrated *in vacuo* to give mesylate **142** (7.95 g, 38.2 mmol, 63%) as a yellow oil.

^{1}H NMR (500 MHz, $CDCl_3$) δ 5.18–5.12 (m, 1H, C***H***(OMs)), 3.04–2.98 (m, 1H, C***H***H), 2.99 (s, 3H, OSO_2C***H***$_3$), 2.62–2.54 (m, 2H, CH***H*** and C***H***$(CH_3)_2$), 1.46 (d, J = 6.4 Hz, 3H, CH(OMs)C***H***$_3$), 1.09 (d, J = 6.9 Hz, 3H, CH(C***H***$_3$)$_2$), 1.08 (d, J = 7.0 Hz, 3H, CH(C***H***$_3$)$_2$).

^{13}C NMR (125 MHz, $CDCl_3$) δ 210.7 (***C***=O), 75.8 (***C***H(OMs)), 46.3 (***C***H_2), 41.2 (***C***$H(CH_3)_2$), 38.0 (OSO_2CH_3), 21.6 (CH(OH)***C***H_3), 17.8 (CH(***C***H_3)$_2$), 17.8 (CH(***C***H_3)$_2$).

(*E*)-2-Methylhex-4-en-3-one (92o):

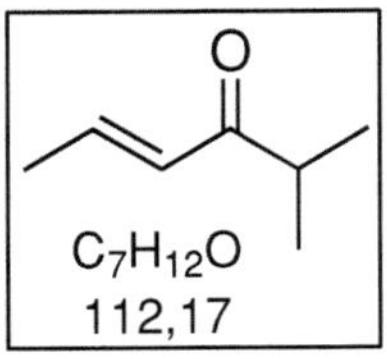

The crude mesylate **142** (7.85 g, 37.7 mmol) was dissolved in Et_2O (40 mL), and triethylamine (7.84 mL, 56.6 mmol, 1.5 equiv) was added. After stirring for 18 h at room temperature, water (80 mL) was added, and the mixture was extracted with Et_2O (3 × 80 mL). The combined organic layers were successively washed with cold aqueous 1% HCl (80 mL), saturated aqueous $NaHCO_3$ solution (80 mL), water (80 mL), and brine (80 mL). The organic phase was dried (Na_2SO_4), filtered, and evaporated. The crude product was purified by flash column chromatography (silica gel, 3–5% Et_2O in pentane) to afford enone **92o** [2.64 g, 23.6 mmol, 62% (*reduced yield due to the high volatility of* ***92o***)] as a pale yellow oil.

^{1}H NMR (500 MHz, CD_2Cl_2) δ 6.85 (dq, J = 15.4, 6.8 Hz, 1H, =C***H***CH_3), 6.17 (dq, J = 15.4, 1.6 Hz, 1H, C***H***=$CHCH_3$), 2.80 (hept, J = 6.8 Hz, 1H, C$\boldsymbol{H}_{al}$), 1.88 (dd, J = 7.0, 1.9 Hz, 3H, =CHC$\boldsymbol{H}_3$), 1.06 (d, J = 6.6 Hz, 6H, CH(C$\boldsymbol{H}_3$)$_2$).

^{13}C NMR (125 MHz, $CDCl_3$) δ 203.7 (***C***=O), 142.4 (= ***C***HCH_3), 130.3 (***C***H=$CHCH_3$), 38.6 (***C***HMe_2), 18.6 (2C, CH(***C***H_3)$_2$), 18.3 (=CH***C***H_3).

GC-MS (GC-EI) *m/z* 112 [M^+].

HRMS (EI-FE) calcd for $C_7H_{12}O$ [M^+] 112.0888, found 112.0888.

7.5.2.4 Synthesis of (Z)-6-phenyl-3-hexen-2-one ((Z)-92b)

(*Z*)-6-Phenyl-3-hexen-2-one ((*Z*)-**92b**) was prepared according to a reaction sequence described by Oare et al. [22].

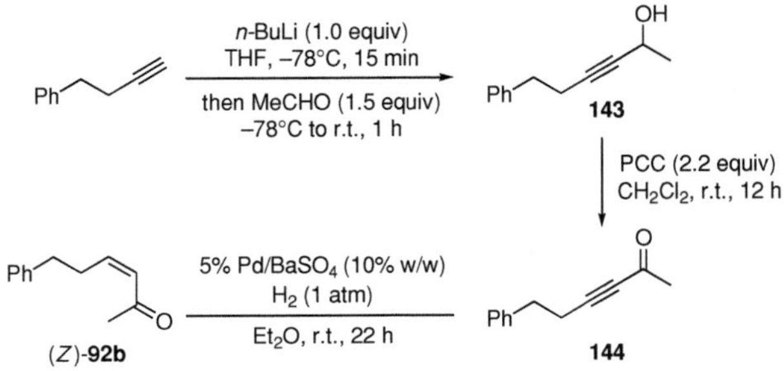

6-Phenylhex-3-yn-2-ol (143):

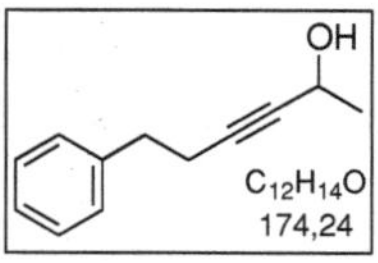

n-BuLi (8.0 mL, 20.0 mmol, 2.5 M in hexanes, 1.0 equiv) was added to a stirred solution of 4-phenylbut-1-yne (2.81 mL, 20.0 mmol, 1.0 equiv) in THF (11 mL) at −78 °C. After 15 min, acetaldehyde (1.70 mL, 30.0 mmol, 1.5 equiv) was added over 1 min. Then the reaction mixture was allowed to warm to room temperature and after 1 h, the reaction was quenched by the addition of aqueous saturated NH_4Cl-solution (20 mL) and repeatedly extracted with Et_2O (3 × 50 mL). The combined organic phases were successively washed with water and brine, dried over Na_2SO_4, filtered, and concentrated under reduced pressure. The crude product was purified by flash column chromatography (silica gel, 20% Et_2O in pentane) and 6-phenylhex-3-yn-2-ol (**143**; 3.13 g, 18.0 mmol, 90%) was obtained as a colorless oil.

^{1}H NMR (500 MHz, CD_2Cl_2) δ 7.32–7.29 (m, 2H, C$\boldsymbol{H}_{Ph,\,m}$), 7.24–7.20 (m, 3H, C$\boldsymbol{H}_{Ph,\,o,\,p}$), 4.48–4.43 (m, 1H, CHO***H***), 2.81 (t, J = 7.5 Hz, 2H, PhC***H***$_2$), 2.49 (td, J = 7.5, 1.9 Hz, 2H, PhCH$_2$C***H***$_2$), 1.92–1.88 (m, 1H, O***H***), 1.37 (d, J = 6.6 Hz, 3H, C***H***$_3$).

^{13}C NMR (125 MHz, CD_2Cl_2) δ 141.4 (***C***q$_{Ph}$), 129.1 (2C, ***C***H$_{Ph}$), 128.8 (2C, ***C***H$_{Ph}$), 126.8 (***C***H$_{Ph,\,p}$), 84.0 (***C***≡C), 83.8 (C≡***C***), 58.9 (***C***OH), 35.5 (Ph***C***H$_2$), 25.1 (***C***H$_3$), 21.4 (PhCH$_2$***C***H$_2$).

GC-MS (GC-EI) *m/z* (%) 173 (1), 156 (6), 141 (4), 129 (21), 115 (9), 102 (2), 91 (100), 77 (4), 65 (16), 51 (7), 43 (9), 29 (7).

HRMS (EI-FE) calcd for $C_{12}H_{15}O$ [(M + H)$^+$] 175.1124, found 175.1123.

6-Phenylhex-3-yn-2-one (**144**):

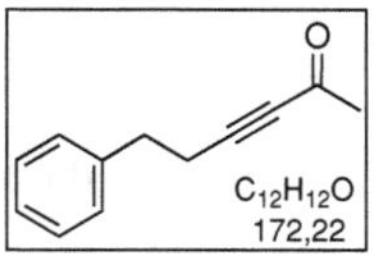

A solution of propargylic alcohol **143** (2.75 g, 15.8 mmol) and PCC (7.50 g, 34.7 mmol, 2.2 equiv) in CH_2Cl_2 (80 mL) was stirred at room temperature for 12 h. Florisil (5 g) was added to the reaction mixture, and stirring was continued for another 15 min. After filtration through a plug of silica gel (eluent: Et_2O), the solvents were removed *in vacuo* giving 6-phenylhex-3-yn-2-one (**144**; 2.42 g, 14.1 mmol, 89%) as a clear oil, which was used in the next step without further purification.

^{1}H NMR (500 MHz, CD_2Cl_2) δ 7.34–7.31 (m, 2H, C$\boldsymbol{H}_{Ph,\,m}$), 7.26–7.23 (m, 3H, C$\boldsymbol{H}_{Ph,\,o,\,p}$), 2.89 (t, J = 7.4 Hz, 2H, PhC***H***$_2$), 2.67 (t, J = 7.4 Hz, 2H, PhCH$_2$C***H***$_2$), 2.26 (s, 3H, C***H***$_3$).

^{13}C NMR (125 MHz, CD_2Cl_2) δ 184.8 (***C***=O), 140.2 (***C***q$_{ar}$), 128.8 (2C, ***C***H$_{ar}$), 128.7 (2C, ***C***H$_{ar}$), 126.9 (***C***H$_{Ph,\,p}$), 92.9 (CH$_2$C≡***C***), 82.0 (CH$_2$***C***≡), 34.2 (Ph***C***H$_2$), 32.9 (***C***H$_3$), 21.3 (PhCH$_2$***C***H$_2$).

MS (EI-DE) *m/z* (%) 172 [M$^+$] (2), 157 (13), 144 (1), 129 (31), 115 (1), 102 (1), 91 (100), 77 (2), 65 (13), 51 (4), 43 (14), 27 (1).

HRMS (CI-FE, *i*-butane) calcd for $C_{12}H_{13}O$ [(M + H)$^+$] 173.0965, found 173.0966.

(*Z*)-6-phenyl-3-hexen-2-one ((*Z*)-**92b**):

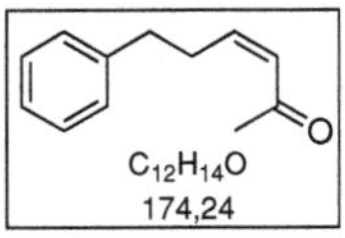

A mixture of ketone **144** (2.0 g, 11.6 mmol), quinoline (20 mg, 1 wt%), 5% Pd/ $BaSO_4$ (200 mg, 10 wt%), and Et_2O (12 mL) as the solvent were placed under a hydrogen atmosphere (1 atm) and stirred at room temperature for 22 h. The catalyst was removed by filtration and the volatiles under reduced pressure. ^{1}H NMR of the crude product showed the desired (*Z*)-enone **92b** contaminated with 25% of the (*E*)-isomer, which was separated by flash column chromatography (silica gel, 10–15% Et_2O in pentane). Pure (*Z*)-enone ((*Z*)-**92b**; 1.40 g, 8.04 mmol, 69%) was obtained as a pale yellow oil.

^{1}H NMR (500 MHz, CD_2Cl_2) δ 7.30–7.27 (m, 2H, C***H***$_{Ph,\ m}$), 7.22–7.17 (m, 3H, C***H***$_{Ph,\ o\ and\ p}$), 6.15 (dt, J = 11.3, 1.3 Hz, 1H, C***H***=CHCH$_2$), 6.08 (dt, J = 11.4, 7.1 Hz, 1H, =C***H***CH$_2$), 2.90 (app. qd, J = 7.6, 1.2 Hz, 2H, =CHC***H***$_2$), 2.74 (t, J = 7.7 Hz, 2H, PhC***H***$_2$), 2.15 (s, 3H, C***H***$_3$).

^{13}C NMR (125 MHz, CD_2Cl_2) δ 198.9 (***C***=O), 146.5 (=***C***HCH$_2$), 141.5 (***C***q$_{Ph}$), 128.5 (2C, ***C***H$_{Ph,\ o}$), 128.4 (2C, ***C***H$_{Ph,\ m}$), 127.6 (***C***H=CHCH$_2$), 126.0 (***C***H$_{Ph,\ p}$), 35.1 (Ph***C***H$_2$), 31.4 (***C***H$_3$), 30.9 (=CH***C***H$_2$).

MS (EI-DE) *m/z* (%) 174 [M$^+$] (19), 159 (3), 141 (2), 131 (33), 117 (7), 104 (19), 91 (100), 83 (2), 77 (3), 65 (12), 51 (3), 43 (21), 39 (4), 27 (2).

HRMS (EI-FE) calcd for $C_{12}H_{14}O$ [M$^+$] 174.1045, found 174.1045.

7.5.2.5 Synthesis of 4,8-dimethylnona-3,7-dien-2-one (100)

4,8-Dimethylnona-3,7-dien-2-one (**100**) was prepared according to a procedure of Usuda et al. [9].

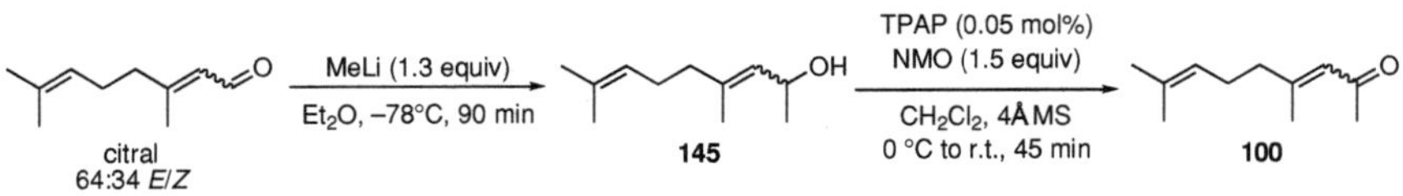

4,8-Dimethylnona-3,7-dien-2-ol (145):

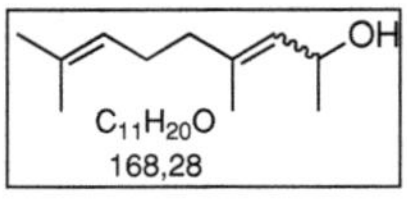

MeLi (24.4 mL, 39.0 mmol, 1.3 equiv; 1.6 M in Et_2O) was added at −78 °C to a solution of citral (5.14 mL, 30.0 mmol, 1.0 equiv; geranial/neral 64:36) in Et_2O (150 mL). After stirring at the same temperature for 1.5 h, 1 N aqueous HCl was added slowly. The phases were separated and the aqueous phase was extracted twice with Et_2O. The combined organic layers were washed with water and brine, dried (Na_2SO_4), and concentrated under reduced pressure. The residue was purified by flash column chromatography (silica gel; 20–30% Et_2O in pentane) to give alcohol **145** (4.55 g, 27.1 mmol, 90%; *E*/*Z* 64:36) as a colorless liquid. *Characterized as a mixture of* 3*E*/3*Z*-*isomers.*

^{1}H NMR (500 MHz, CD_2Cl_2) δ 5.21–5.17 (m, 2H, C***H***$_{ol}$), 5.14–5.08 (m, 2H, C***H***$_{ol}$), 4.56–4.47 (m, 2H, C***H***OH), 2.11–1.97 (m, 8H, C***H***$_2$), 1.71–1.66 (m, 12H, Cq$_{ol}$C***H***$_3$), 1.61 (s, 6H, Cq$_{ol}$C***H***$_3$), 1.42–1.40 (m, 2H, O***H***), 1.19–1.16 (m, 6H, CHC***H***$_3$).

^{13}C NMR (125 MHz, CD_2Cl_2) 3*E*-4,8-*dimethylnona*-3,7-*dien*-2-*ol*: δ 137.4 (CH$_2$***C***q$_{ol}$), 131.9 (Me$_2$***C***q$_{ol}$), 129.8 (***C***H$_{ol}$CH), 124.3 (***C***H$_{ol}$CH$_2$), 64.9 (***C***HOH), 39.8 (***C***H$_2$Cq$_{ol}$), 26.8 (CH$_{ol}$***C***H$_2$), 25.7 (***C***H$_3$), 23.8 (***C***H$_3$), 17.7 (***C***H$_3$), 16.4 (***C***H$_3$); 3*Z*-4,8-*dimethylnona*-3,7-*dien*-2-*ol*: δ 137.7 (CH$_2$***C***q$_{ol}$), 132.6 (Me$_2$***C***q$_{ol}$), 130.7 (***C***H$_{ol}$CH), 124.3 (***C***H$_{ol}$CH$_2$), 64.4 (***C***HOH), 32.5 (***C***H$_2$Cq$_{ol}$), 26.9 (CH$_{ol}$***C***H$_2$), 25.7 (***C***H$_3$), 23.8 (***C***H$_3$), 23.3 (***C***H$_3$), 17.7 (***C***H$_3$).

GC-MS (GC-EI) 3*E*-4,8-*dimethylnona*-3,7-*dien*-2-*ol*: *m*/*z* (%) 168 [M$^+$] (trace), 150 (4), 135 (9), 123 (4), 107 (53), 91 (13), 79 (20), 69 (100), 53 (11), 43 (34), 41 (72), 29 (8); 3*Z*-4,8-*dimethylnona*-3,7-*dien*-2-*ol*: *m*/*z* (%) 168 [M$^+$] (trace), 150 (6), 121 (5), 107 (72), 93 (18), 82 (28), 69 (100), 65 (6), 59 (2), 53 (13), 41 (85), 39 (19), 29 (9).

HRMS (EI-DE) calcd for $C_{11}H_{20}O$ [M$^+$] 168.1513, found 168.1514.

4,8-Dimethylnona-3,7-dien-2-one (100):

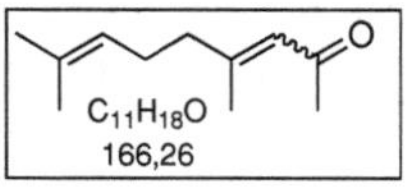

To a suspension of alcohol **145** (2.20 g, 13.1 mmol) and powered 4 Å MS (6.54 g) in CH_2Cl_2 (40 mL) were added successively at 0 °C NMO (2.67 g, 19.7 mmol, 1.5 equiv) and TPAP (233 mg, 0.66 mmol, 0.05 mol%). After stirring for 45 min at ambient temperature, the mixture was filtered through a short pad of silica gel (eluent: EtOAc). The filtrate was concentrated under reduced pressure and flash column chromatography (silica gel, 8% Et_2O in pentane) afforded pure fractions of 3*E*- and 3*Z*-isomers as colorless oils [3*E*-4,8-dimethylnona-3,7-dien-2-one ((*E*)-**100**; 450 mg, 2.71 mmol, 21%); 3*Z*-4,8-dimethylnona-3,7-dien-2-one ((*Z*)-**100**; 303 mg, 1.82 mmol, 14%); 3*Z*/*E*-4,8-dimethylnona-3,7-dien-2-one (1.17 g, 7.04 mmol, 54%; *E*/*Z* 73:27)].

(3*E*)-4,8-Dimethylnona-3,7-dien-2-one (*E*-**100**):

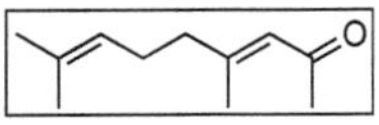

^{1}H NMR (500 MHz, CD_2Cl_2) δ 6.06 (s, 1H, =C***H***C(=O)), 5.11–5.07 (m, 1H, =C***H***CH_2), 2.17–2.11 (m overlapped, 4H, C$\boldsymbol{H_2}$), 2.13 (s overlapped, 3H, C$\boldsymbol{H_3}$C(=O)), 2.09 (d, J = 1.2 Hz, 3H, C$\boldsymbol{H_3}$), 1.68 (s, 3H, C$\boldsymbol{H_3}$), 1.61 (s, 3H, C$\boldsymbol{H_3}$).

^{13}C NMR (125 MHz, CD_2Cl_2) δ 198.8 (***C***=O), 158.1 (CH_2***C***q_{ol}), 132.8 (Me_2***C***q_{ol}), 124.0 (***C***H_{ol}), 123.5 (***C***H_{ol}), 41.4 (***C***H_2Cq_{ol}), 31.9 (***C***H_3C(=O)), 26.5 (CH_{ol}***C***H_2), 25.7 (***C***H_3), 19.2 (***C***H_3), 17.7 (***C***H_3).

GC-MS (GC-EI) *m*/*z* (%) 166 [M^+] (5), 151 (8), 133 (2), 123 (24), 108 (21), 98 (19), 93 (7), 83 (56), 69 (100), 53 (10), 41 (80), 39 (16), 27 (6).

HRMS (EI-FE) calcd for $C_{11}H_{18}O$ [M^+] 166.1358, found 166.1355.

(3*Z*)-4,8-Dimethylnona-3,7-dien-2-one (*Z*-**100**)

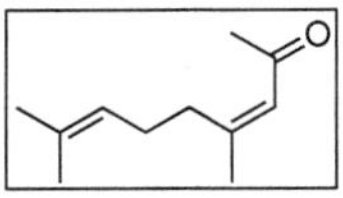

^{1}H NMR (500 MHz, CD_2Cl_2) δ 6.06 (s, 1H, =C***H***C(=O)), 5.15–5.12 (m, 1H, =C***H***CH_2), 2.55 (dd, J = 7.9 Hz, 2H, C$\boldsymbol{H_2}$Cq_{ol}), 2.14–2.09 (m overlapped, 2H, =CHC$\boldsymbol{H_2}$), 2.11 (s overlapped, 3H, C$\boldsymbol{H_3}$C(=O)), 1.86 (d, J = 1.3 Hz, 3H, C$\boldsymbol{H_3}$), 1.68 (s, 3H, C$\boldsymbol{H_3}$), 1.62 (s, 3H, C$\boldsymbol{H_3}$).

^{13}C NMR (125 MHz, CD_2Cl_2) δ 198.2 (***C***=O), 158.7 (CH_2***C***q_{ol}), 132.4 (Me_2***C***q_{ol}), 124.5 (***C***H_{ol}), 124.1 (***C***H_{ol}), 33.9 (***C***H_2Cq_{ol}), 31.8 (***C***H_3C(=O)), 27.1 (CH_{ol}***C***H_2), 25.7 (***C***H_3), 25.5 (***C***H_3), 17.7 (***C***H_3).

GC-MS (GC-EI) *m*/*z* (%) 166 [M^+] (6), 151 (10), 133 (4), 123 (34), 108 (34), 98 (23), 83 (75), 69 (100), 65 (5), 59 (5), 55 (16), 41 (90), 39 (21), 27 (8).

HRMS (EI-FE) calcd for $C_{11}H_{18}O$ [M^+] 166.1358, found 166.1356.

7.6 Catalyst Synthesis

7.6.1 Synthesis of 9-Amino(9-deoxy) Cinchona Alkaloid Derivatives

Cinchona alkaloid-derived primary amines described in the following paragraphs were prepared according to literature procedures reported by Brunner et al. [23] and Vakulya et al. [24].

9-Amino(9-deoxy)*epi*quinine (9-NH$_2$-*epi*Q; 13)

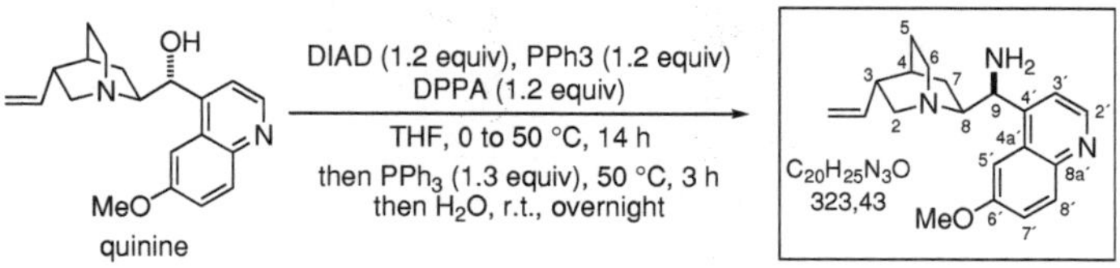

9-NH$_2$-*epi*Q (**13**)

Reference [24] Quinine (**Q**; 6.48 g, 20.0 mmol) and triphenylphosphine (6.30 g, 24.0 mmol, 1.2 equiv) were dissolved in THF (100 mL) and the solution was cooled to 0 °C. Then diisopropyl azodicarboxylate (4.64 mL, 24.0 mmol, 1.2 equiv) was added all at once followed by the dropwise addition of a solution of diphenyl phosphoryl azide (5.16 mL, 24.0 mmol, 1.2 equiv) in THF (40 mL). Then the reaction mixture was allowed to warm to room temperature, stirred overnight (12 h), and the resulting solution was further heated to 50 °C for additional 2 h. Next, triphenylphosphine (6.82 g, 26.0 mmol, 1.3 equiv) was added and the heating (50 °C) was maintained until the gas evolution has ceased (3 h). Then, the solution was cooled to room temperature, and water (2 mL) was added. After stirring for 12 h, the solvents were removed under reduced pressure, and the residue was extracted with CH_2Cl_2 and 10% aqueous HCl (1:1, 200 mL). The phases were separated, and the aqueous phase was repeatedly washed with CH_2Cl_2 (4 × 100 mL). Then the aqueous phase was made alkaline with excess aqueous ammonia at 0 °C and subsequently extracted with CH_2Cl_2 (4 × 100 mL). The combined organic phases were dried (Na_2SO_4), filtered, and concentrated *in vacuo*. The residue was purified by flash column chromatography (silica gel, methanol/EtOAc 50:50 with 1% aq. NH_4OH) affording the title compound as a white semi-solid (4.93 g, 15.3 mmol, 76%). The analytical data were identical in all respects to those previously reported [23].

^{1}H NMR (400 MHz, $CDCl_3$) δ 8.73 (d, $J = 4.5$ Hz, 1H, C$\boldsymbol{H}^{2'}$), 8.01 (d, $J = 9.2$ Hz, 1H, C$\boldsymbol{H}^{8'}$), 7.64 (br s, 1H, C$\boldsymbol{H}^{5'}$), 7.43 (br d, $J = 4.2$ Hz, 1H, C$\boldsymbol{H}^{3'}$), 7.36 (dd, $J = 9.1$, 2.8 Hz, 1H, C$\boldsymbol{H}^{7'}$), 5.78 (ddd, $J = 17.3$, 10.1, 7.3 Hz, 1H, C***H***=CH$_2$), 5.00–4.93 (m, 2H, CH=C$\boldsymbol{H}_2$), 4.57 (br d, $J = 9.7$ Hz, 1H, C$\boldsymbol{H}^9$NH$_2$), 3.94 (s, 3H, OC$\boldsymbol{H}_3$), 3.26 (dd, $J = 13.8$, 10.0 Hz, 1H, C***H***H^2), 3.21–3.15 (m, 1H, C***H***H^6), 3.11–3.02 (m, 1H, C$\boldsymbol{H}^8$), 2.82–2.75 (m, 2H, CH$\boldsymbol{H}^6$ and CH$\boldsymbol{H}^2$), 2.30–2.23 (m, 1H, C$\boldsymbol{H}^3$), 1.94 (br s, 2H, N$\boldsymbol{H}_2$), 1.62–1.58 (m, 1H, C$\boldsymbol{H}^4$), 1.57–1.50 (m, 2H, C$\boldsymbol{H}_2^5$), 1.45–1.36 (m, 1H, C***H***H^7), 0.75 (ddt, $J = 13.6$, 7.5, 1.9 Hz, 1H, CH$\boldsymbol{H}^7$).

^{13}C NMR (100 MHz, $CDCl_3$) δ 157.6 (***C***q$^{6'}$OMe), 147.8 ($\boldsymbol{C}^{2'}$H), 147.0 (***C***q), 144.7 (***C***q), 141.7 (***C***H=CH$_2$), 131.8 ($\boldsymbol{C}^{8'}$H), 128.7 (***C***q$^{4a'}$), 121.2 ($\boldsymbol{C}^{7'}$H), 119.9 ($\boldsymbol{C}^{3'}$H), 114.3 (CH=***C***H$_2$), 102.0 ($\boldsymbol{C}^{5'}$H), 61.9 ($\boldsymbol{C}^8$H), 56.3 ($\boldsymbol{C}^2$H$_2$), 55.5 (O***C***H$_3$), 52.5 ($\boldsymbol{C}^9$HNH$_2$), 40.9 ($\boldsymbol{C}^6$H$_2$), 39.8 ($\boldsymbol{C}^3$H), 28.2 ($\boldsymbol{C}^5$H$_2$), 27.5 ($\boldsymbol{C}^4$H), 26.0 ($\boldsymbol{C}^7$H$_2$).

MS (EI-DE) *m/z* (%) 323 [M$^+$] (2), 199 (1), 187 (15), 160 (2), 136 (100), 108 (6), 95 (3), 82 (8), 70 (5), 56 (4), 42 (4).

HRMS (ESI+) calcd for $C_{20}H_{25}N_3NaO$ [(M + Na)$^+$] 346.1889, found 346.1890.

9-Amino(9-deoxy)*epi*quinidine (9-NH$_2$-*epi*QD; 67) [23]:

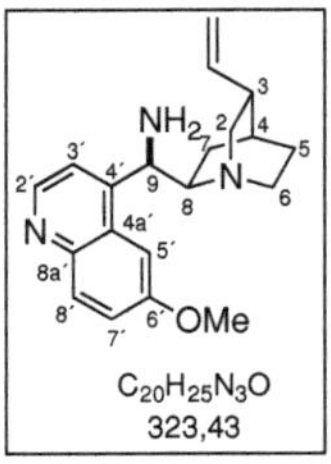

Amine **67** was prepared starting from quinidine (**QD**; 1.95 g, 6.0 mmol) following the same procedure described for the synthesis of 9-amino(9-deoxy)*epi*quinine (**13**). After hydrolyzing the reaction overnight (12 h), the solvents were removed under reduced pressure, and the residue was extracted with CH_2Cl_2 and 10% aqueous HCl (1:1, 200 mL). The phases were separated, the aqueous phase was concentrated *in vacuo*, and the residue was crystallized from methanol. The white precipitate was collected and dissolved in water. Then the aqueous phase was made alkaline by the addition of K_2CO_3, and repeatedly extracted with CH_2Cl_2 (3 × 50 mL). The combined organic phases were dried (Na_2SO_4) and evaporated under reduced pressure affording the pure title compound (820 mg, 2.54 mmol, 42%) as a pale yellow viscous oil.

^{1}H NMR (400 MHz, $CDCl_3$) δ 8.72 (d, J = 4.6 Hz, 1H, C$\boldsymbol{H}^{2'}$), 8.00 (d, J = 9.1 Hz, 1H, C$\boldsymbol{H}^{8'}$), 7.60 (br s, 1H, C$\boldsymbol{H}^{5'}$), 7.50 (br s, 1H, C$\boldsymbol{H}^{3'}$), 7.34 (dd, J = 9.2, 2.7 Hz, 1H, C$\boldsymbol{H}^{7'}$), 5.86 (ddd, J = 17.0, 10.6, 6.4 Hz, 1H, C$\boldsymbol{H}$=CH$_2$), 5.07–5.01 (m, 2H, CH=C$\boldsymbol{H}_2$), 4.64 (br d, J = 9.1 Hz, 1H, C$\boldsymbol{H}^9$NH$_2$), 3.94 (s, 3H, OC$\boldsymbol{H}_3$), 3.04–2.88 (m, 5H, C$\boldsymbol{H}^8$, C$\boldsymbol{H}^6_2$and C$\boldsymbol{H}^2_2$), 2.24 (app. br q, J = 8.0 Hz, 1H, C$\boldsymbol{H}^3$), 2.02 (br s, 2H, N$\boldsymbol{H}_2$), 1.60–1.56 (m, 1H, C$\boldsymbol{H}^4$), 1.54–1.48 (m, 2H, C$\boldsymbol{H}^5_2$), 1.11 (dd, J = 13.4, 8.8 Hz, 1H, C$\boldsymbol{H}$H^7), 0.96–0.88 (m, 1H, CH$\boldsymbol{H}^7$).

^{13}C NMR (100 MHz, $CDCl_3$) δ 157.6 (**C**q$^{6'}$OMe), 147.8 (**C**$^{2'}$H), 147.5 (**C**q), 144.7 (**C**q), 140.7 (**C**H=CH$_2$), 131.8 (**C**$^{8'}$H), 128.7 (**C**q$^{4a'}$), 121.6 (**C**$^{7'}$H), 119.9 (**C**$^{3'}$H), 114.4 (CH=**C**H$_2$), 101.4 (**C**$^{5'}$H), 62.4 (**C**^{8}H), 55.4 (O**C**H$_3$), 51.6 (**C**^{9}HNH$_2$), 49.5 (**C**^{6}H$_2$), 47.4 (**C**^{2}H$_2$), 39.4 (**C**^{3}H), 27.6 (**C**^{4}H), 26.7 (**C**^{5}H$_2$), 25.0 (**C**^{7}H$_2$).

MS (EI-DE) *m/z* (%) 323 [M$^+$] (63), 306 (10), 282 (3), 265 (2), 240 (2), 200 (8), 187 (100), 160 (11), 137 (67), 122 (11), 108 (33), 95 (10), 82 (43), 70 (15), 56 (11), 42 (7).

HRMS (ESI+) calcd for $C_{20}H_{26}N_3O$ [(M + H)$^+$] 324.2067, found 324.2070.

9-Amino(9-deoxy)*epi*dihydroquinine (9-NH$_2$-*epi*DHQ; 72) [24]:

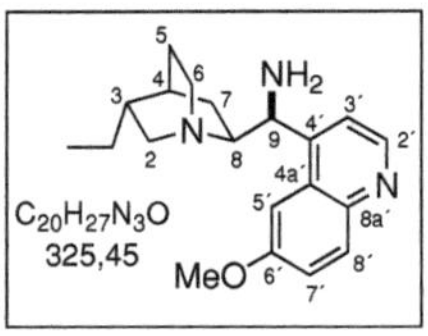

Amine **72** was prepared starting from dihydroquinine (**DHQ**; 1.95 g, 3.86 mmol) following the same procedure described for the synthesis of 9-amino

(9-deoxy)*epi*quinine (**13**). Purification by flash column chromategraphy (silica gel, methanol/EtOAc 50:50 with 1% aq. NH_4OH) gave the title compound (842 mg, 2.59 mmol, 67%) as pale yellow viscous oil.

^{1}H NMR (400 MHz, $CDCl_3$) δ 8.70 (d, $J = 4.5$ Hz, 1H, $CH^{2'}$), 7.99 (d, $J = 9.1$ Hz, 1H, $CH^{8'}$), 7.62 (br s, 1H, $CH^{5'}$), 7.43 (br d, $J = 4.0$ Hz, 1H, $CH^{3'}$), 7.34 (dd, $J = 9.1, 2.8$ Hz, 1H, $CH^{7'}$), 4.55 (br d, $J = 9.6$, 1H, CH^9NH_2), 3.92 (s, 3H, OCH_3), 3.21 (dd, $J = 13.6, 9.8$ Hz, 1H, CHH^2), 3.17–3.10 (m, 1H, CHH^6), 3.05–2.98 (m, 1H, CH^8), 2.77–2.69 (m, 1H, CHH^6), 2.47 (ddd, $J = 13.6, 4.6, 2.3$ Hz, 1H, CHH^2), 2.09 (br s, 2H, NH_2), 1.54–1.19 (m, 7H, CH^4, CH^5_2, CHH^7, CH^3, and CH_2CH_3), 0.78 (t, $J = 7.3$ Hz, 3H, CH_2CH_3), 0.70 (dd, $J = 13.6, 7.6$ Hz, 1H, CHH^7).

^{13}C NMR (100 MHz, $CDCl_3$) δ 157.5 ($Cq^{6'}OMe$), 147.8 ($C^{2'}H$), 147.2 (Cq), 144.7 (Cq'), 131.7 ($C^{8'}H$), 128.8 ($Cq^{4a'}$), 121.1 ($C^{7'}H$), 119.9 ($C^{3'}H$), 102.0 ($C^{5'}H$), 61.7 (C^8H), 57.9 (C^2H_2), 55.5 (OCH_3), 54.5 (C^9HNH_2), 41.0 (C^6H_2), 37.5 (C^3H), 28.9 (C^5H_2), 27.6 (CH_2CH_3), 25.8 (C^7H_2), 25.2 (C^4H), 12.0 (CH_2CH_3).

MS (EI-DE) *m/z* (%) 325 [M^+] (47), 308 (9), 279 (1), 251 (1), 201 (4), 187 (69), 160 (9), 139 (69), 110 (97), 96 (3), 82 (100), 70 (5), 55 (15), 41 (7).

HRMS (ESI+) calcd for $C_{20}H_{27}N_3NaO$ [$(M + Na)^+$] 348.2046, found 348.2046.

9-Amino(9-deoxy)*epi*cinchonidine (9-NH_2-*epi*CD; 69) [24]:

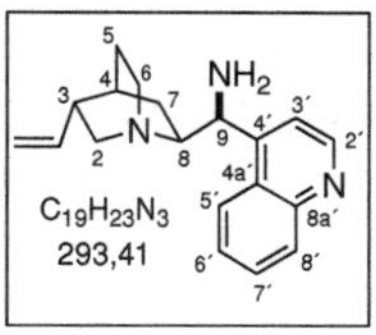

Amine **69** was prepared starting from cinchonidine (**CD**; 1.77 g, 6.0 mmol) following the same procedure described for the synthesis of 9-amino(9-deoxy)*epi*quinine (**13**). Purification by flash column chromatography (silica gel, methanol/EtOAc 50:50 with 1% aq. NH_4OH) gave the title compound as colorless viscous oil (1.23 g, 4.19 mmol, 70%).

^{1}H NMR (400 MHz, $CDCl_3$) δ 8.87 (d, $J = 4.6$ Hz, 1H, $CH^{2'}$), 8.33 (br s, 1H, $CH^{5'}$), 8.11 (dd, $J = 8.6, 0.8$ Hz, 1H, $CH^{8'}$), 7.68 (ddd, $J = 8.3, 6.9, 1.3$ Hz, 1H, $CH^{7'}$), 7.56 (ddd, $J = 8.5, 6.9, 1.3$ Hz, 1H, $CH^{6'}$), 7.49 (br d, $J = 4.1$ Hz, 1H, $CH^{3'}$), 5.77 (ddd, $J = 17.2, 10.1, 7.3$ Hz, 1H, $CH{=}CH_2$), 4.99–4.91 (m, 2H, $CH{=}CH_2$), 4.67 (br d, $J = 9.1$, 1H, CH^9NH_2), 3.24 (dd, $J = 13.9, 10.1$ Hz, 1H, CHH^2), 3.20–3.13 (m, 1H, CHH^6), 3.04 (app. br q, $J = 8.7$ Hz, 1H, CH^8), 2.81–2.73 (m, 2H, CHH^2 and CHH^6), 2.27–2.21 (m, 1H, CH^3), 2.01 (br s, 2H, NH_2), 1.59–1.50 (m, 3H, CH^4 and CH^5_2), 1.38 (app. br t, $J = 11.6$ Hz, 1H, CHH^7), 0.71 (ddt, $J = 13.6, 7.5, 1.9$ Hz, 1H, CHH^7).

^{13}C NMR (100 MHz, $CDCl_3$) δ 150.3 ($C^{2'}H$), 148.7 ($Cq^{4'}$), 148.6 ($Cq^{8a'}$), 141.8 ($CH{=}CH_2$), 130.4 ($C^{8'}H$), 128.9 ($C^{7'}H$), 127.8 ($Cq^{4a'}$), 126.4 ($C^{6'}H$), 123.3 ($C^{5'}H$), 119.6 ($C^{3'}H$), 114.2 ($CH{=}CH_2$), 61.9 (C^8H), 56.3 (C^2H_2), 51.8 (C^9HNH_2), 40.9 (C^6H_2), 39.8 (C^3H), 28.1 (C^5H_2), 27.5 (C^4H), 26.0 (C^7H_2).

MS (EI-DE) *m/z* (%) 293 [M$^+$] (1), 196 (1), 181 (1), 169 (3), 157 (12), 136 (100), 108 (6), 95 (4), 81 (9), 70 (4), 56 (4), 42 (6), 30 (2).

HRMS (ESI+) calcd for $C_{19}H_{24}N_3$ [(M + H)$^+$] 294.1964, found 294.1965.

9-Amino(9-deoxy)*epi*cinchonine (9-NH$_2$-*epi*C; 70) [23]:

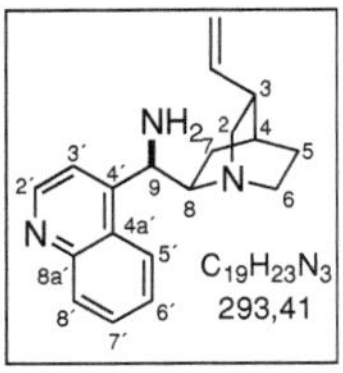

Amine **70** was prepared starting from cinchonine (**C**; 1.77 g, 6.0 mmol; *contains* ~10% *of dihydrocinchonine*) following the same procedure described for the synthesis of 9-amino(9-deoxy)*epi*quinine (**13**). After hydrolyzing the reaction overnight (12 h), the solvents were removed under reduced pressure, and the residue was extracted with CH_2Cl_2 and 10% aqueous HCl (1:1, 200 mL). The phases were separated, the aqueous phase was concentrated *in vacuo*, and the residue was crystallized from methanol. The white precipitate was collected and dissolved in water. Then the aqueous phase was made alkaline by the addition of K_2CO_3, and repeatedly extracted with CH_2Cl_2 (3 × 50 mL). The combined organic phases were dried (Na_2SO_4) and evaporated under reduced pressure affording the pure title compound (808 mg, 2.75 mmol, 46%; *contaminated by 9-amino(9-deoxy)epidihydrocinchonine* (~10%)) as colorless viscous oil.

^{1}H NMR (300 MHz, $CDCl_3$) δ 8.87 (d, J = 4.5 Hz, 1H, C$\boldsymbol{H}^{2'}$), 8.33 (br d, J = 8.0 Hz, 1H, C$\boldsymbol{H}^{5'}$), 8.11 (dd, J = 8.5, 0.9 Hz, 1H, C$\boldsymbol{H}^{8'}$), 7.69 (ddd, J = 8.3, 7.0, 1.3 Hz, 1H, C$\boldsymbol{H}^{7'}$), 7.59–7.53 (m, 2H, C$\boldsymbol{H}^{3'}$ and C$\boldsymbol{H}^{6'}$), 5.84 (ddd, J = 16.6, 11.1, 6.6 Hz, 1H, C***H***=CH$_2$), 5.07–5.01 (m, 2H, CH=C$\boldsymbol{H}_2$), 4.73 (br d, J = 9.8, 1H, C$\boldsymbol{H}^9$NH$_2$), 3.07–2.87 (m, 5H, C$\boldsymbol{H}^8$, C$\boldsymbol{H}_2^6$and C$\boldsymbol{H}_2^2$), 2.25 (app. br q, J = 8.4 Hz, 1H, C$\boldsymbol{H}^3$), 1.99 (br s, 2H, N$\boldsymbol{H}_2$), 1.57–1.48 (m, 3H, C$\boldsymbol{H}^4$ and C$\boldsymbol{H}_2^5$), 1.09 (dd, J = 13.3, 8.8 Hz, 1H, C***H***H^7), 0.97–0.88 (m, 1H, CH$\boldsymbol{H}^7$).

^{13}C NMR (75 MHz, $CDCl_3$) δ 150.3 ($\boldsymbol{C}^{2'}$H), 149.0 ($\boldsymbol{C}$q$^{4'}$), 148.6 ($\boldsymbol{C}$q$^{8a'}$), 140.6 (***C***H=CH$_2$), 130.4 ($\boldsymbol{C}^{8'}$H), 129.0 ($\boldsymbol{C}^{7'}$H), 127.9 ($\boldsymbol{C}$q$^{4a'}$), 126.3 ($\boldsymbol{C}^{6'}$H), 123.3 ($\boldsymbol{C}^{5'}$H), 119.7 ($\boldsymbol{C}^{3'}$H), 114.5 (CH=***C***H$_2$), 62.2 ($\boldsymbol{C}^8$H), 51.4 ($\boldsymbol{C}^9$HNH$_2$), 49.5 ($\boldsymbol{C}^6$H$_2$), 47.4 ($\boldsymbol{C}^2$H$_2$), 39.7 ($\boldsymbol{C}^3$H), 27.7 ($\boldsymbol{C}^4$H), 26.7 ($\boldsymbol{C}^5$H$_2$), 25.0 ($\boldsymbol{C}^7$H$_2$).

MS (EI-DE) *m/z* (%) 293 [M$^+$] (87), 276 (13), 252 (6), 235 (3), 211 (3), 183 (9), 169 (14), 157 (90), 136 (100), 122 (14), 115 (4), 108 (46), 95 (18), 82 (70), 70 (23), 56 (20), 42 (21).

HRMS (ESI+) calcd for $C_{19}H_{23}N_3Na$ [(M + Na)$^+$] 316.1785, found 316.1784.

9-Amino(9-deoxy)-*epi*-6′-isopropoxycinchonidine (71) [24]:

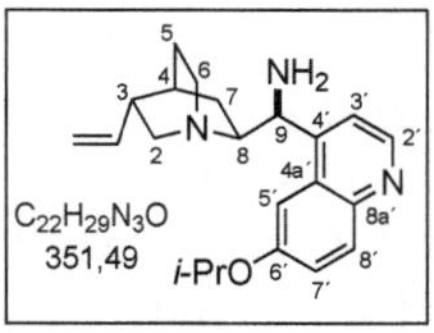

Amine **71** was prepared starting from 6′-isopropoxy-cinchoni-dine (**146**; 650 mg, 1.84 mmol) following the same procedure described for the synthesis of 9-amino(9-deoxy)*epi*quinine (**13**). Purification by flash column chromatography (silica gel, methanol/EtOAc 50:50 with 1% aq. NH_4OH) gave the title compound (498 mg, 1.41 mmol, 77%) as pale yellow solid.

^{1}H NMR (300 MHz, $CDCl_3$) δ 8.71 (d, $J = 4.4$ Hz, 1H, C$H^{2'}$), 8.01 (d, $J = 9.1$ Hz, 1H, C$H^{8'}$), 7.66 (br s, 1H, C$H^{5'}$), 7.42 (br s, 1H, C$H^{3'}$), 7.34 (dd, $J = 9.3$, 2.7 Hz, 1H, C$H^{7'}$), 5.78 (ddd, $J = 17.3$, 9.9, 7.6 Hz, 1H, C***H***=CH$_2$), 5.00–4.93 (m, 2H, CH=CH_2), 4.72 (hept, $J = 6.1$ Hz, 1H, C***H***Me$_2$), 4.55 (br s, 1H, CH^9NH$_2$), 3.26 (dd, $J = 13.9$, 10.1 Hz, 1H, C***H***H^2), 3.22–3.15 (m, 1H, C***H***H^6), 3.10–3.01 (m, 1H, CH^8), 2.81–2.75 (m, 2H, CHH^6 and CHH^2), 2.28–2.24 (m, 1H, CH^3), 1.89 (br s, 2H, NH_2), 1.62–1.60 (m, 1H, CH^4), 1.56–1.51 (m, 2H, CH^5_2), 1.43–1.40 (m, overlapped, 1H, C***H***H^7), 1.42 (d overlapped, $J = 6.3$ Hz, 3H, CH(CH_3)$_2$), 1.41 (d overlapped, $J = 6.3$ Hz, 3H, CH(CH_3)$_2$), 0.78–0.73 (m, 1H, CHH^7).

^{13}C NMR (125 MHz, $CDCl_3$) δ 155.8 (*C*q$^{6'}$O*i*-Pr), 147.8 ($C^{2'}$H), 146.8 (*C*q$^{4'}$), 144.5 (*C*q$^{8a'}$), 141.6 (***C***H=CH$_2$), 131.8 ($C^{8'}$H), 128.7 (*C*q$^{4a'}$), 122.1 ($C^{7'}$H), 119.8 ($C^{3'}$H), 114.4 (CH=***C***H$_2$), 104.7 ($C^{5'}$H), 70.2 (O***C***HMe$_2$), 61.9 (C^8H), 56.2 (C^2H$_2$), 51.5 (C^9HNH$_2$), 40.9 (C^6H$_2$), 39.7 (C^3H), 28.0 (C^5H$_2$), 27.5 (C^4H), 26.0 (C^7H$_2$), 22.1 (CH(***C***H$_3$)$_2$), 21.8 (CH(***C***H$_3$)$_2$).

MS (EI-DE) *m*/*z* (%) 351 [M^+] (3), 336 (1), 308 (2), 291 (1), 215 (12), 173 (12), 146 (7), 136 (100), 108 (7), 95 (4), 81 (10), 56 (4), 43 (14).

HRMS (EI-DE) calcd for $C_{22}H_{29}N_3O$ [M^+] 351.2312, found 351.2311.

9-Amino(9-deoxy)quinine (9-NH$_2$-Q; 73) [24]:

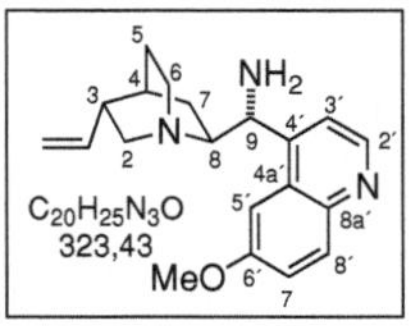

Amine **73** was prepared starting from 9-*epi*quinine (**147**; 1.24 g, 3.82 mmol) following the same procedure described for the synthesis of 9-amino (9-deoxy)*epi*quinine (**13**). Purification by flash column chromatography (silica gel, methanol/EtOAc 50:50 with 1% aq. NH_4OH) gave the title compound as yellowish viscous oil (861 mg, 2.66 mmol, 70%). The analytical data were identical in all respects to those previously reported [25].

^{1}H NMR (500 MHz, $CDCl_3$) δ 8.70 (d, J = 4.7 Hz, 1H, C$H^{2'}$), 7.98 (d, J = 9.1 Hz, 1H, C$H^{8'}$), 7.40 (d, J = 2.5 Hz, 1H, C$H^{5'}$), 7.34 (d overlapped, J = 4.7 Hz, 1H, C$H^{3'}$), 7.32 (dd overlapped, J = 9.1, 2.5 Hz, 1H, C$H^{7'}$), 5.90 (ddd, J = 17.3, 10.2, 7.3 Hz, 1H, C**H**=CH_2), 5.05–5.00 (m, 2H, CH=CH_2), 4.60 (d, J = 9.5, 1H, CH^9NH_2), 3.93 (s, 3H, OCH_3), 3.18 (app. q, J = 8.6 Hz, 1H, CH^8), 3.01 (dd, J = 13.9, 10.1 Hz, 1H, C**H**H^2), 2.99–2.91 (m, 1H, C**H**H^6), 2.62 (ddd, J = 13.8, 4.2, 2.4 Hz, 1H, CHH^2), 2.53 (ddd, J = 14.2, 10.6, 4.1 Hz, 1H, CHH^6), 2.28–2.23 (m, 1H, CH^3), 2.15–2.08 (m, 1H, C**H**H^7), 2.00–1.77 (br s, 2H, NH_2), 1.89–1.85 (m, 1H, CH^4), 1.69–1.60 (m, 1H, C**H**H^5), 1.54–1.46 (m, 2H, CHH^5 and CHH^7).

^{13}C NMR (125 MHz, $CDCl_3$) δ 157.6 ($Cq^{6'}$OMe), 149.2 ($Cq^{4'}$), 147.9 ($C^{2'}$H), 144.8 ($Cq^{8a'}$), 141.9 (**C**H=CH_2), 131.9 ($C^{8'}$H), 127.6 ($Cq^{4a'}$), 121.1 ($C^{7'}$H), 118.3 ($C^{3'}$H), 114.4 (CH=CH_2), 101.2 ($C^{5'}$H), 60.6 (C^8H), 56.2 (C^2H_2), 55.6 (OCH_3), 53.8 (C^9HNH_2), 41.9 (C^6H_2), 39.7 (C^3H), 27.8 (C^5H_2), 27.7 (C^4H), 26.5 (C^7H_2).

MS (EI-DE) *m/z* (%) 323 [M^+] (2), 213 (1), 199 (1), 187 (15), 160 (3), 136 (100), 108 (7), 95 (4), 81 (12), 70 (6), 56 (6), 42 (8).

HRMS (ESI+) calcd for $C_{20}H_{25}N_3NaO$ [$(M + Na)^+$] 346.1886, found 346.1890.

7.6.2 Synthesis of 9-epiquinine (147)

9-*Epi*quinine (**147**) was prepared according to the protocol of Franz et al. [26].

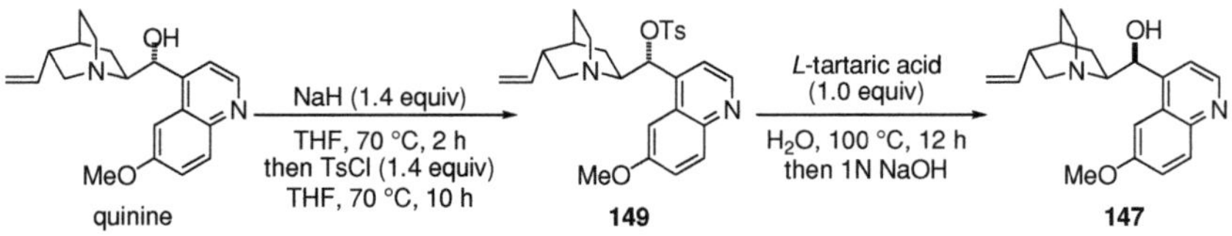

***O*-Tosyl quinine (149)**:

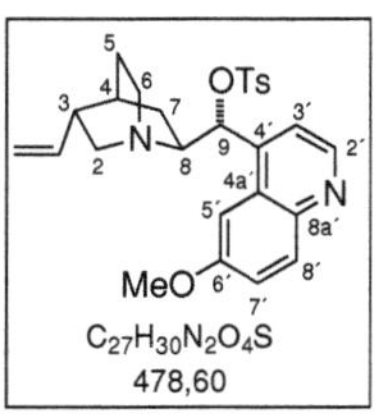

Sodium hydride (207 mg, 8.62 mmol, 1.4 equiv; 60 wt% suspension in mineral oil) was added all at once to a solution of quinine (2.0 g, 6.16 mmol) in THF (35 mL), and the resulting suspension was heated to 70 °C for 2 h. Then, the reaction mixture was cooled to 0 °C, and a solution of tosyl chloride (1.64 g, 8.62 mmol, 1.4 equiv) in THF (10 mL) was added dropwise. Once the addition was complete, heating (70 °C) was continued for 10 h. The solvents were removed under reduced pressure, and the residue was extracted with Et_2O and 1 N aqueous HCl (1:1, 60 mL). The phases were separated, and the aqueous phase was repeatedly washed with Et_2O (3 × 20 mL). Then the aqueous phase was made alkaline (pH ~11) by the addition of 2 N aqueous NaOH, and subsequently extracted with Et_2O (3 × 60 mL). The combined organic phases were washed with brine, dried (Na_2SO_4), filtered, and concentrated *in vacuo*. The crude product was filtered through a short plug of silica gel (eluent: 10% MeOH in Et_2O) affording *O*-tosylated quinine **149** as a white solid (1.73 g, 3.62 mmol, 59%).

^{1}H NMR (300 MHz, $CDCl_3$) δ 8.49 (d, J = 4.5 Hz, 1H, C$\boldsymbol{H}^{2'}$), 7.85 (d, J = 9.0 Hz, 1H, C$\boldsymbol{H}^{8'}$), 7.30–7.08 (m, 5H, C$\boldsymbol{H}^{3'}$, C$\boldsymbol{H}^{5'}$, C$\boldsymbol{H}^{7'}$, and 2 × C$\boldsymbol{H}_{tolyl}$), 6.75 (d, J = 7.5 Hz, 2H, C$\boldsymbol{H}_{tolyl}$), 6.07 (br s, 1H, C$\boldsymbol{H}^9$OTs), 5.80 (ddd, J = 17.5, 10.0, 7.3 Hz, 1H, C***H***=CH_2), 5.01–4.94 (m, 2H, CH=C$\boldsymbol{H}_2$), 3.92 (s, 3H, OC$\boldsymbol{H}_3$), 3.38–3.20 (m, 1H), 3.15–3.01 (m, 1H), 2.97–2.88 (m, 1H), 2.64–2.43 (m, 2H), 2.28–2.19 (m, 1H), 2.14 (s, 3H, C_6H_4C$\boldsymbol{H}_3$), 2.10–1.97 (m, 1H), 1.91–1.85 (m, 1H), 1.78–1.60 (m, 2H), 1.57–1.45 (m, 1H).

MS (EI-DE) *m/z* 478 [M^+] (1), 323 (1), 306 (15), 251 (4), 225 (4), 172 (6), 136 (100), 81 (4), 55 (2).

9-*Epi*quinine (**147**):

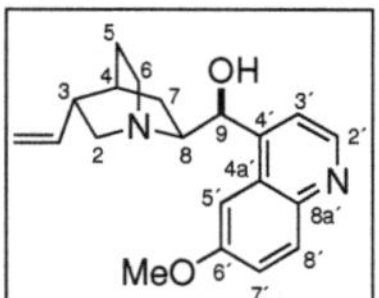

$C_{20}H_{24}N_2O_2$
324,42

O-Tosyl quinine (**149**; 1.70 g, 3.55 mmol) and *L*-tartaric acid (533 mg, 3.55 mmol) were dissolved in water (34 mL). The reaction mixture was heated to 100 °C for 12 h; then cooled to room temperature and neutralized with 1 N aqueous NaOH. The aqueous phase was separated and repeatedly extracted with CH_2Cl_2. The combined organic phases were dried (Na_2SO_4), filtered, and concentrated. Purification of the crude product by flash column chromatography (silica gel, methanol/EtOAc 20:80 with 0.5% aq. NH_4OH) gave 9-*epi*quinine (**147**; 1.15 g, 3.55 mmol, quant.) as a white solid.

^{1}H NMR (300 MHz, $CDCl_3$) δ 8.70 (d, J = 4.5 Hz, 1H, C$\boldsymbol{H}^{2'}$), 8.02 (d, J = 9.4 Hz, 1H, C$\boldsymbol{H}^{8'}$), 7.64 (d, J = 2.6 Hz, 1H, C$\boldsymbol{H}^{5'}$), 7.39 (d, J = 4.9 Hz, 1H, C$\boldsymbol{H}^{3'}$), 7.36 (dd, J = 9.0, 2.9 Hz, 1H, C$\boldsymbol{H}^{7'}$), 5.73 (ddd, J = 17.1, 9.9, 7.1 Hz, 1H, C***H***=CH_2), 5.02–4.92 (m, 3H, CH=C$\boldsymbol{H}_2$ and C$\boldsymbol{H}^9$OH), 3.93 (s, 3H, OC$\boldsymbol{H}_3$), 3.26 (dd, J = 13.8, 10.0 Hz, 1H, C***H***H^2), 3.22–3.04 (m, 2H, C***H***H^6 and C$\boldsymbol{H}^8$), 2.83–2.74 (m, 2H, CH$\boldsymbol{H}^6$ and CH$\boldsymbol{H}^2$), 2.36–2.27 (m, 1H, C$\boldsymbol{H}^3$), 1.74–1.69 (m, 1H, C$\boldsymbol{H}^4$), 1.63–1.57 (m, 2H, C$\boldsymbol{H}_2^5$), 1.51–1.41 (m, 1H, C***H***H^7), 0.96 (ddt, J = 13.6, 7.9, 1.8 Hz, 1H, CH$\boldsymbol{H}^7$).

MS (EI-DE) *m/z* (%) 324 [M$^+$] (1), 189 (4), 160 (2), 136 (100), 117 (3), 95 (2), 81 (6), 67 (1), 55 (3), 42 (4).

HRMS (ESI+) calcd for $C_{20}H_{25}N_2O_2$ [(M + H)$^+$] 325.1913, found 325.1911.

7.6.3 Synthesis of 6′-isopropoxycinchonidine (146)

6′-Isopropoxycinchonidine (146) was prepared through a synthetic route reported by Berkessel et al. [27].

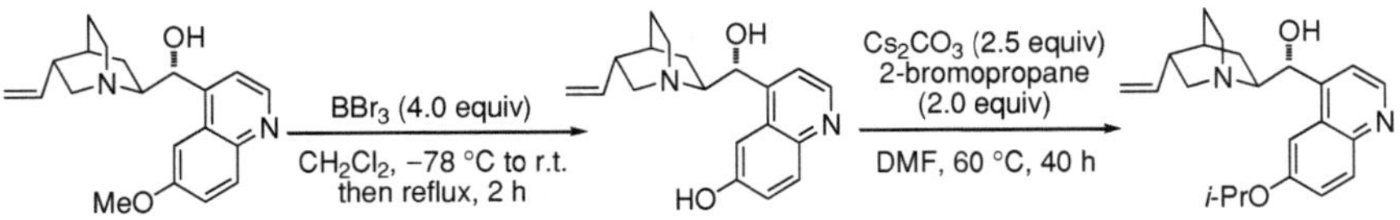

quinine **148** **146**

6′-Hydroxycinchonidine (148):

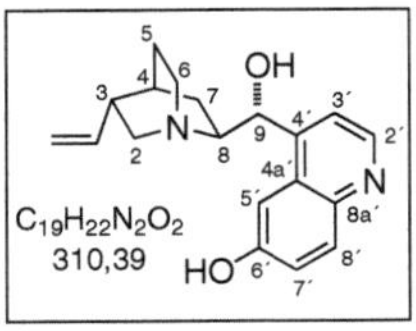

Under vigorous stirring, BBr_3 (2.93 mL, 30.8 mmol, 4.0 equiv) in CH_2Cl_2 (30 mL) was slowly added to a solution of quinine (2.5 g, 7.71 mmol) in CH_2Cl_2 (250 mL) at –78 °C. The reaction mixture was allowed to warm to room temperature, and then refluxed for 2 h. The reaction was quenched by the addition of 10% aqueous NaOH (130 mL) at 0 °C, and stirring was maintained for 1 h. The phases were separated, and the aqueous phase was washed with CH_2Cl_2. Then, 2 M aqueous HCl (80 mL) was added dropwise until a colorless solid precipitated (pH ~8). Extraction with $CHCl_3$, drying (Na_2SO_4) of the organic phase, filtering, and evaporating to dryness *in vacuo* afforded the title compound (1.30 g, 4.19 mmol, 54%) as a white solid.

^{1}H NMR (400 MHz, CD_3OD) δ 8.61 (d, J = 4.6 Hz, 1H, C$\boldsymbol{H}^{2'}$), 7.93(d, J = 9.1 Hz, 1H, C$\boldsymbol{H}^{8'}$), 7.64 (d, J = 4.5 Hz, 1H, C$\boldsymbol{H}^{3'}$), 7.38–7.34 (m, 2H, C$\boldsymbol{H}^{5'}$ and C$\boldsymbol{H}^{7'}$), 5.73 (ddd, J = 17.2, 10.1, 7.3 Hz, 1H, C$\boldsymbol{H}$=CH_2), 5.58 (d, J = 3.0 Hz, 1H, C$\boldsymbol{H}^9$OH), 4.97 (dt, J = 17.2, 1.4 Hz, 1H, =C$\boldsymbol{H}_{trans}$H), 4.90 (dt, J = 10.4, 1.4 Hz, 1H, =CH$\boldsymbol{H}_{cis}$), 3.79–3.70 (m, 1H, C$\boldsymbol{H}$H^6), 3.15–3.07 (m, 2H, C$\boldsymbol{H}$H^2 and C$\boldsymbol{H}^8$), 2.77–2.65 (m, 2H, CH$\boldsymbol{H}^2$ and CH$\boldsymbol{H}^6$), 2.39–2.31 (m, 1H, C$\boldsymbol{H}^3$), 1.93–1.83 (m, 2H), 1.82–1.76 (m, 1H), 1.65–1.54 (m, 1H), 1.47–1.39 (m, 1H).

^{13}C NMR (100 MHz, CD_3OD) δ 158.3 ($\boldsymbol{C}$q$^{6'}$OH), 150.0 ($\boldsymbol{C}$q$^{4'}$), 147.7 ($\boldsymbol{C}^{2'}$H), 144.2 ($\boldsymbol{C}$q$^{8a'}$), 142.8 ($\boldsymbol{C}$H=CH_2), 131.7 ($\boldsymbol{C}^{8'}$H), 128.7 ($\boldsymbol{C}$q$^{4a'}$), 123.7 ($\boldsymbol{C}^{7'}$H), 120.1

($\boldsymbol{C}^{3'}$H), 115.3 (CH=$\boldsymbol{C}H_2$), 105.5 ($\boldsymbol{C}^{5'}$H), 72.4 ($\boldsymbol{C}^{9}$HOH), 61.2 ($\boldsymbol{C}^{8}$H), 57.8 ($\boldsymbol{C}^{2}H_2$), 44.5 ($\boldsymbol{C}^{6}H_2$), 41.1 ($\boldsymbol{C}^{3}$H), 29.4 ($\boldsymbol{C}^{4}$H), 28.4 ($\boldsymbol{C}^{5}H_2$), 21.9 ($\boldsymbol{C}^{7}H_2$).

MS (EI-DE) *m/z* (%) 310 [M^+] (1), 175 (5), 158 (2), 136 (100), 95 (3), 81 (5), 67 (1), 55 (2), 42 (2).

HRMS (ESI+) calcd for $C_{19}H_{22}N_2NaO_2$ [$(M + Na)^+$] 333.1576, found 333.1573.

6′-Isopropoxycinchonidine (146):

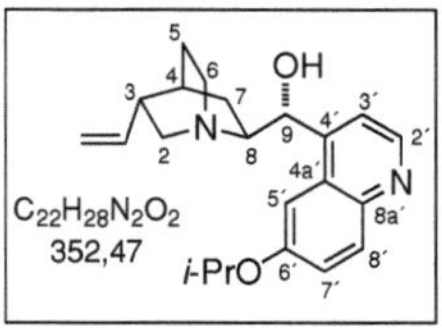

Cs_2CO_3 (2.36 g, 7.25 mmol, 2.5 equiv) was added to a stirred solution of 6-hydroxycinchonidine (**148**; 900 mg, 2.90 mmol) in dry DMF (145 mL) and stirred at room temperature for 10 min. 2-Bromopropane (0.55 mL, 5.80 mmol, 2.0 equiv) was added, and the reaction mixture was heated to 60 °C for 40 h. Then, the solvent was removed under reduced pressure, and the resulting solid was purified by flash column chromatography (silica gel, 10% MeOH in $CHCl_3$) to obtain 6′-isopropoxycinchonidine (**146**; 733 mg, 2.08 mmol, 72%) as a white solid.

^{1}H NMR (500 MHz, $CDCl_3$) δ 8.59 (d, $J = 4.4$ Hz, 1H, C$\boldsymbol{H}^{2'}$), 7.91 (d, $J = 9.1$ Hz, 1H, C$\boldsymbol{H}^{8'}$), 7.47 (d, $J = 4.4$ Hz, 1H, C$\boldsymbol{H}^{3'}$), 7.47–7.21 (m, 2H, C$\boldsymbol{H}^{5'}$ and C$\boldsymbol{H}^{7'}$), 5.69 (ddd, $J = 17.3, 10.1, 7.4$ Hz, 1H, C$\boldsymbol{H}$=CH_2), 5.60 (br s, 1H, C$\boldsymbol{H}^{9}$OH), 4.93 (dt, $J = 17.0, 1.3$ Hz, 1H, CH=C$\boldsymbol{H}_{trans}$H), 4.89 (dt, $J = 10.4$, 1.3 Hz, 1H, CH=CH$\boldsymbol{H}_{cis}$), 4.67 (sept, $J = 6.1$ Hz, 1H, OC$\boldsymbol{H}$$Me_2$), 3.94 (br s, 1H, O$\boldsymbol{H}$), 3.57–3.48 (m, 1H, C$\boldsymbol{H}$$H^6$), 3.16–3.06 (m, 2H, C$\boldsymbol{H}$$H^2$ and C$\boldsymbol{H}^{8}$), 2.71–2.64 (m, 2H, CH$\boldsymbol{H}^{2}$ and CH$\boldsymbol{H}^{6}$), 2.32–2.25 (m, 1H, C$\boldsymbol{H}^{3}$), 1.83–1.80 (m, 1H, C$\boldsymbol{H}^{4}$), 1.78–1.71 (m, 2H, C$\boldsymbol{H}$$H^5$ and C$\boldsymbol{H}$$H^7$), 1.53–1.45 (m, 2H, CH$\boldsymbol{H}^{5}$ and CH$\boldsymbol{H}^{7}$), 1.33 (dd, $J = 6.0$ Hz, 6H, OCH(C$\boldsymbol{H}_3$)$_2$).

^{13}C NMR (125 MHz, $CDCl_3$) δ 155.9 ($\boldsymbol{C}q^{6'}$O*i*-Pr), 147.4 ($\boldsymbol{C}^{2'}$H), 147.0 ($\boldsymbol{C}q^{4'}$), 143.9 ($\boldsymbol{C}q^{8a'}$), 141.4 ($\boldsymbol{C}$H=CH_2), 131.5 ($\boldsymbol{C}^{8'}$H), 126.5 ($\boldsymbol{C}q^{4a'}$), 122.7 ($\boldsymbol{C}^{7'}$H), 118.4 ($\boldsymbol{C}^{3'}$H), 114.6 (CH=$\boldsymbol{C}H_2$), 103.5 ($\boldsymbol{C}^{5'}$H), 71.6 ($\boldsymbol{C}^{9}$HOH), 70.1 (O$\boldsymbol{C}$HMe_2), 60.0 ($\boldsymbol{C}^{8}$H), 56.9 ($\boldsymbol{C}^{2}H_2$), 43.3 ($\boldsymbol{C}^{6}H_2$), 39.6 ($\boldsymbol{C}^{3}$H), 27.7 ($\boldsymbol{C}^{4}$H), 27.3 ($\boldsymbol{C}^{5}H_2$), 22.0 (OCH($\boldsymbol{C}H_3$)$_2$), 21.5 (OCH($\boldsymbol{C}H_3$)$_2$), 21.4 ($\boldsymbol{C}^{7}H_2$).

MS (EI-DE) *m/z* (%) 352 [M^+] (2), 309 (4), 217 (2), 200 (1), 186 (1), 174 (4), 158 (2), 136 (100), 117 (2), 95 (3), 81 (7), 67 (2), 55 (4), 43 (10).

HRMS (EI-DE) calcd for $C_{22}H_{28}N_2O_2$ [M^+] 352.2154, found 352.2151.

7.6.4 *Synthesis of 9-amino(9-deoxy)epi-6′-hydroxycinchonidine (68)*

A literature-known procedure by the *Chen* group was followed [28].

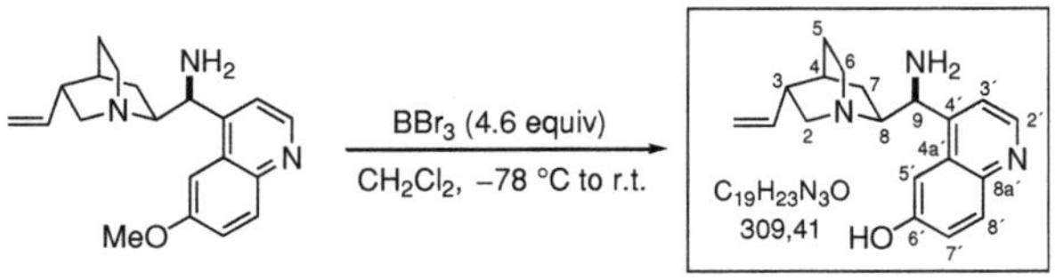

9-NH$_2$-*epi*Q (**13**) **68**

BBr_3 (23 mL, 23.0 mmol, 1.0 M in CH_2Cl_2, 4.6 equiv) was added to a solution of amine **13** (1.62 g, 5.01 mmol) in CH_2Cl_2 (15 mL) at −78 °C. The mixture was slowly warmed to room temperature and stirred for 12 h. The reaction was quenched by the addition of water (50 mL) at 0 °C, and further stirred for 4 h to ensure complete hydrolysis. The phases were separated, and the aqueous phase was washed with CH_2Cl_2 (3 × 50 mL). Subsequently the aqueous phase was neutralized with excess aqueous ammonia at 0 °C, and repeatedly extracted with EtOAc (3 × 50 mL). The combined organic phases were dried (Na_2SO_4), filtered, and concentrated. The residue was purified by flash column chromatography (silica gel, methanol/EtOAc 50:50 with 1% aq. NH_4OH) affording the title compound as a yellow solid (936 mg, 3.03 mmol, 60%). The analytical data were identical in all respects to those previously reported [28].

^{1}H NMR (500 MHz, $CDCl_3$) δ 8.65 (d, J = 4.4 Hz, 1H, C$\boldsymbol{H}^{2'}$), 7.93 (d, J = 9.2 Hz, 1H, C$\boldsymbol{H}^{8'}$), 7.60 (br s, 1H, C$\boldsymbol{H}^{5'}$), 7.34 (br s, 1H, C$\boldsymbol{H}^{3'}$), 7.27 (dd, J = 8.9, 2.2 Hz, 1H, C$\boldsymbol{H}^{7'}$), 5.75 (ddd, J = 17.3, 9.6, 7.4 Hz, 1H, C***H***=CH$_2$), 4.98–4.93 (m, 2H, CH=C***H***$_2$), 4.90–4.63 (br s, 2H, N***H***$_2$), 4.48 (br s, 1H, C$\boldsymbol{H}^9$NH$_2$), 3.20 (dd, J = 13.9, 10.1 Hz, 1H, C***H***H^2), 3.14–3.06 (m, 1H, C$\boldsymbol{H}^8$), 3.05–2.97 (m, 1H, C***H***H^6), 2.78–2.65 (m, 2H, CH$\boldsymbol{H}^6$ and CH$\boldsymbol{H}^2$), 2.26–2.20 (m, 1H, C$\boldsymbol{H}^3$), 1.57–1.53 (m, 1H, C$\boldsymbol{H}^4$), 1.49–1.40 (m, 2H, C***H***H^5 and C***H***H^7), 1.37–1.29 (m, 1H, CH$\boldsymbol{H}^5$), 0.72–0.63 (m, 1H, CH$\boldsymbol{H}^7$).

^{13}C NMR (125 MHz, $CDCl_3$) δ 156.6 (**C**q$^{6'}$OH), 146.5 ($\boldsymbol{C}^{2'}$H), 146.4 (**C**q$^{4'}$), 143.4 (**C**q$^{8a'}$), 141.4 (**C**H=CH$_2$), 131.1 ($\boldsymbol{C}^{8'}$H), 129.2 (**C**q$^{4a'}$), 122.7 ($\boldsymbol{C}^{7'}$H), 118.9 ($\boldsymbol{C}^{3'}$H), 114.7 (CH=**C**H$_2$), 105.1 ($\boldsymbol{C}^{5'}$H), 61.7 ($\boldsymbol{C}^8$H), 56.0 ($\boldsymbol{C}^2$H$_2$), 50.7 ($\boldsymbol{C}^9$HNH$_2$), 40.8 ($\boldsymbol{C}^6$H$_2$), 39.6 ($\boldsymbol{C}^3$H), 27.7 ($\boldsymbol{C}^5$H$_2$), 27.4 ($\boldsymbol{C}^4$H), 26.1 ($\boldsymbol{C}^7$H$_2$).

MS (ESI+) *m/z* (%) 309 [M$^+$] (trace), 294 (2), 225 (1), 212 (1), 199 (1), 185 (1), 173 (14), 159 (1), 136 (100), 122 (3), 108 (9), 95 (5), 82 (11), 56 (6), 44 (10).

HRMS (ESI+) calcd for $C_{19}H_{24}N_3O$ [(M + H)$^+$] 310.1914, found 310.1914.

7.6.5 Synthesis of 2-(α-aminobenzyl)quinuclidine (86a)

2-(α-Hydroxybenzyl)quinuclidine (**150**) was prepared according to a procedure of Kessar et al. [29].

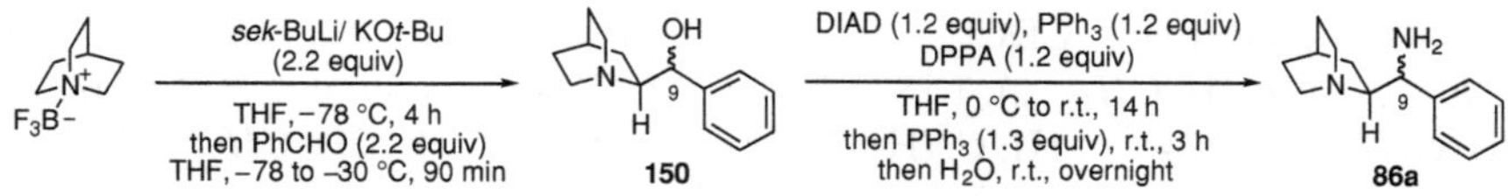

2-(α-Hydroxybenzyl)quinuclidine (150) [29]:

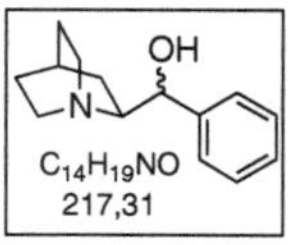

Quinuclidine (1.11 g, 10.0 mmol, 1.0 equiv) was dissolved in THF (60 mL) and treated with $BF_3 \cdot Et_2O$ (1.38 mL, 11.0 mmol, 1.1 equiv) at 0 °C for 40 min. In the meantime, *s*-BuLi (15.7 mL, 22.0 mmol, 2.2 equiv; 1.4 M in cyclohexane) was added to a solution of potassium *tert*-butoxide (2.47 g, 22.0 mmol, 2.2 equiv) in THF (60 mL) at −78 °C, and the resulting solution was stirred for 30 min. Then, the solution of the quinuclidine –BF_3 complex was added slowly to the flask containing the *Schlosser* base solution. After stirring for 4 h at −78 °C, freshly distilled benzaldehyde (2.22 mL, 22.0 mmol) dissolved in THF (20 mL) was added dropwise. The temperature was maintained at −78 °C for 30 min and then allowed to rise to −30 °C over a period of 1 h. The reaction was quenched with 10% aqueous HCl (50 mL) and repeatedly extracted with 10% aqueous HCl (2 × 40 mL). The phases were separated. The aqueous phase was made alkaline with excess aqueous NH_4OH solution, and was then repeatedly extracted with Et_2O (3 × 100 mL) and EtOAc (3 × 100 mL). The combined Et_2O/EtOAc layers were washed with brine (60 mL), dried (Na_2SO_4), filtered, and concentrated under reduced pressure. The crude product contained mainly the *threo* isomer of **150** (669 mg, 3.08 mmol, 31%) which was purified by flash column chromatography (silica gel, 15–25% methanol in EtOAc with 1% aq. NH_4OH). Afterwards, the original reaction mixture was extracted again with 10% aqueous HCl (2 × 50 mL). The phases were separated and the aqueous phase was made alkaline with excess aqueous NH_4OH solution. Backextraction of the aqueous phase II with EtOAc (3 × 100 mL) gave—after washing of the combined EtOAc layers with brine (60 mL), drying (Na_2SO_4), filtering, and evaporating—a crude product which contained predominantly the *erythro* isomer of **150**. Purification by flash column chromatography (silica gel, 15–25% methanol in EtOAc with 1% aq. NH_4OH) and subsequent recrystallization from Et_2O gave pure *erythro*-**150** (555 mg, 2.55 mmol, 26%) as a white solid. The physical data were identical in all respects to those previously reported [30].

threo-**150:**

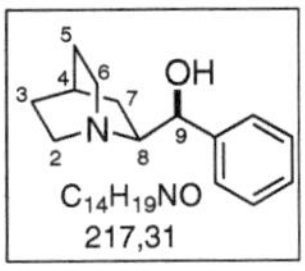

^{1}H NMR (500 MHz, $CDCl_3$) δ 7.39 (d, $J = 7.3$ Hz, 2H, C***H***$_{Ph,\ o}$), 7.32 (t, $J = 7.3$ Hz, 2H, C***H***$_{Ph,\ m}$), 7.28–7.25 (m, 1H, C***H***$_{Ph,\ p}$), 4.36 (d, $J = 9.8$ Hz, 1H, C***H***^{9}OH), 3.10–3.04 (m, 1H, C***H***H), 2.95 (app. t, $J = 7.7$ Hz, 2H, C***H***$_2$), 2.80–2.73 (m, 2H, CH***H*** and C***H***^{8}), 1.97 (br s, 1H, O***H***), 1.78–1.74 (m, 1H, C***H***^{4}), 1.53–1.42 (m, 4H, C***H***$^3_2$ and C***H***$^5_2$), 1.34–1.28 (m, 1H, C***H***H^7), 1.11 (ddt, $J = 13.1$, 8,4, 1.9 Hz, 1H, CH***H***^{7}).

^{13}C NMR (125 MHz, $CDCl_3$) δ 141.1 (***C***q$_{Ph}$), 128.3 (2C, ***C***H$_{Ph,\ m}$), 127.8 (***C***H$_{Ph,\ p}$), 127.3 (2C, ***C***H$_{Ph,\ o}$), 74.4 (***C***^{9}HOH), 62.8 (***C***^{8}H), 49.5 (***C***H$_2$), 41.4 (***C***H$_2$), 29.2 (***C***^{7}H$_2$), 26.6 (***C***H$_2$), 25.6 (***C***H$_2$), 21.5 (***C***^{4}H).

MS (EI-DE) *m/z* (%) 217 [M$^+$] (51), 200 (25), 188 (13), 176 (15), 158 (7), 140 (5), 120 (4), 110 (69), 105 (11), 98 (23), 91 (17), 82 (100), 77 (27), 68 (9), 55 (89), 42 (34), 29 (23).

HRMS (EI-DE) calcd for $C_{14}H_{19}NO$ [M$^+$] 217.1465, found 217.1467.

erythro-**150:**

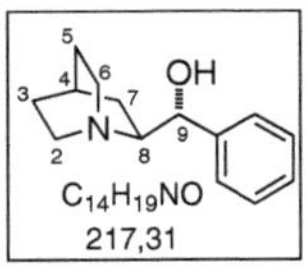

^{1}H NMR (500 MHz, $CDCl_3$) δ 7.35–7.30 (m, 4H, C***H***$_{Ph}$), 7.27–7.22 (m, 1H, C***H***$_{Ph,\ p}$), 4.85 (d, $J = 6.3$ Hz, 1H, C***H***^{9}OH), 3.45 (br s, 1H, O***H***), 3.29–3.22 (m, 1H, C***H***H^2), 3.02 (dd, $J = 16.4$, 8.8 Hz, 1H, C***H***^{8}), 2.90–2.76 (m, 2H, C***H***$^6_2$), 2.69–2.63 (m, 1H, CH***H***^{2}), 1.87–1.84 (m, 1H, C***H***^{4}), 1.71–1.66 (m, 1H, C***H***H^7), 1.65–1.59 (m, 1H, CH***H***^{7}), 1.55–1.43 (m, 4H, C***H***$^3_2$ and C***H***$^5_2$).

^{13}C NMR (125 MHz, CD_2Cl_2) δ 143.8 (***C***q$_{Ph}$), 128.2 (2C, ***C***H$_{Ph}$), 127.4 (***C***H$_{Ph,\ p}$), 126.4 (2C, ***C***H$_{Ph}$), 76.0 (***C***^{9}HOH), 61.4 (***C***^{8}H), 50.3 (***C***^{6}H$_2$), 43.3 (***C***^{2}H$_2$), 28.0 (***C***^{7}H$_2$), 26.4 (***C***H$_2$), 25.5 (***C***H$_2$), 21.9 (***C***^{4}H).

MS (EI-DE) *m/z* (%) 217 [M$^+$] (44), 200 (30), 188 (13), 176 (18), 158 (18), 140 (5), 120 (5), 110 (95), 98 (25), 91 (19), 82 (93), 77 (30), 68 (9), 55 (100), 42 (36), 29 (26).

HRMS (EI-DE) calcd for $C_{14}H_{19}NO$ [M$^+$] 217.1465, found 217.1467.

***erythro*-2-(α-Aminobenzyl)quinuclidine** (*erythro*-**86**):

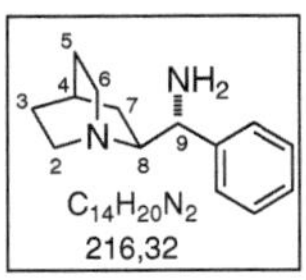

erythro-**86** was prepared starting from *threo*-**150** (200 mg, 920 μmol) following the same procedure described for the synthesis of 9-amino(9-deoxy)*epi*quinine

(**13**). The reaction was always kept between 0 °C and room temperature. Purification by flash column chromatography (silica gel, methanol/EtOAc 40:60 with 1% aq. NH_4OH) gave *erythro*-**86** as a white semi-solid (90.3 mg, 418 μmol, 45%). The enantiomers were separated by chiral HPLC (Chiralcel OD, 20 μm, 250 × 20 mm BIAX column; 10% *i*-PrOH in *i*-hexane with 0.1% diethylamine, 10 mL/min, 1.3 MPa, 308 K).

Optical rotation $[\alpha]_D^{25} = +61.0$ (c = 0.7, $CHCl_3$); $[\alpha]_D^{25} = -63.0$ (c = 0.6, $CHCl_3$).

^{1}H NMR (500 MHz, CD_2Cl_2) δ 7.31–7.26 (m, 4H, C***H***$_{Ph}$), 7.23–7.19 (m, 1H, C***H***$_{Ph,\ p}$), 3.87 (d, *J* = 9.5 Hz, 1H, C***H***^{9}NH$_2$), 3.05–2.97 (m, 1H, C***H***H^2), 2.87 (dd, *J* = 17.4, 8.9 Hz, 1H, C***H***^{8}), 2.76–2.73 (m, 2H, C***H***$_2^6$), 2.61–2.55 (m, 1H, CH***H***^{2}), 1.95–1.89 (m, 1H, C***H***H^7), 1.87–1.83 (m, 1H, C***H***^{4}), 1.79 (br s, 2H, N***H***$_2$), 1.54–1.43 (m, 5H, C***H***$_2^5$, C***H***$_2^3$ and CH***H***^{7}).

^{13}C NMR (125 MHz, CD_2Cl_2) δ 147.1 (***C***q$_{Ph}$), 128.5 (2C, ***C***H$_{Ph}$), 127.2 (2C, ***C***H$_{Ph}$), 127.0 (***C***H$_{Ph,\ p}$), 62.7 (***C***^{8}H), 60.1 (***C***^{9}HNH$_2$), 50.3 (***C***^{6}H$_2$), 42.5 (***C***^{2}H$_2$), 32.6 (***C***^{7}H$_2$), 27.2 (***C***H$_2$), 26.0 (***C***H$_2$), 22.6 (***C***^{4}H).

MS (EI-DE) *m/z* (%) 216 [M$^+$] (6), 199 (1), 158 (1), 145 (2), 132 (3), 106 (48), 91 (4), 82 (100), 79 (10), 69 (4), 55 (11), 42 (8), 28 (12).

HRMS (EI-DE) calcd for $C_{14}H_{20}N_2$ [M$^+$] 216.1625, found 216.1626.

***threo*-2-(α-Aminobenzyl)quinuclidine** (*threo*-**86**):

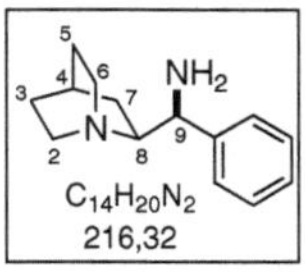

threo-**86** was prepared starting from *erythro*-**150** (200 mg, 920 μmol) following the same procedure described for the synthesis of 9-amino(9-deoxy)*epi*quinine (**13**). The reaction was always kept between 0 °C and room temperature. Purification by flash column chromatography (silica gel, methanol/EtOAc 40:60 with 1% aq. NH_4OH) gave *threo*-**86** as a white semi-solid (121 mg, 561 μmol, 61%).

^{1}H NMR (500 MHz, CD_2Cl_2) δ 7.38 (d, *J* = 7.6 Hz, 2H, C***H***$_{Ph,\ o}$), 7.31 (t, *J* = 7.6 Hz, 2H, C***H***$_{Ph,\ m}$), 7.24 (t, *J* = 7.1 Hz, 1H, C***H***$_{Ph,\ p}$), 3.79 (d, *J* = 10.1 Hz, 1H, C***H***^{9}NH$_2$), 3.07–3.00 (m, 1H, C***H***H^2), 2.93 (app. t, *J* = 7.7 Hz, 2H, C^6***H***$_2$), 2.75 (dd overlapped, *J* = 18.6, 9.5 Hz, 1H, C***H***^{8}), 2.74–2.68 (m overlapped, 1H, CH***H***^{2}), 1.92 (br s, 2H, N***H***$_2$), 1.66–1.64 (m, 1H, C***H***^{4}), 1.47–1.41 (m, 4H, C***H***$_2^3$ and C***H***$_2^5$), 1.16–1.10 (m, 1H, C***H***H^7), 0.98–0.94 (m, 1H, CH***H***^{7}).

^{13}C NMR (125 MHz, CD_2Cl_2) δ 144.1 (***C***q$_{Ph}$), 128.6 (2C, ***C***H$_{Ph,\ o}$), 128.5 (2C, ***C***H$_{Ph,\ m}$), 127.5 (***C***H$_{Ph,\ p}$), 62.5 (***C***^{8}H), 58.9 (***C***^{9}HNH$_2$), 50.3 (***C***^{6}H$_2$), 41.8 (***C***^{2}H$_2$), 31.0 (***C***^{7}H$_2$), 27.3 (***C***H$_2$), 26.1 (***C***H$_2$), 22.3 (***C***^{4}H).

MS (EI-DE) *m/z* (%) 216 [M$^+$] (3), 158 (1), 146 (5), 130 (2), 118 (2), 106 (42), 91 (4), 82 (100), 79 (11), 68 (5), 55 (14), 42 (12), 28 (23).

HRMS (EI-DE) calcd for $C_{14}H_{20}N_2$ [M$^+$] 216.1628, found 216.1626.

7.6.6 Synthesis of Amino Acid-Derived Diamines

Amino acid-derived diamines were prepared according to standard peptide coupling protocols followed by amide reduction [31, 32].

Synthesis of (*S*)-phenylglycine-*N*,*N*-dimethylamide (154a)

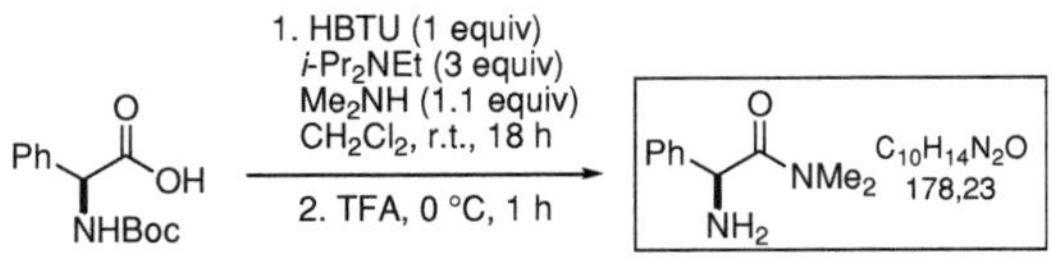

154a

To a solution of *N*-Boc-(*S*)-phenylglycine (1.51 g, 6.00 mmol, 1.0 equiv) in CH_2Cl_2 (48 mL) was added *O*-(benzotriazol-1-yl)-tetramethyluronium hexafluorophosphate (HBTU, 2.28 g, 6.00 mmol, 1.0 equiv), and the white suspension was stirred for 5 min, followed by the sequential addition of diisopropylethylamine (3.06 mL, 18.0 mmol, 3.0 equiv) and dimethylamine (0.33 mL, 6.60 mmol, 1.1 equiv). After 18 h at room temperature, the reaction mixture was combined with CH_2Cl_2 (50 mL) and water (50 mL). The organic layer was separated, repeatedly washed with aqueous 1 N HCl (3 × 30 mL), dried (Na_2SO_4), filtered, and concentrated under reduced pressure. The crude *N*-Boc-protected *N*,*N*-dimethylamide was treated with trifluoroacetic acid (TFA, 6 mL) at 0 °C. After 1 h, the mixture was diluted with CH_2Cl_2, basified (pH 8 ~ 9) at 0 °C with saturated aqueous $NaHCO_3$ solution, and then extracted with CH_2Cl_2 (3 × 50 mL). The combined organic layers dried (Na_2SO_4), filtered, and evaporated. L-phenylglycine-*N*,*N*-dimethylamide (**154a**; 739 mg, 4.15 mmol, 69%) was obtained as a pale yellow solid by flash column chromatography (silica gel, 10–25% methanol in EtOAc).

^{1}H NMR (500 MHz, $CDCl_3$) δ 7.27–7.17 (m, 5H, $C_6\boldsymbol{H}_5$), 4.65 (s, 1H, C***H***NH_2), 2.89 (s, 3H, C$\boldsymbol{H}_3$), 2.75 (s, 3H, C$\boldsymbol{H}_3$), 2.26 (br s, 2H, N$\boldsymbol{H}_2$).

^{13}C NMR (125 MHz, $CDCl_3$) δ 172.7 (***C***=O), 141.0 (***C***q_{Ph}), 129.1 (2C, ***C***H_{Ph}), 127.8 (***C***$H_{Ph,\ p}$), 127.1 (2C, ***C***H_{Ph}), 56.8 (***C***HNH_2), 36.6 (***C***H_3), 36.1 (***C***H_3).

Synthesis of (*S*)-*tert*-leucine-*N*,*N*-dimethylamide (154b):

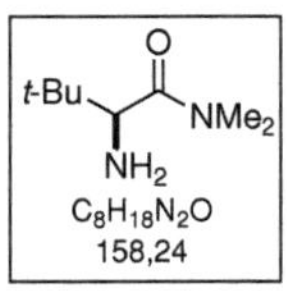

N,*N*-dimethylamide **154b** was obtained from (*S*)-*N*-Boc-*tert*-leucine by the method described for *N*,*N*-dimethylamide **154a**. Flash column chromatography (silica gel, 10–25% methanol in EtOAc) gave *N*,*N*-dimethylamide **154b** as a white solid (949 mg, 6.00 mmol, 95% yield).

^{1}H NMR (500 MHz, $CDCl_3$) δ 3.56 (s, 1H, C***H***), 3.06 (s, 3H, N(C$\boldsymbol{H}_3$)$_2$), 2.95 (s, 3H, N(C$\boldsymbol{H}_3$)$_2$), 2.11 (br s, 2H, N$\boldsymbol{H}_2$), 0.96 (s, 9H, C(C$\boldsymbol{H}_3$)$_3$).

^{13}C NMR (125 MHz, $CDCl_3$) δ 174.2 (***C***=O), 57.6 (***C***HNH_2), 38.1 (***C***qMe_3), 35.6 (***C***H_3), 35.3 (***C***H_3), 26.3 (3C, Cq(***C***H_3)$_3$).

Synthesis of (*S*)-tryptophane-*N*,*N*-dimethylamide (154c):

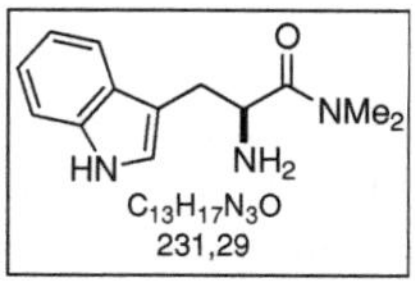

N,*N*-dimethylamide **154c** was obtained from (*S*)-*N*-Boc-tryptophane by the method described for *N*,*N*-dimethylamide **154a**. Flash column chromatography (silica gel, 10–30% methanol in EtOAc) gave *N*,*N*-dimethylamide **154c** as a pale yellow solid (505 mg, 2.16 mmol, 36%).

^{1}H NMR (500 MHz, $CDCl_3$) δ 8.28 (br s, 1H, N***H***), 7.52 (d, $J = 7.8$ Hz, 1H, C***H***$_{indol}$), 7.34 (d, $J = 8.0$ Hz, 1H, C***H***$_{indol}$), 7.16 (app. t, $J = 7.6$ Hz, 1H, C***H***$_{indol}$), 7.10–7.07 (m, 2H, C***H***$_{indol}$), 4.08–4.05 (m, 1H, C***H***NH_2), 3.11 (dd, $J = 14.6$, 6.4 Hz, 1H, C***H***H), 2.94 (dd, $J = 14.6$, 7.7 Hz, 1H, CH***H***), 2.84 (s, 3H, C***H***$_3$), 2.76 (s, 3H, C***H***$_3$), 2.40 (br s, 2H, N***H***$_2$).

Synthesis of (*S*)-phenylalanine amide (152)

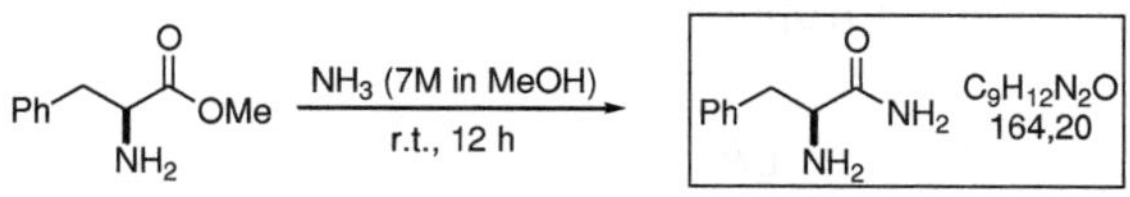

152

(*S*)-Phenylalanine methyl ester (3.30 g, 18.4 mmol) was dissolved in a methanolic solution of ammonia (13.1 mL, 92.0 mmol, 5.0 equiv; 7 M in methanol), and the resulting solution was stirred at room temperature overnight (12 h). The volatiles were removed *in vacuo* and (*S*)-phenylalanine amide (**152**; white solid; 3.02 g, >99% yield) was used in the next step without further purification.

^{1}H NMR (500 MHz, $CDCl_3$) δ 7.31 (t, $J = 7.4$ Hz, 2H, C***H***$_{Ph,\ m}$), 7.25–7.23 (m overlapped, 1H, C***H***$_{Ph,\ p}$), 7.22 (d overlapped, $J = 7.6$ Hz, 2H, C***H***$_{Ph,\ o}$), 7.11 (br s, 1H, C(=O)N***H***H), 6.00 (br s, 1H, C(=O)NH***H***), 3.60 (dd, $J = 9.5$, 4.1 Hz, 1H, C***H***NH_2), 3.26 (dd, $J = 13.7$, 4.2 Hz, 1H, C***H***H), 2.70 (dd, $J = 13.7$, 9.6 Hz, 1H, CH***H***), 1.38 (br s, 2H, CHN***H***$_2$).

^{13}C NMR (125 MHz, $CDCl_3$) δ 177.5 (***C***=O), 137.8 (***C***q_{Ph}), 129.2 (2C, ***C***H_{Ph}), 128.7 (2C, ***C***H_{Ph}), 126.8 (***C***$H_{Ph,\ p}$), 56.5 (***C***HNH_2), 40.9 (***C***H_2).

MS (EI-DE) *m*/*z* (%) 164 [M^+] (trace), 147 (2), 131 (1), 120 (100), 103 (14), 91 (18), 77 (9), 73 (31), 65 (7), 51 (4), 42 (4), 28 (6).

HRMS (CI-FE, *i*-butane) calcd for $C_9H_{13}N_2O$ [$(M + H)^+$] 165.1027, found 165.1028.

Synthesis of (*S*)-phenylalanine-*N*-methylamide (153)

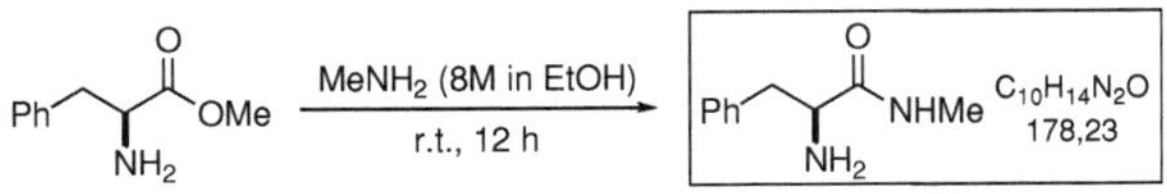

153

(*S*)-Phenylalanine methyl ester (3.59 g, 20.0 mmol) was dissolved in an ethanolic solution of methylamine (12.5 mL, 100 mmol, 5.0 equiv; 8 M in ethanol), and the resulting solution was stirred at room temperature overnight (12 h). The volatiles were removed *in vacuo* and (*S*)-phenylalanine-*N*-methylamide (**153**; white solid; 3.56 g, >99% yield) was used in the next step without further purification.

^{1}H NMR (500 MHz, $CDCl_3$) δ 7.29–7.17 (m, 6H, $C_6\boldsymbol{H}_5$ and N***H***Me), 3.57 (dd, $J = 9.5$, 3.8 Hz, 1H, C***H***NH_2), 3.25 (dd, $J = 13.8$, 4.0 Hz, 1H, C***H***H), 2.78 (d, $J = 5.0$ Hz, 3H, C$\boldsymbol{H}_3$), 2.64 (dd, $J = 14.0, 9.9$ Hz, 1H, CH***H***), 1.41 (br s, 2H, N$\boldsymbol{H}_2$).

^{13}C NMR (125 MHz, $CDCl_3$) δ 174.7 (***C***=O), 137.9 (***C***q_{Ph}), 129.2 (2C, ***C***H_{Ph}), 128.6 (2C, ***C***H_{Ph}), 126.7 (***C***$H_{Ph,\ p}$), 56.4 (***C***HNH_2), 40.9 (***C***H_2), 25.8 (***C***H_3).

MS (EI-DE) *m*/*z* (%) 178 [M^+] (1), 161 (2), 131 (2), 120 (100), 103 (11), 91 (13), 87 (29), 77 (7), 69 (5), 58 (4), 51 (2), 42 (7), 30 (3).

HRMS (CI-FE, *i*-butane) calcd for $C_{10}H_{15}N_2O$ [$(M + H)^+$] 179.1183, found 179.1184.

Synthesis of (*S*)-3-Phenylpropane-1,2-diamine (87)

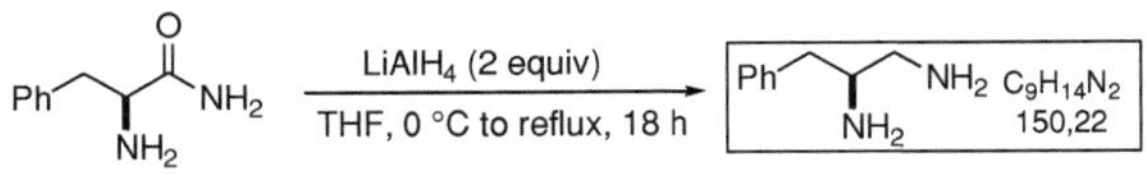

152 **87**

$LiAlH_4$ (228 mg, 6.00 mmol, 2.0 equiv) was added in portions to a solution of amide **152** (493 mg, 3.00 mmol) in THF (21 mL) at 0 °C. The reaction was heated to reflux until TLC analysis indicated complete disappearance of the starting material (18 h), and was then carefully quenched by the sequential addition of water, 15% aqueous NaOH, and water. The reaction mixture was repeatedly extracted with Et_2O, and the combined organic layers were dried (Na_2SO_4), filtered, and evaporated. Purification of the residue by short path distillation (bp 50–55 °C at 3.1^{-2} mbar) gave pure diamine **87** as a colorless liquid (255 mg, 1.55 mmol, 52%).

^{1}H NMR (500 MHz, $CDCl_3$) δ 7.28 (t, $J = 7.6$ Hz, 2H, C$\boldsymbol{H}_{Ph,\ m}$), 7.21 (d, $J = 7.6$ Hz, 1H, C$\boldsymbol{H}_{Ph,\ p}$), 7.18 (d, $J = 7.3$ Hz, 2H, C$\boldsymbol{H}_{Ph,\ o}$), 2.99–2.94 (m, 1H, C***H***NH_2), 2.80 (dd, $J = 12.6$, 4.1 Hz, 1H, C***H***HNH_2), 2.77 (dd, $J = 13.3$, 4.8 Hz, 1H, PhC***H***H), 2.54 (dd, $J = 12.6$, 7.9 Hz, 1H, CH***H***NH_2), 2.49 (dd, $J = 13.6$, 8.5 Hz, 1H, PhCH***H***), 1.56–1.46 (br m, 4H, N$\boldsymbol{H}_2$).

^{13}C NMR (125 MHz, $CDCl_3$) δ 139.1 (***C***q_{Ph}), 129.2 (2C, ***C***H_{Ph}), 128.5 (2C, ***C***H_{Ph}), 126.3 (***C***$H_{Ph,\ p}$), 55.0 (***C***HNH_2), 48.0 (***C***H_2NH_2), 42.3 (Ph***C***H_2).

MS (EI-DE) *m*/*z* (%) 150 [M^+] (trace), 132 (1), 120 (100), 103 (17), 91 (19), 77 (11), 65 (8), 59 (54), 51 (5), 42 (11), 30 (51), 28 (6).

HRMS (CI-FE, *i*-butane) calcd for $C_9H_{15}N_2$ [$(M + H)^+$] 151.1236, found 151.1235.

Synthesis of (*S*)-*N*[1]-methyl-3-phenylpropane-1,2-diamine (88):

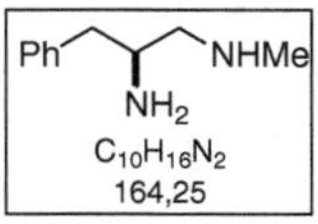

Diamine **88** was obtained from the corresponding *N*-methylamide **153** according to the procedure described for the synthesis diamine **87** [$LiAlH_4$ (2.0 equiv); 4 h at reflux]. Short path distillation (bp 38–39 °C at $4.2–4.4^{-2}$ mbar) gave pure diamine **88** as a colorless liquid (218 mg, 1.33 mmol, 44%).

^{1}H NMR (500 MHz, $CDCl_3$) δ 7.28 (t, $J = 7.4$ Hz, 2H, $C\boldsymbol{H}_{Ph,\ m}$), 7.21 (d, $J = 7.6$ Hz, 1H, $C\boldsymbol{H}_{Ph,\ p}$), 7.18 (d, $J = 7.6$ Hz, 2H, $C\boldsymbol{H}_{Ph,\ o}$), 3.10 (ddd, $J = 12.8$, 8.6, 4.3 Hz, 1H, C***H***NH_2), 2.77 (dd, $J = 13.4$, 4.9 Hz, 1H, PhC***H***H), 2.65 (dd, $J = 11.8$, 3.9 Hz, 1H, C***H***HNHMe), 2.49 (dd, $J = 13.4$, 8.7 Hz, 1H, PhCH***H***), 2.44 (dd overlapped, $J = 11.7$, 8.5 Hz, 1H, CH***H***NHMe), 2.43 (s overlapped, 3H, $C\boldsymbol{H}_3$), 1.47 (br s, 3H, $N\boldsymbol{H}_2$ and N***H***Me).

^{13}C NMR (125 MHz, $CDCl_3$) δ 139.1 ($\boldsymbol{C}q_{Ph}$), 129.2 (2C, $\boldsymbol{C}H_{Ph}$), 128.5 (2C, $\boldsymbol{C}H_{Ph}$), 126.3 ($\boldsymbol{C}H_{Ph,\ p}$), 58.3 ($\boldsymbol{C}H_2$NHMe), 52.3 (***C***HNH_2), 42.9 (Ph$\boldsymbol{C}H_2$), 36.5 ($\boldsymbol{C}H_3$).

MS (EI-DE) *m*/*z* (%) 164 [M^+] (trace), 132 (1), 120 (100), 103 (12), 91 (12), 77 (7), 73 (28), 65 (5), 56 (3), 44 (68), 30 (11).

HRMS (CI-FE, *i*-butane) calcd for $C_{10}H_{17}N_2$ [$(M + H)^+$] 165.1392, found 165.1392.

Synthesis of (*S*)-*N*[1],*N*[1]-dimethyl-3-phenylpropane-1,2-diamine (89d):

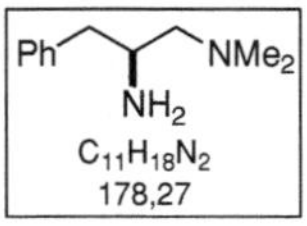

Diamine **89d** was obtained from the corresponding *N*,*N*-dimethylamide hydrochloride salt according to the procedure described for diamine **87** [$LiAlH_4$ (3.0 equiv); 12 h at room temperature]. Short path distillation (bp 52 °C at 2.2^{-1} mbar) gave pure diamine **89d** as a colorless liquid (315 mg, 1.77 mmol, 44%).

^{1}H NMR (500 MHz, CD_2Cl_2) δ 7.29 (t, $J = 7.6$ Hz, 2H, $C\boldsymbol{H}_{Ph,\ m}$), 7.22–7.18 (m, 3H, $C\boldsymbol{H}_{Ph,\ o\ \mathrm{and}\ p}$), 3.10 (app. sept, $J = 4.5$ Hz, 1H, C***H***NH_2), 2.72 (dd, $J = 13.2, 4.4$ Hz, 1H, PhC***H***H), 2.42 (dd, $J = 13.3$, 8.5 Hz, 1H, PhCH***H***), 2.20 (s, 6H, $C\boldsymbol{H}_3$), 2.19 (dd, $J = 11.7$, 9.3 Hz, 1H, CH***H***NMe_2), 2.17 (dd, $J = 12.1$, 4.5 Hz, 1H, C***H***HNMe_2), 1.27 (br s, 2H, $N\boldsymbol{H}_2$).

^{13}C NMR (125 MHz, CD_2Cl_2) δ 140.3 ($\boldsymbol{C}q_{Ph}$), 129.7 (2C, $\boldsymbol{C}H_{Ph}$), 128.6 (2C, $\boldsymbol{C}H_{Ph}$), 126.3 ($\boldsymbol{C}H_{Ph,\ p}$), 67.0 ($\boldsymbol{C}H_2NMe_2$), 50.5 (***C***HNH_2), 46.0 (2C, $\boldsymbol{C}H_3$), 42.7 (Ph$\boldsymbol{C}H_2$).

MS (EI-DE) *m/z* (%) 178 [M$^+$] (4), 161 (1), 132 (1), 120 (65), 103 (7), 91 (12), 77 (4), 70 (3), 65 (4), 58 (100), 51 (1), 42 (9), 30 (5).

HRMS (EI-FE) calcd for $C_{11}H_{18}N_2$ [M$^+$] 178.1469, found 178.1470.

Synthesis of (*S*)-*N*^{1},*N*^{1}-dimethyl-2-phenylethane-1,2-diamine (89a):

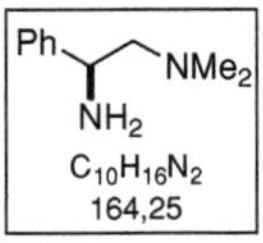

Diamine **89a** was obtained from the corresponding *N*,*N*-dimethylamide **154a** according to the procedure described for diamine **87** [$LiAlH_4$ (3.0 equiv); 4 h at room temperature]. Short path distillation (bp 95–96 °C at 1.2–1.4^{-1} mbar) gave pure diamine **89a** as a colorless liquid (396 mg, 2.41 mmol, 71%).

^{1}H NMR (500 MHz, CD_2Cl_2) δ 7.38 (d, J = 7.3 Hz, 2H, C***H***$_{Ph,\ o}$), 7.31 (t, J = 7.6 Hz, 2H, C***H***$_{Ph,\ m}$), 7.24–7.21 (m, 1H, C***H***$_{Ph,\ p}$), 4.05 (dd, J = 10.2, 3.9 Hz, 1H, C***H***NH$_2$), 2.39 (dd, J = 12.0, 10.4 Hz, 1H, C***H***H), 2.26 (s, 6H, C***H***$_3$), 2.19 (dd, J = 12.0, 4.1 Hz, 1H, CH***H***), 1.67 (br s, 2H, N***H***$_2$).

^{13}C NMR (125 MHz, CD_2Cl_2) δ 145.5 (***C***q$_{Ph}$), 128.5 (2C, ***C***H$_{Ph}$), 127.2 (***C***H$_{Ph,\ p}$), 127.1 (2C, ***C***H$_{Ph}$), 68.8 (***C***H$_2$), 53.6 (***C***HNH$_2$), 45.8 (2C, ***C***H$_3$).

GC-MS (GC-EI) *m/z* (%) 118 (1), 106 (10), 91 (1), 79 (5), 58 (100), 51 (2), 42 (5), 30 (4).

HRMS (CI-FE, *i*-butane) calcd for $C_{10}H_{17}N_2$ [(M + H)$^+$] 165.1391, found 165.1392.

Synthesis of (*S*)-*N*^{1},*N*^{1}-dimethyl-2-*tert*-butylethane-1,2-diamine (89b):

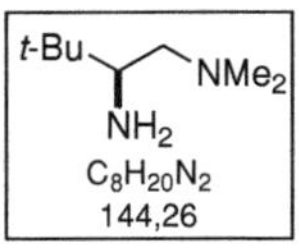

Diamine **89b** (clear liquid; 228 mg, 1.39 mmol, 46%) was obtained from the corresponding *N*,*N*-dimethylamide **154b** according to the procedure described for diamine **87** [$LiAlH_4$ (3.0 equiv) in Et_2O; 4 h at room temperature].

^{1}H NMR (500 MHz, CD_2Cl_2) δ 2.55 (dd, J = 9.6, 4.0 Hz, 1H, C***H***), 2.18 (s, 6H, N(C***H***$_3$)$_2$), 2.13–2.07 (m, 2H, C***H***$_2$), 1.30 (br s, 2H, N***H***$_2$), 0.86 (s, 9H, Cq(C***H***$_3$)$_3$).

^{13}C NMR (125 MHz, CD_2Cl_2) δ 62.1 (***C***H$_2$), 57.2 (***C***H), 45.9 (2C, N(***C***H$_3$)$_2$), 33.3 (***C***q), 26.3 (3C, Cq(***C***H$_3$)$_3$).

Synthesis of (*S*)-*N*[1],*N*[1]-dimethyl-3-(1*H*-indol-3-yl)-propane-1,2-diamine (89c):

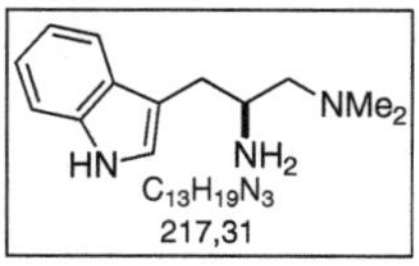

Diamine **89c** was obtained from the corresponding *N,N*-dimethylamide **154c** according to the procedure described for diamine **87** [$LiAlH_4$ (3.0 equiv); 3 h at room temperature]. Flash column chromatography (silica gel, methanol/EtOAc 50:50 with 1% aq. NH_4OH) gave pure diamine **89c** as a pale yellow solid (260 mg, 1.22 mmol, 69%).

^{1}H NMR (500 MHz, $CDCl_3$) δ 8.22 (br s, 1H, N***H***), 7.61 (d, J = 7.3 Hz, 1H, CH_{ar}, 7.34 (d, J = 8.2 Hz, 1H, CH_{ar}), 7.18 (app. t, J = 8.0 Hz, 1H, CH_{ar}), 7.10 (app. t, J = 7.7 Hz, 1H, CH_{ar}), 7.06 (s, 1H, CH_{ar}NH), 3.25 (app. sept, J = 4.6 Hz, 1H, C***H***NH_2), 2.91 (dd, J = 14.2, 4.1 Hz, 1H, Cq_{ar}C***H***H), 2.64 (dd, J = 14.2, 8.5 Hz, 1H, Cq_{ar}CH***H***), 2.31 (dd, J = 12.0, 9.5 Hz, 1H, C***H***HNMe_2), 2.25–2.23 (m overlapped, 1H, CH***H***NMe_2), 2.23 (s ovelapped, 6H, CH_3), 2.03 (br s, 2H, NH_2).

^{13}C NMR (125 MHz, $CDCl_3$) δ 136.4 (***C***q_{ar}), 127.7 (***C***q_{ar}), 122.7 (***C***H_{ar}), 122.0 (***C***H_{ar}), 119.3 (***C***H_{ar}), 118.9 (***C***H_{ar}), 113.0 (***C***q_{ar}), 111.2 (***C***H_{ar}), 66.5 (***C***H_2NMe_2), 48.8 (***C***HNH_2), 45.9 (2C, ***C***H_3), 31.5 (Cq_{ar}***C***H_2).

*7.6.7 Synthesis of (1***R***,2***R***)-***N***-iso-propyl-1,2-diphenylethylene-1,2-diamine (61)*

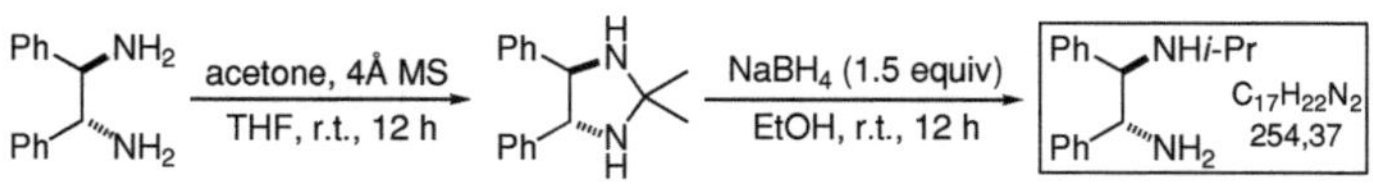

155 **61**

Reference [33, 34] (1*R*,2*R*)-DPEN (200 mg, 0.94 mmol) and 4 Å molecular sieves (500 mg) were placed in a flask, and THF (2 mL) and acetone (0.3 mL) were successively added. After stirring overnight (12 h), GC/MS analysis indicated full conversion to the corresponding acetonide. The mixture was filtered through Celite (eluent: Et_2O) and the filtrate was concentrated to give the crude acetonide **155**, which was used without further purification in the next step.

^{1}H NMR (300 MHz, $CDCl_3$) δ 7.29–7.19 (m, 10H, C_6***H***$_5$), 4.24 (s, 2H, C***H***C***H***), 2.04 (br s, 2H, N***H***), 1.53 (s, 6H, Cq(C***H***$_3$)$_2$).

The crude acetonide **155** was dissolved in ethanol (7 mL) and treated with $NaBH_4$ (54 mg, 1.41 mmol, 1.5 equiv). After stirring overnight, water and 1 N aqueous NaOH (2:1, 7 mL), and the reaction mixture was repeatedly extracted with Et_2O (3 × 20 mL). The combined organic layers were washed with brine, dried (Na_2SO_4), filtered, and concentrated to give analytically pure (1*R*,2*R*)-*N*-isopropyl-1,2-diphenylethylene-1,2-diamine (**61**; 407 mg, 1.60 mmol, 63%) as a colorless oil. The analytical data were identical in all respects to those previously reported [35].

^{1}H NMR (400 MHz, CD_2Cl_2) δ 7.21–7.09 (m, 10H, $C_6\boldsymbol{H}_5$), 3.93 (d, $J = 7.6$ Hz, 1H, C***H***CH), 3.78 (d, $J = 7.6$ Hz, 1H, CHC***H***), 2.51 (sept, $J = 6.1$ Hz, 1H, C***H***Me$_2$), 1.72 (br s, 3H, N$\boldsymbol{H}_2$ and N***H***), 0.95 (dd, $J = 6.3$ Hz, 6H, CH(C$\boldsymbol{H}_3$)$_2$).

^{13}C NMR (100 MHz, CD_2Cl_2) δ 144.5 (***C***q$_{Ph}$), 143.0 (***C***q$_{Ph}$), 128.2 (2C, ***C***H$_{Ph}$), 128.2 (2C, ***C***H$_{Ph}$), 128.1 (2C, ***C***H$_{Ph}$), 127.6 (2C, ***C***H$_{Ph}$), 127.0 (***C***H$_{Ph,\ p}$), 126.9 (***C***H$_{Ph,\ p}$), 67.3 (***C***HCH), 62.3 (CH***C***H), 46.1 (***C***HMe$_2$), 24.5 (CH(***C***H$_3$)$_2$), 22.0 (CH(***C***H$_3$)$_2$).

MS (EI-DE) *m/z* (%) 254 [M$^+$] (trace), 196 (1), 178 (1), 165 (1), 148 (100), 132 (2), 118 (1), 106 (47), 91 (3), 79 (13), 51 (1), 43 (3).

HRMS (ESI+) calcd for $C_{17}H_{22}N_2Na$ [(M + Na)$^+$] 277.1673, found 277.1675.

7.6.8 Synthesis of (S)-3,3′-bis(triphenylsilyl)-1,1′-binaphthyl-2,2′-diol phosphate ((S)-7b)

(*S*)-3,3′-Bis(triphenylsilyl)-1,1′-binaphthyl-2,2′-diol phosphate ((*S*)-**7b**) was prepared according to literature-known procedures [35, 36].

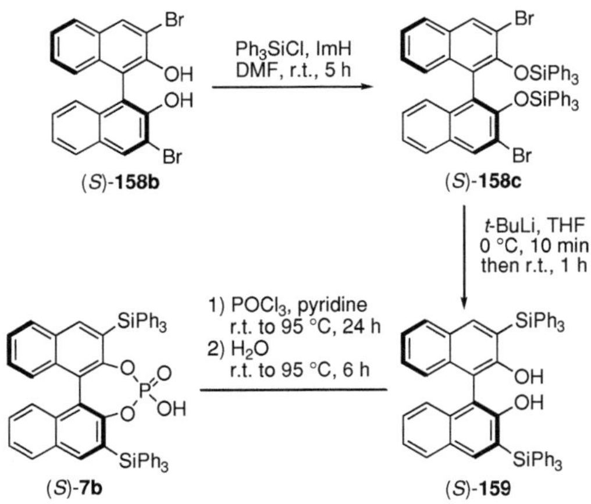

(*S*)-(3,3′-Dibromo-1,1′-binaphthyl-2,2′-diyl)bis(oxy)bis(triphenylsilane) ((*S*)-157c):

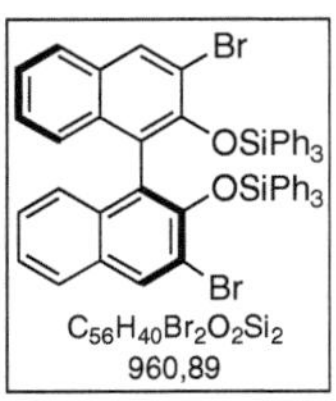

(*S*)-3,3′-Dibromo-1,1′-binaphthyl-2,2′-diol (**158b**) (3.50 g, 7.88 mmol) was dissolved in DMF (47 mL) and then imidazole (1.64 g, 24.0 mmol) followed by triphenylsilyl chloride (5.90 g, 20.0 mmol) was added. The mixture was stirred at room temperature for 5 h, then poured into saturated aqueous $NaHCO_3$ solution, and extracted with CH_2Cl_2. The organic layers were washed again with saturated aqueous $NaHCO_3$ solution, dried (Na_2SO_4), filtered, and concentrated *in vacuo*. The residue was purified by flash column chromatography (silica gel, 1:2 CH_2Cl_2/hexane) to furnish bis(triphenyl)silyl ether **158c** (7.15 g, 7.44 mmol, 94%) as a colorless solid.

^{1}H NMR (300 MHz, $CDCl_3$) δ 7.57 (s, 2H), 7.38–7.30 (m, 4H), 7.21–6.95 (m, 32H), 6.74 (d, $J = 8.3$ Hz, 2H).

^{13}C NMR (75 MHz, $CDCl_3$) δ 147.2 (*C*q), 134.4, 134.2 (*C*q), 133.6, 133.3, 132.8, 132.2, 131.8, 128.9 (*C*q), 128.7 (*C*q), 128.4 (*C*q), 128.2, 126.9 (*C*q), 126.7, 126.2, 126.0, 125.0, 124.9, 123.2 (*C*q), 122.5 (*C*q), 116.1 (*C*q).

MS (EI-DE) *m*/*z* 960 [M^+] (27), 880 (1), 259 (100), 181 (6), 105 (1).

HRMS (ESI+) calcd for $C_{56}H_{44}N_1O_2Si_2Br_2$ [$(M + NH_4)^+$] 976.1267, found 976.1272.

(*S*)-3,3′-Bis(triphenylsilyl)-1,1′-binaphthyl-2,2′-diol ((*S*)-159):

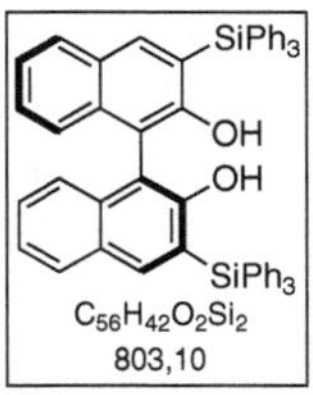

To a solution of **158c** (6.60 g, 6.87 mmol) in dry THF (100 mL) was added dropwise *t*-BuLi (11.7 mL, 19.9 mmol; 1.7 M in pentane) over a period of 10 min at 0 °C. The mixture was stirred at room temperature for 1 h, poured into saturated aqueous NH_4Cl solution, and extracted with CH_2Cl_2. The combined extracts were dried (Na_2SO_4), filtered, and concentrated. Pure diol (*S*)-**159** (4.66 g, 5.81 mmol, 85%) was obtained after flash column chromatography (silica gel, 1:2 CH_2Cl_2/hexane).

^{1}H NMR (300 MHz, $CDCl_3$) δ 7.84 (s, 2H), 7.62–7.56 (m, 8H), 7.35–7.18 (m, 30H), 5.22 (s, 2H).

^{13}C NMR (75 MHz, $CDCl_3$) δ 155.5 (*C*q), 141.0, 135.3, 133.7 (*C*q), 133.3 (*C*q), 128.5, 128.2, 128.0, 127.2, 126.8 (*C*q), 122.9, 122.8 (*C*q), 122.6 (*C*q), 109.6 (*C*q).

MS (EI-DE) *m*/*z* 802 [M^+] (7), 724 (8), 646 (100), 567 (5), 491 (4), 429 (5), 369 (2), 323 (6), 284 (27), 245 (9), 78 (1).

HRMS (ESI+) calcd for $C_{56}H_{46}N_1O_2Si_2$ [$(M + NH_4)^+$] 820.3069, found 820.3062.

(*S*)-3,3′-Bis(triphenylsilyl)-1,1′-binaphthyl-2,2′-diol phosphate ((*S*)-7b):

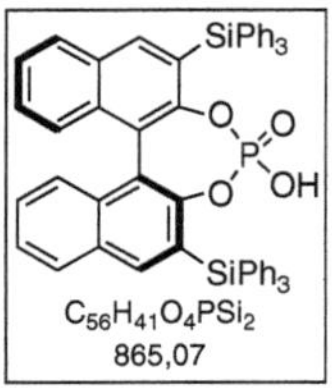

Diol **159** (2.20 g, 2.74 mmol) was suspended in pyridine (10 mL) and phosphorus oxychloride (0.5 mL, 5.48 mmol) was added dropwise at room temperature with rapid stirring. The resulting suspension was heated to 95 °C for 24 h. Then, the reaction mixture was cooled to room temperature and water (2 mL) was added. The resulting biphasic suspension was heated to 95 °C for an additional 6 h. The reaction mixture was diluted with CH_2Cl_2 and repeatedly extracted with 1 N aqueous HCl. The combined organic phases were dried (Na_2SO_4), filtered, and concentrated under reduced pressure. Purification by flash column chromatography (silica gel, 2% methanol in CH_2Cl_2) provided phosphoric acid derivative (*S*)-**7b** (2.18 g, 2.52 mmol, 92%) as a white solid. The analytical data were identical in all respects to those previously reported [35].

^{1}H NMR (400 MHz, $(CD_3)_2SO$) δ 7.94 (s, 2H), 7.80 (d, J = 7.5 Hz, 2H), 7.60–7.57 (m, 12H), 7.37–7.28 (m, 22H), 7.06 (d, J = 8.0 Hz, 2H).

^{13}C NMR (75 MHz, $(CD_3)_2SO$) δ 152.1 (*C*q), 152.1 (*C*q), 141.1 (*C*q), 136.3, 134.0 (*C*q), 133.5 (*C*q), 129.8, 129.5, 128.8, 127.7 (*C*q), 126.2 (*C*q), 125.8, 125.4 (*C*q), 121.2 (*C*q).

^{31}P NMR (161 MHz, $(CD_3)_2SO$) δ 2.49 (s).

MS (ESI−) *m/z* 863 [(M-H)$^-$].

HRMS (ESI−) calcd for $C_{56}H_{40}O_4Si_2P_1$ [(M-H)] 863.2222, found 863.2208.

References

1. Ouyang X-Y, Jiang H-F, Huang J-M (2006) Chin J Chem 24:1480
2. Wynberg H, Marsman B (1980) J Org Chem 45:158
3. Carlone A, Bartoli G, Bosco M, Pesciaioli F, Ricci P, Sambri L, Melchiorre P (2007) Eur J Org Chem 33:5492
4. Wang X, Reisinger CM, List B (2008) J Am Chem Soc 130:6070
5. Rickards RW, Rodwell JL, Schmalzl KJ (1977) J Chem Soc Chem Commun 849
6. Lv J, Zhang J, Lin Z, Wang Y (2009) Chem Eur J 15:972
7. Wipf P, Smitrovich JH, Moon CW (1992) J Org Chem 57:3178
8. Reisinger CM, Wang X, List B (2008) Angew Chem Int Ed 47:8112
9. Usuda H, Kuramochi A, Kanai M, Shibasaki M (2004) Org Lett 6:4387
10. Sato T, Kikuchi T, Tsujita H, Kaetsu A, Sootome N, Nishida K-i, Tachibana K, Murayama E (1991) Tetrahedron 47:3281–3304
11. Scheller ME, Frei B (1992) Helv Chim Acta 75:69
12. Miyashita M, Suzuki T, Yoshikoshi A (1987) Tetrahedron Lett 28:4293
13. Dussault PH, Liu X (1999) Org Lett 1:1391

14. Woods GF, Tucker IW (1948) J Am Chem Soc 70:2174
15. Woods GF, Griswold PH, Armbrecht BH, Blumenthal DI, Plapinger R (1949) J Am Chem Soc 71:2028
16. Coote SC, O'Brien P, Whitwood AC (2008) Org Biomol Chem 6:4299
17. Uyanik M, Fukatsu R, Ishihara K (2009) Org Lett 11:3470
18. Larock RC, Hightower TR, Kraus GA, Hahn P, Zheng D (1995) Tetrahedron Lett 36:2423
19. Yu J-Q, Corey EJ (2002) Org Lett 4:2727
20. Cahiez G, Avedissian H (1998) Synthesis 1199
21. Zhou J, Wakchaure V, Kraft P, List B (2008) Angew Chem Int Ed 47:7656
22. Oare DA, Henderson MA, Sanner MA, Heathcock CH (1990) J Org Chem 55:132
23. Brunner H, Bügler J, Nuber B (1995) Tetrahedron Asymmetry 6:1699
24. Vakulya B, Varga S, Csámpai A, Soós T (1967) Org Lett 2005:7
25. Brunner H, Bügler J (1997) Bull Soc Chim Belg 106:77
26. Franz MH, Röper S, Wartchow R, Hoffmann HMR (2004) J Org Chem 69:2983
27. Berkessel A, Guixá M, Schmidt F, Neudörfl JM, Lex J (2007) Chem Eur J 13:4483
28. Chen W, Du W, Duan Y-Z, Wu Y, Yang S-Y, Chen Y-C (2007) Angew Chem Int Ed 46:7667
29. Kessar SV, Singh P, Singh KN, Singh SK (1999) Chem Commun 19:1927
30. Lygo B, Crosby J, Lowdon T, Wainwright PG (1999) Tetrahedron 55:2795
31. Stotter PL, Friedman MD (1985) J Org Chem 50:29
32. Andrés JM, Manzano R, Pedrosa R (2008) Chem Eur J 14:5116
33. Martin NJA, Ozores L, List B (2007) J Am Chem Soc 129:8976
34. Kim KH, Lee S, Lee D-W, Ko D-H, Ha D-C (2005) Tetrahedron Lett 46:5991
35. Storer RI, Carrera DE, Ni Y, MacMillan DWC (2005) J Am Chem Soc 128:84
36. Maruoka K, Itoh T, Araki Y, Shirasaka T, Yamamoto H (1988) Bull Chem Soc Jpn 61:2975

Appendix

Curriculum Vitae

Corinna Marie Reisinger
19.07.1981, Bretten, Germany

01.2006–03.2010	Ph.D. candidate with Prof. Dr. Benjamin List, Max-Planck-Institut für Kohlenforschung, Mülheim an der Ruhr, Germany 06.2010 Ph.D. (summa cum laude) in organic chemistry Title of the Thesis: "Organocatalytic asymmetric epoxidations and hydroperoxidations of α,β-unsaturated ketones" 05.2010 Laureate of the 25th DSM Science & Technology Awards (North), Vaals, Netherlands 07.2006–06.2008 Kekulé Fellowship, Fonds der Chemischen Industrie, Germany
10.2000–09.2005	Chemistry student, University of Ulm, Germany 09.2005 Diploma in chemistry, grade: "with honors" (1.0) 03.2005–09.2005 Diploma thesis with Prof. Dr. Dr.h.c. Bernhard Rieger, University of Ulm, Germany Title of the Thesis: "Novel late transition metal complexes employing anionic chelating [P,O] ligands based on phosphoric acid derivatives" 05.2006 Diploma Award, Dr. Barbara Mez-Starck-Stiftung 01.2003 Pre-diploma Award, University of Ulm, Germany 10.2000–09.2002 Undergraduate fellowship, Fonds der Chemischen Industrie, Germany
09.1991–06.2000	Edith-Stein-Gymnasium, Bretten, Germany 06.2000 Abitur, grade: "very good" (1.0) 06.2000 Award, Fonds der Chemischen Industrie, Germany

C. Reisinger, *Epoxidations and Hydroperoxidations of α,β-Unsaturated Ketones*,
Springer Theses, DOI: 10.1007/978-3-642-28118-1,

MIX
Papier aus verantwortungsvollen Quellen
Paper from responsible sources
FSC® C105338

If you have any concerns about our products,
you can contact us on
ProductSafety@springernature.com

In case Publisher is established outside the EU,
the EU authorized representative is:
Springer Nature Customer Service Center GmbH
Europaplatz 3, 69115 Heidelberg, Germany

Printed by Libri Plureos GmbH
in Hamburg, Germany